Restoring America's Rivers

Richard M. Robinson

Restoring America's Rivers

The Movement for Dam Deconstructions and Rehabilitations

palgrave
macmillan

Richard M. Robinson
School of Business
State University of New York
Fredonia, NY, USA

ISBN 978-3-031-81757-1 ISBN 978-3-031-81758-8 (eBook)
https://doi.org/10.1007/978-3-031-81758-8

This Palgrave Macmillan imprint is published by the registered company Springer Nature Switzerland AG
The registered company address is: Gewerbestrasse 11, 6330 Cham, Switzerland

This book is dedicated to Dr. Maryann Robinson who assisted with reviewing some of the more difficult material in the thirteen chapters of this volume. It is also dedicated to the environmental advocacy organizations American Rivers, The Nature Conservancy, and The Environmental Defense Fund because of their leadership in restoring America's rivers.

Preface

Rivers and Dams as Common Property

This book describes an American renaissance, a movement to return our rivers to being closer to their natural state of healthy ecology rather than stagnant ecologically vacuous water bodies. One essential element of this renaissance is the reformation of our federal procedures for water project evaluation, a process designed to decide which projects to undertake or reject, and which existing facilities to alter or remove. These procedures are now well established and properly documented in our federal law and agency manuals. They present our society's moral declarations as to how we should proceed and decide about some of the most important physical elements of our existence—our rivers, lakes, bays, and other water bodies. These statements of "what to consider and how to evaluate" are moral declarations now aimed at recognizing the ethical conundrums of environmental justice, economic infrastructure design, and resource allocations that shape our society and its future. In the same way that ethics shape our personal lives, our society's ethical declarations also shape our socio-economic dynamics. Over the previous few decades, we have reformed these moral declarations concerning water project evaluation procedures, and the effects of this reformation and resource allocations are reviewed here.

In the last twenty-five years, the US has removed thousands of its dams, both large and small but mostly low-head derelict structures on smaller streams.[1] These smaller structures were dangerous to recreators and disrupted the natural ecology of our watersheds. The reservoirs behind many of these obstructions stored the poisonous sediments of defunct industries. Their removals allowed us to once again interact with some of our more essential natural assets. We suspect that a healthy psychology benefits from interacting with nature, and nonpoisonous streams can facilitate those beneficial interactions.

Our era of big dam construction is over, and some of our larger dams are being deconstructed on our larger rivers. (Northern California's Klamath River Dams and Washington's Elwha River Dams are prime examples.) Consequently, some of our more significant rivers are being renewed. These larger dam deconstructions are sometimes motivated by the pursuit of environmental justice for indigenous tribes, but they are also motivated by our battle to adjust to climate change, and our struggle to preserve species. We now know that our previous attempts to replumb our great watersheds by constructing some of our largest dams were not based on rational analysis. We were ignorantly sacrificing some of our most important natural assets.

This book documents that our current watershed restorations are properly motivated by ecological pursuits, species preservation, desires for healthy recreation, and environmental justice. Our dam removals and river restorations pose a significant movement that involves both effective government agencies and effective *environmental advocacy organizations*. This book reviews the reforms in our society's approaches to remedying environmental disasters that previously government either caused or allowed. These reforms pose very significant parts of our overall political environmental movement.

Dams as Monuments

In the 1950s, America's new Interstate Highway System was being constructed. "See the USA" was a clarion call for families with new Detroit made autos that seemed perfectly designed for the highway road trip. "Boomers" can remember those road trips. Those highway tours typically visited the great American monuments of the twentieth century:

[1] See "Free Rivers: The State of Dam Removal in the United States," *American Rivers*, at www.americanrivers.org/wp-content/uploads/2023/02/DamList2021_Report_02172022_FINAL3,pdf.

Mount Rushmore, the Golden Gate bridge, Disneyland, a Hollywood movie studio, the Grand Canyon, Monument Valley, the Saint Louis Arch, Hoover Dam, and others. Ask any American to name a dam and it will likely be "Hoover Dam." It has become one of America's monuments, a permanent testament to American engineering accomplishment. Hoover Dam appears to tourists to be indestructibly fixed in place. Its sleek design and massive size gives the appearance that it will last forever as if it were a natural part of the Rocky Mountains or Grand Canyon. A generation of boomer children were taken on these highway tours to view Hoover Dam and other all-American monuments. Since Las Vegas is nearby, a Hoover Dam trip has been "a sight worth seeing" for countless visitors to that electrified City. (Las Vegas' electric power comes from Hoover's generators.)

One can still watch the well-made documentary about Hoover Dam's construction (now on *You Tube*), during which one marvels about the construction problems overcome for pouring, and curing, and even making that massive volume of concrete, plus the placement of the turbines, the arch design of the dam, and the construction of the desert housing necessary for the workers.[2] Construction workers even made their own hardhats from tar. Building this dam was a 1930s "Depression Era" accomplishment that seemed even bigger than the constructions of the Interstate Highway System in the 1950s.

As a ten-year old in the late 1950s, I was taken on one of those family highway trips, and I saw that engineering marvel of Hoover Dam. Thirteen years after that cross-country trip, I moved to Oregon. Upon arrival, I purchased a used Chevrolet, and drove east from Portland to see one of America's finest natural vistas, the Columbia River Gorge with its Multnomah Falls and Horsetail Falls. A little farther east I saw the Bonneville Dam and the Dalles Dam, both gigantic structures. They also were monuments to American engineering aimed at controlling one of North America's biggest arteries—the Columbia River. They were my second and third views of big concrete dams. Hoover is a big hard-rock deep-canyon dam. One is impressed by its considerable depth, but it is also relatively narrow. Bonneville and the Dalles Dams are very wide but not as high. They had fish passage ramps for the salmon and steelhead trout on their upriver migrations. You can watch fish pass on the ramps

[2] See www.youtube.com/watch2v=nGy_nPpwSU.

by looking through glass panels. But soon after, I talked to a biologist friend who worked for the US Fish and Wildlife Department. He indicated that few fish made the passage past these dams. There was a crisis in salmon and steelhead survival. I observed all of this in 1971; the problem has only gotten worse since.

This problem with salmon and steelhead did not surprise me. Between the first time I saw the Hoover Dam and the time I moved to Oregon, my opinion of dams deteriorated. As a youngster, my fishing buddy and I explored the Ipswich River and its tributaries looking for good spots to fish. We rode bicycles. We did find three different low-head dams along the Ipswich.[3] They were ugly, close to the road, and with many broken beer bottles smashed against the concrete and nearby rocks. Worst of all, we could find no signs of fish. One of these dams had clearly been constructed to provide mechanical power to a currently broken-down and abandoned small mill. The remains of the wrecked abandoned building remained next to the derelict dam.

The American East is filled with old structures; some of these have been restored. The American West is much newer, filled with new structures that are much bigger, perhaps to match the size of the Western landscape. When I first viewed the Bonneville and Dalles Dams, they were relatively new, both constructed within the previous couple of decades. But I was biased and envisioned them as I thought they might be in a distant future, i.e. derelict structures responsible for destroying our salmon and steelhead runs. I was sure this would be the result. In recent decades, however, we have begun to seriously address this problem for the Columbia and other rivers. This book in part concerns these efforts.

This task of restoration is currently so large that it appears to pose a crisis. But over the last twenty-five years the record of dam deconstructions and river renewals has been substantial, and the public's demand for restoration remains vigorous. As an example, a local government agency in Ojai, California, recently deconstructed the Matilija Dam, a 168-foot high concrete structure on Matilija Creek. This Dam is upstream of the Town of Ojai, which is twelve miles inland from the coastal town of Ventura, a short distance north of Los Angeles. The Dam was constructed in 1948 for flood control, but after its first fifteen years, it had become heavily silted. The sediment buildup was so high that the

[3] "Lowhead dams" are those less than 15 feet in height. See Chapters 7 and 8.

Dam had little water storage capacity left. The Matilija flows into the Pacific Ocean at a beach in Ventura created by the sandy sediment the Creek once carried downstream. After the Dam's construction, the sandy beach disappeared. The Dam also prevented the endangered steelhead trout from reaching their spawning beds. In addition, the Dam's concrete had severely deteriorated so that the risk of failure was substantial. The Dam was deconstructed in 2023. The flood control problem is now being handled by some downstream levees and floodplain areas. We are now learning that dams do not often provide the answer to managing our river problems, particularly when ecological destruction is the cost.

The Saranac River's Dams

Our history of building America's dams demonstrates a cavalier attitude toward the natural environment. In the nineteenth century, American society ignored the apparent environmental degradations because of America's natural abundance. Why worry about destroying the ecology of a single river when other streams were healthy? These single degraded rivers then became numerous. We became familiar with converting our Nation's streams into "working rivers" for industrial reasons. (See Chapter 1 for examples.)

In 1822, the naturalist and at that time the future Governor of New York, Dewitt Clinton, commented on the extraordinary salmon runs he observed in the early 1790s on the Saranac River near where it empties into Lake Champlain.[4] But in 1796, Zephaniah Platt—the founder of Plattsburgh—constructed the Saranac River's first dam three miles upstream from Lake Champlain at the site of what was to become the Imperial Mills Dam. (See Figure 10.4 in Chapter 10.) In 1800, however, New York established a law to protect salmon spawning runs by requiring fish passages "for any and all dams." The Imperial Mills Dam did not have a fish passage. Consequently, New York State sued Platt to restore the salmon runs. Unfortunately, in 1819, Platt won the case on the grounds that the dam was entirely on private property. As a result, the barrier remained without fish passage, and the Atlantic salmon ceased

[4] See Dewitt Clinton's (1822), "Letters on the Natural History and Internal Resources of the State of New York," E, Bliss & E. White, New York, NY.

migrating to their upstream spawning grounds on the Saranac. After two-hundred and thirty years, however, the *Lake Champlain Basin Program* now seeks to restore salmon and other fish to the upper Saranac. Consequently, the NYS Department of Environmental Conservation (DEC) is now installing a modern fish ladder at the Imperial Mills Dam. Its planned completion year is 2026. As a result, combined with the removal of both the upstream Indian Rapids Dam and the Fredenburgh Falls Dam, plus a recently installed fish ladder at Treadwell Mills Dam (all on the Saranac), for the first time since 1796 the upper Saranac River might soon be opened to salmon and other migrating fish.[5]

Some fishermen and other naturalists are consequently pleased by these restoration prospects; but not all are pleased. Trout Unlimited, the *environmental advocacy organization* (EAO), is arguing that the total removal of the Imperial Mills Dam is preferred.[6] Since 2022, New York State has classified this dam as "high hazard" and "unsound." It follows that the Dam is unacceptably dangerous to the City of Plattsburgh, and that it is also a barrier to fish migration despite the installation of a new fish ladder.

At a minimum, therefore, the Imperial Mills Dam will have substantial safety upgrades over the next few years. This is also true of the upstream Bartlett Carry Dam. In addition, prior to 2027, four of the eight hydroelectric dams on the Saranac—Lake Flower, Franklin Falls, Union Falls, and Treadwell Mills—are all up for *Federal Energy Regulatory Commission* (FERC) relicensing review. Four other Saranac River dams have been relicensed since 2000. But the upcoming relicensing examinations do pose possibilities for substantial changes in the Saranac's ecology. Waiting for safety improvements to be approved and financed, especially at the Imperial Mills Dam, is not within the spirit of the standards of the *Association of State Dam Safety Officials* (ASDSO), i.e. the body of safety requirements we now apply to dams.[7] In Particular, waiting means asking the citizens of Plattsburg to endure the danger of a disastrous dam failure over whatever time is required to bring this dam up to safety code.

[5] It should be noted that since the 1970s, the US Fish and Wildlife Service have been stocking some land-locked salmon from Lake Sebago into Lake Champlain.

[6] See Lynch, Mike (2023, September 11), "Work underway to remove remnant dams for salmon," *Adirondack Explorer*, at www.adirondackexplorer.org/stories/work-underway-to-removw-remnant-dams-for-salmon.

[7] This Association has organized our current safety standards for dams. See Chapter 10.

An interesting question, however, concerns the 1819 New York State Court decision which declared that a dam that spanned the Saranac River was on private property, and that therefore it was legal without any prior public approval before construction, i.e. State approval. The Court's vision was that the Saranac River was not a public common property subject to public control. This was clearly a pernicious view of society's natural assets.

Rivers are common property. Historically they have been widely used for public water, irrigation, power generation, and unfortunately pollution dissipation. All of these uses have the potential for degrading a river; the pollution dissipation use always has a degrading effect. Public control is therefore necessary for avoiding the tragedy of the commons phenomena that afflicts most of our rivers. In their natural states, rivers are ecological assets that are weakened to some extent by any of the above listed uses, especially when these uses are in combination. Dams not only facilitate these uses, they also interrupt aquatic migrations, they segment watersheds, and they prevent silt from flowing downstream to the rivers' natural deltas. They destroy the ecology that makes them society's valuable natural assets.

Rivers can also cause floods, and dams are often justified as necessary for flood control. But no dam lasts forever! They become filled by silt; they age and deteriorate, and if allowed, they fail. That is how particularly dangerous floods occur, the ones that wash away towns and sweep people and structures downriver. The average age of America's dams is over fifty years, and dams that old often need serious maintenance.[8] Safety concerns are justifiably on the rise.[9] Identifying dangerous structures, and either removing or rehabilitating them is of paramount importance.

But this is now the age of the hydrologist. Our ability to analyze a watershed as a whole entity to decide where dams might be best placed, and what their optimal flood control design should be, and where floodplains should be established, are now our tasks. For example, the Penobscot River Watershed of Maine had large numbers of dams constructed at sites unrelated to flood control as though the sites were selected independently without consideration of other dams.[10] Currently,

[8] See footnote 1.

[9] See Chapter 10.

[10] See Chapter 1.

these obstructions are either being removed or modified since their functions are now outdated. Flood control requires a coordinated watershed hydrological approach. This is especially essential given our age of climate change and current experience with storms of greater precipitation. ("Hurricane Helene" of late September, 2024, is an example.)

Dams were constructed on our rivers for each of the purposes listed above, and frequently for combinations of these purposes. They were also constructed in part for public employment projects. They were not constructed with ecological consideration in mind. We are now in the age of general ecological renaissance, so it is natural that we deconstruct our outdated, dangerous, and ecologically destructive dams. Today, given the rapid rise of wind power and solar-cell energy, and our new large-scale electric lithium-ion battery storage, we often deconstruct even hydroelectric dams with little opposition. As a result, there is a dynamic currently working, i.e. when people begin to engage in natural interactions, they demand more. They now demand to live and recreate in cleaner environments. After two centuries of industrial degradations, our society is moving rapidly in another cleaner direction.

The Structure of the Book

Six chapters in this volume (1–3, 5, 6, and 10) review the construction, modification, rehabilitation, and deconstruction of big dams. These removals tend to be in our over engineered watersheds such as California's Central Valley, the Tennessee River Valley, and the Columbia River Watershed. Other chapters focus on small "low-head dams" that were mostly constructed during the nineteenth and early twentieth centuries. These are most often derelict and abandoned dams that seriously upset the ecology of our watersheds. They are now being removed, but the problems they pose remain massive. The Northeastern US still has many thousands of these old "low-head" structures. Their pernicious effects, and their political difficulties in removal, are reviewed in chapters 7–9. Some other chapters focus on the ecological renewal and preservation of our rivers for the purpose of promoting our natural interactions, recreation, and cleaner healthier domiciles. This relatively new recreational demand has also fed our political movement for dam removals as reviewed throughout this book, but particularly in Chapter 11.

This book cannot review all the deconstructions and reconstructions currently ongoing in the US. This renewal movement is too large for a complete examination. But by reviewing the selected rivers and their histories, the reader may obtain a broader understanding of this very broad movement, e.g. an understanding of the politics of dam removal, particularly of the low-head dam aspects, the flood control aspects, the irrigation aspects, and the hydropower aspects. This volume reviews the recent increase in recreational demands for cleaner, ecologically sound, and unobstructed rivers, and also the changes in demand for our rivers to be largely just transportation arteries. All of this represents a considerable component of our greater environmental movement. The structure and nature of this component is what this book seeks to communicate.

We know that this era of climate change poses the most significant challenges for our environmental movement. The effects of this global conundrum on our rivers and their restoration must be, and is a consistent focus of the examinations presented in the thirteen chapters of this volume. Most notably, I review the political movement within our government agencies for a significantly reformed evaluation methodology for water projects be they dam removals or rehabilitations or other efforts that affect our rivers. This latter reform effort has roots going back to the formation of the *Federal Water Resources Council* in the mid-1960s, and it marks a search for correct philosophical foundations for society's water project evaluation methodology that are not built on raw utilitarianism or the mere exercise of the power of vested interests. This philosophical review might at first appear to be simplistic, but I firmly believe it is not. Ethical philosophy is an exercise in organizing one's thinking, and when properly organized, it appears simple. But it is a substantive task to achieve this organization. Having instructed courses in business ethics and environmental ethics for years, I know that this organizational journey is worthwhile and satisfying.[11] I also believe that our procedures for water project evaluation are indeed moving in the moral direction.

For particular acknowledgment, I certainly thank the efforts of Palgrave Macmillan's very competent and creative editor Bronwyn Geyer. Editing a volume of this size and content is a difficult task, but Bronwyn is diligent and persistent. Her efforts made this book significantly better

[11] See Robinson, Richard M. (2024), *Business Ethics: Kant, Virtue, and the Nexus of Duty*, Second Edition, Springer Texts in Business and Economics. This text includes substantial material concerning the philosophical foundations of ethics, including chapters concerning environmental ethics.

than it would otherwise have been. Nonetheless, its defects are my responsibility only.

The figures presented at the end of each chapter provide rough mappings of the rivers and dams addressed in each. These are meant to only indicate the relative positions of the water bodies and structures referred to, and are not meant to accurately indicate distances or shapes. They might, however, help the reader to understand the locations of the bodies addressed in each chapter.

North East, PA, USA Richard M. Robinson
February, 2025

Contents

About the Author

Dr. Richard M. Robinson is a long serving Professor of Business at SUNY Fredonia. He received his Ph.D. from the University of Oregon (Economics) in 1981. This is his fourth recent book published with Palgrave Macmillan: *Environmental Advocacy and Local Restorations* (2023), *Environmental Organizations and Reasoned Discourse* (2021), and *The Imperfect Duties of Management* (2019). He also recently authored *Business Ethics: Kant, Virtue, and the Nexus of Duty* in 2022 (Second Edition in 2024) with Springer Texts in Business and Economics. He has published numerous peer-reviewed articles in scholarly journals, many of which include environmental concerns. He is former editor of the *Journal of Economics and Finance,* a Springer journal. He instructs courses in business ethics, environmental ethics, and statistics. His web page is at "profrichardrobinson.com,"

Abbreviations

ACE	US Army Corps of Engineers
APWC	Applegate Partnership Watershed Council
ASDSO	Association of State Dam Safety Officials
BPA	Bonneville Power Administration
BOR	Bureau of Reclamation
CAFO	Concentrated Animal Feeding Operations
CAP	Central Arizona Project
CBA	Cost-Benefit Analysis
CEP	US Council on Environmental Policy
COPCO	California Oregon Power Corporation
Corps	US Army Corps of Engineers
CPR	Common Property Resource
CRWA	Charles River Watershed Association
CVP	Central Valley Project (California)
CWA	Clean Water Act
DEC	NYS Department of Environmental Conservation
DEEP	Connecticut Department of Energy and Environmental Protection
DNRC	Montana Department of Natural Resources and Conservation
DWR	California Department of Water Resources
EAO	Environmental Advocacy Organization
EDF	Environmental Defense Fund
EIS	Environmental Impact Statement
EPA	US Environmental Protection Administration
EQ	Environmental Quality Account Measures
ESA	Endangered Species Act
ESU	Evolutionary Significant Unit

EWC	Edwards Manufacturing Group
FEMA	Federal Emergency Management Agency
FERC	Federal Energy Regulatory Commission
FishPass	Fish Pass Technology
FWS	US Fish and Wildlife Service
GE	General Electric
GLCA	Garnet Lake Conservation Association
GLRI	Great lakes Restoration Initiative
GTB	Grand Traverse Band (Chippewa or Ojibwe)
HHU	High Hazard Unsound Dams
HRCCC	Housatonic River Conservation Coordinating Committee
HVA	Housatonic Valley Association
IID	Imperial Valley Irrigation District
IRWA	Ipswich River Watershed Association
KHDG	Kennebec Hydro Development Group
MDEQ	Michigan Department of Environmental Quality
MDER	Massachusetts Department of Ecological Restoration
MDMR	Main Department of Marine Resources
MDNR	Michigan Department of Natural Resources
MEPA	Massachusetts Environmental Protection Administration
MICSA	Maine Indian Claims Settlement Act
MPPA	Mill Pond Preservation Association
NED	National Economic Development Benefits
NEPA	National Environmental Policy Act
NFWF	Northwest Fish and Wildlife Foundation
NGO	Nongovernment Organization
NID	US National Inventory of Dams
NMFS	US National Marine Fisheries Service
NOAA	US National Oceanographic and Atmospheric Administration
NPS	US National Park Service
NRCM	Natural Resources Council of Maine
ODEQ	Oregon Department of Environmental Quality
OPEC	Organization of Petroleum Exporting Countries
ORV	Outstanding Remarkable Value (an NPS Designation)
P&G	US Army Corps of Engineers "Principles and Guidelines" for CBA
PAH	Polycyclic Aromatic Hydrocarbons
PCB	Polychlorinated Biphenyls
PDEP	Pennsylvania Department of Environmental Protection
PGE	Pacific General Electric
PPL	Penobscot Power and Light
PRRP	Penobscot River Restoration Project
PRRT	Penobscot River Restoration Trust
REA	US Rural Electrification Administration

RNA	Regulated Navigation Area (US Coast Guard)
SOLC	Southern Oregon Land Conservancy
SMUD	Sacramento Municipal Utility District
SWP	California State Water Project
SWRCB	California State Water Resources Control Board
TVA	Tennessee Valley Authority
UNESCO	United Nations Educational, Scientific and Cultural Organization
USDA	US Department of Agriculture
USGS	US Geological Survey
VDEC	Vermont Department of Environmental Conservation
VPUC	Vermont Public Utility Commission
WPA	US Works Progress Administration (programs of the 1930s)
WRC	US Water Resources Council

List of Figures

List of Tables

CHAPTER 1

An Environmental Movement Begins: Maine's Rivers

The Festival of the Kennebec and the First Undamming

A small festival formed in the early light of July 1, 1999, when more than a thousand spectators gathered on the banks of the Kennebec River near the Edwards Dam in Augusta Maine. (Figure 1.1 shows a schematic map of Maine's river locations.) The gathering's purpose was to celebrate the dam's imminent destruction. Located at the "run-of-the-tide" of the Kennebec's estuary, it was erected in considerable environmental controversy in 1837.[1] To use the descriptive language of the age when this dam was constructed, the Kennebec was transformed into "a working river." Its purpose was to power riverside sawmills that were eventually destroyed by fire in 1968. But before that, in 1882 the dam was transformed to power a large textile mill that actually operated until the early 1980s. The Edwards Dam was even converted a third time in 1913 to produce hydroelectric power. But in that early July morning of 1999, the dam was finally to be removed, and the river would run free after more than 160 years of obstruction.

Just prior to its deconstruction, the Edwards Dam provided one-hundredth of one percent of the total electrical power of Maine, but it

[1] The "run-of-the-tide" indicates the farthest point upstream reached by the estuary's tide.

R. M. Robinson, *Restoring America's Rivers*,
https://doi.org/10.1007/978-3-031-81758-8_1

charged exorbitant rates to the consumers (several times above Maine's going market price). The ecological benefits to be garnered from the dam's removal were judged by federal and state regulatory authorities to far exceed the minor amount of electrical-power generation the dam provided in 1999, and so on July 1, the Edwards Dam was to be breached and removed.

On the day of removal, on the southwest side of the dam a large diesel-powered backhoe maneuvered to be ready to create the breach. On the northeast side, a small stage was constructed close by the river to accommodate speeches from dignitaries—principally Maine's Governor Angus King, and US Secretary of the Interior Bruce Babbitt.[2] A small acoustic band provided festive music through amplifying speakers, and a helicopter hovered over the area providing noise slightly louder than the flow of the river. Television news cameras stood by to record the historic event. The first US hydroelectric dam to be decommissioned was to be deconstructed thereby providing the rebirth of a significant river.

Governor King's speech was expansive.[3] He reviewed the lengthy process and negotiations that led to this historic event for the environmental movement. He was clearly optimistic about the future of river restorations. He knew that more was about to occur to restore Maine's rivers. Secretary Babbitt spoke of the day's historic event as showing "respect for God's creation!" But Babbitt was reticent to claim this undamming foretold future dam removals, particularly among the large dams of the American West that were then the focal point of America's emerging river restoration movement. Babbitt focused his remarks on that day's historic event; not on the potential for initiating Western US dam removals as anticipated by the environmental advocates who brought about the undamming of the Kennebec. July 1, 1999, was still very early in the dam removal movement. There were many more dam removals to come.

At 9:30 AM, the backhoe dragged a large section of the dam from its old foundation. As though the river came alive with celebration, the Kennebec flowed through with a roar louder than the helicopter above,

[2] Other dignitaries included the US Department of Commerce Assistant Secretary Terry Garcia, and the heads of the four environmental organization that composed the Kennebec Coalition, i.e. the alliance that orchestrated the political movement for the dam's removal.

[3] See "Angus King at Edwards Dam," at www.bing.com/videos/search?q=video+of+angus+king+at+Edwards+Dam.

and louder than the cheers of the streamside celebrators. Shortly thereafter, fish runs returned quickly to seventeen miles of the upper river. Within a few days, this first significant undamming began demonstrating the river's resilience in restoring its natural health while avoiding all the disasters that the nay-sayers predicted. This first undamming did provide momentum for future restorations. The environmental movement drew energized direction from this event.

The Kennebec River and Its Dams

The Kennebec flows from Moosehead Lake in the central interior of Maine. The river flows 150 miles with hundreds of tributaries adding water before it enters Merrymeeting Bay below the City of Augusta. This freshwater bay was once—and after restoration it is now again—one of the most alive estuaries on the Atlantic coast (see Figs. 1.1 and 1.2).

The Kennebec's watershed drains approximately one-fifth of Maine's area. Prior to the damming, Kennebec's abundant anadromous fish runs included Atlantic salmon, rainbow smelt, blueback and alewife herring, American shad, striped bass, and both short-nosed and Atlantic sturgeons. The American eel, which is born and spawned in the ocean, but spends its life in freshwater rivers, was also abundant in the Kennebec. The Edwards Dam, along with sixty-nine other dams within the watershed, eliminated these runs. They also eliminated the water's aeration and its natural downstream flow of nutrients and sediments. During the nineteenth century, intensive farming within the watershed, plus industrial development, all helped to destroy the river's ecological health.

In the late eighteenth century, Maine's shipbuilding utilized the watershed's tall and straight-mast pines which fueled the Kennebec's timber industry. Within a decade of the original dam's completion in 1837, seven sawmills were constructed close to the dam. A gristmill and a machine shop were also hydro-powered by the dam. In 1882, a large textile mill was constructed next to the dam site. This mill employed 700 workers mostly from the nearby French-Canadian community. The majority of the workers were young females who worked the looms. This factory was owned by the Edwards Manufacturing Company, and thereafter the dam was called the Edwards Dam. An ice business was also established behind the dam, one that produced over a million tons cut from the river in the year 1894. The ice blocks were shipped in boats from Augusta to the Boston metropolitan area.

Prior to all this industrialization, an acrimonious political battle concerning whether to even build the dam began in 1834. Multiple constituencies petitioned Maine's Legislature in Augusta concerning whether to construct this obstruction of the Kennebec's natural flow. The logging industry wanted the proposed dam to power its sawmills. The area's strong fishing industry opposed it. The opponents lost. The dam was constructed in 1837, and Augusta became Maine's center of industrialization. But Augusta's new urbanization meant raw sewerage was dumped into the river. The log drives covered the river's bed with bark and wood. The mills filled the river with dioxins.[4] The untreated sewerage filled the river with human effluent. Bacteria spread and depleted the oxygen; the fish and other life could therefore not survive in the river's waters which smelled of effluent and degradation. Until the 1970s, falling into the river posed a significant health risk.[5]

The dam was constructed of logs and rock, and stood 917 feet long and 24 feet above its downstream water level. It was breached several times during flood seasons, the most recent being in the winter of 1974 when Thomas Squiers, a fish biologist with the *Maine Department of Marine Resources* (MDMR), argued that the breach should be allowed and maintained so that fish runs could move upstream. The breach, however, was repaired. But in the 1970s, the American environmental movement was beginning to question the economic-environmental rationale for dams.

The political impetus for cleaning our rivers began with the environmental movement of the 1960s. In particular, the demand for cleaner waters was accelerated with the *Clean Water Act of 1972* (CWA) from which federal funding for water treatment began. Between 1972 and 1990, millions in federal dollar funding were spent by Maine on sewerage treatment which substantially cleansed the Kennebec. The river began to move toward ecological health.[6] Consequently, the political question for Maine became, "Why leave an old ineffectual dam system in place that prevents our fishery's restoration?".

[4] For a review of dioxins and their effects, see the Appendix to Chapter 7.

[5] See Steven Brooke's interview in Grossman (2002, pp. 12–13).

[6] For example, Federal wastewater development projects for Maine in 2024 were over $40 million. See www.collins.senate.gov/newsroom/congress-approves-more-for-maine-wastewater-infrastructure-secured-by-senator-collins.

In 1985, the MDMR offered a plan for significant fish restoration in the Kennebec—*The Strategic Plan for the Restoration of Shad and Alewives to the Kennebec River above Augusta*. The MDMR's "Plan" asked the owners of the Kennebec's dams to fund fish traps, trucks, and release programs to move shad and alewives above critical dams. The Kennebec Hydro Developers Group (KHDG)—representing the owners of the dams in the watershed—all signed agreements to support the *Plan* with the exception of the Edwards Manufacturing Company (EMC) who owned the Edwards Dam. But this was the most important dam in the watershed for promoting the revival of the river. The EMC assumed that government would continue to support narrow-vested business interests over the broader economic and ecological interests of the area. But the EMC's assumption was incorrect; the politics of dam support was changing. A coalition of sportsmen and environmentalists began to declare their intention to support removal of the Edwards and other Kennebec dams.

Because it was a hydroelectric dam, the Edwards Dam needed relicensing from the *Federal Energy Regulatory Commission* (FERC). Those hydroelectric dams that are not owned by the federal government are licensed and regulated by FERC—a *Commission within the US Department of Energy*. The licensing is for either 30 or 50 years. To be relicensed, a dam must meet safety standards, and also the standards of the *Clean Water Act* (CWA), the *Endangered Species Act*, the *Electric Consumers Protection Act*, and the *National Environmental Policy Act*. Relicensing requires that the dam must be evaluated by FERC as "in the public's interest." In 1986, Congress required that FERC's hydropower licensing process must give equal consideration to the dam's effects on fish, wildlife, and recreation valuation. This requires that FERC evaluate environmental and recreational effects of the dam as well as the electric power benefits. Is the dam detrimental to water quality and ecological health? EMC reapplied for relicensing in 1993.

Dam owners must declare their intention to apply for relicensing five years prior to expiration. The required information submitted to FERC (safety information plus economic and environmental information) must be available to the public. Interested parties can then file a "Motion to Intervene" with FERC, which means the interveners can then submit additional information and analyses to be considered by FERC. The environmentally oriented *Kennebec Coalition* opposed the relicensing. Its members included *The Natural Resources Council of Maine*, *The Atlantic Salmon Federation*, *Trout Unlimited*, and *American Rivers*.

In 1990, Maine Governor McKernan announced he supported removal of the Edwards Dam even if the state had to use eminent domain to accomplish the task. In March, 1990, the environmental organization *American Rivers* announced its support for McKernan's plan for negotiating the Dam's turnover, and it asked FERC to deny relicensing. The Governor then requested that FERC deny relicensing after EMC appeared to cease "negotiating in good faith." The removal effort then turned political with the Kennebec *Coalition* emphasizing public education about river restoration, rather than only focusing on dam removal. The *Coalition* emphasized the historical fish numbers and wide variety of species negatively affected by the dam. The *Coalition* also emphasized the potential return of eagles and other animals that would feed on the fish provided the dam was removed. The *Coalition* provided 7,000 pages of well-researched reports of environmental analyses to FERC, and it also organized large turnouts at FERC's public hearings on relicensing. The turnouts were largely from members of *Trout Unlimited* who were well prepared, respectful, and patient.

FERC's initial draft of its *Environmental Impact Statement* (EIS) called for maintaining the dam but with fish passages, a solution that failed in the past because either the passages were inadequate or not installed despite assurances that they would be. This disappointing draft EIS motivated the *Coalition* to hire consultants to work with federal agencies in order to pose stronger arguments for dam removal. The dam supporters' counter argued that the removal would be too costly, would cause riverbank erosion, and an increase in flooding.

The *Coalition* emphasized that the dam was very old and unstable, that it only produced 3.5 megawatts of power, and that it severely prevented the Kennebec's potentially valuable fisheries from thriving. The benefits of restoring the river outweighed the small amount of electrical power generated. The Augusta City Council supported relicensing since the City derived tax revenues from the Dam. The Council thought it was supporting business. There were also other concerns expressed by the City Manager, namely the rights of the riverside landowners along the upper seventeen miles above the dam. Would the water levels continue to reach the borders of their property? Removing the dam would be unprecedented, and it frightened the City Government.

In his 1994 campaign for Governor of Maine, Angus King committed to remove the Edwards Dam. Governor King (elected in 1994) and his Planning Director had allies in the *Kennebec Coalition*. They recognized that the Edwards Dam did not make economic or ecological sense. The state of Maine provided reports to FERC stating these claims, and the *Coalition* also provided its reports with similar analyses. The City of Augusta continued its opposition until its new City Manager (Bill Bridgeo) posed a vision of downtown development that relied on the restored Kennebec. Eliminating the Dam was the central component of this revitalization. The removal's opponents, however, posed dire predictions about the possible negative effects:

- Substantial industrial contaminants accumulated in the silt built up behind the dam, and these would be released by the removal.
- Removal of the dam would create mudflats along the drained reservoir, and the lowered river level behind the dam's current position would decrease the value of riverside homes.
- Removal would also facilitate downstream flooding.
- Access to the river for the communities that constructed piers and docks into the reservoir would be eliminated by dam removal.

But public education played the critical role that culminated in FERC's two-to-one vote against relicensing the Dam. Nonetheless, the removal of the dam was yet to be accomplished.

Getting to Removal

The FERC decision did not by itself remove the Edwards Dam. Together with the National Hydropower Association and the Edison Electric Institute—both national trade organizations—Edwards Manufacturing Company (EMC) appealed the FERC decision. They challenged in federal court trying to force FERC to either vacate its decision, or short of that, to arrange compensation for "taking the property." Other parties joined the appeal because they feared the precedent for future FERC relicensing procedures given the environmental concerns. The various dam owners feared they would be trapped with the removal costs if they did not obtain relicensing.

In additional support of the hydroelectric industry, four conservative Republican members of the US Senate's "Energy and Natural Resources Committee" challenged the FERC decision. All four were opponents of the environmental movement. Their claim was that the FERC decision comprised "a federal taking." Such a claim ignores the dams' taking of public property without compensation (as was the case of the Edwards dam), and that the dam's generation of negative externalities is also "a taking" when forced on the public.

Achieving a timely removal of the Edwards Dam would require a solution that benefitted all affected parties. To accomplish this, the *Kennebec Coalition* and the state government offered a solution that would avoid further lengthy litigation. Maine's Planning Director Evan Richert announced immediately after the FERC decision that the state wanted to "find ways" for EMC "to meet their obligations." (Cheever, 1997.) The state and *Coalition* sought to expedite the restoration. Governor King appeared before Augusta's City Council to present the State's "plan." The agreement between Augusta, the State of Maine, and EMC was signed on May 8, 1998. It protected EMC from legal liability during the removal and utilized funds from a settlement with a large shipbuilding company (the downriver Bath Iron Works) to finance the removal. The agreement also required a last payment of $250,000 from EMC to the City, and provided some state funding for Augusta's revitalization. With this offer, all parties agreed to the undamming.

The Removal Day Again

In Augusta, at 9:30 on the morning of July 1, 1999, Governor Angus King rang a bell to start the undamming process. Bells from a nearby Saint Augustine's church then began ringing as the backhoe dug a sixty-foot breach in the dam. The church belonged to the old French-Canadian immigrant community that provided the workers who built the dam and who still represented a large segment of Augusta's population. The author John McPhee described the festivities:

> Schools of striped bass had been seen by construction workers nosing up to Edwards Dam that morning. A big Atlantic salmon broached, and two feet out of water it just hung there watching. A high-school bell was mounted on a post near the microphones, and Babbitt, King, and others – on schedule with the speeches and the downstream recession of the morning

> tide – began to ring it, and ring it. Evan Richert of the Maine State Planning Office shouted into the microphones, "Set the fishery free!" (McPhee, 2002, p. 77)

As Interior Secretary, Bruce Babbitt had already endured a continuum of dam-related conflicts and failures. Reflecting on the significant press coverage gathered for the day's event, Babbitt commented,

> They're not just coming here to celebrate good news. I've got to tell you that's not what the American press is all about. ... They're here because they know that this is the beginning of something that is going to affect the entire nation. (Paulson, 1999, 5 July. "SeattleP-I.com")

Secretary Babbitt then refused to respond further to the reporters' inquiries concerning the implications of the Edwards Dam's removal for other potential dam deconstructions. How would the FERC decision affect other removal possibilities, especially those involving the large dams of the American West? The answer would in part depend upon the subsequent environmental success of the Edwards removal for restoration of the Kennebec. But considering the effects in Augusta, Mayor Roger Katz lauded the effects of the dam removal on the city. He particularly pointed out the development of a park and river access at the old mill site. This now popular park features a summer carnival, a weekly farmers' market, a wooded riverfront nature trail, and access for paddle recreation to the river.

The Healthy Kennebec

Since the removal of the Edwards Dam in 1999, the fish have returned to the seventeen miles of the Kennebec between Augusta and Waterville, Maine. But four dams (Waterville's Lockwood Dam and Hydro-Kennebec Dam, Fairfield's Shawmut Dam and Skowhegan's Weston Dam) blocked access of Atlantic Salmon, striped bass, shad, and herring from their prime spawning grounds of the Sandy River, a tributary of the upper Kennebec (as depicted by Fig. 1.2). These four dams were owned by the Canadian company Brookfield Renewable Energy. They provided six percent of Maine's hydroelectric power. But the Sandy was the only remaining essential spawning ground for the Atlantic salmon, a species on the brink of extinction.[7] In 1998, the Kennebec Coalition reached a legal agreement with Brookfield to provide adequate fish passages past the dams so as to facilitate rapid restoration of the salmon and other fish. But in 2020, the anadromous fish still had no access above the Lockwood Dam because of the failure of Brookfield to provide these fish passages.

In 2021, Brookfield was a $60 billion (US $s) multinational corporation with 6,000 power generation facilities worldwide, of which 222 were hydropower facilities. Its four dams between Waterville and Skowhegan were in violation of its federal operating permit since 2019 due to their killing of the endangered Atlantic Salmon. Brookfield had not made adequate remediation attempts. In 2021, the *Atlantic Salmon Federation, Conservation Law Foundation, Maine Rivers,* and the *Natural Resource Council of Maine* (NRCM) sued Brookfield for repeated violations of the *Endangered Species Act*. A year earlier, federal agencies had already rejected Brookfield's proposed "Species Protection Plan" for the salmon.

A newly proposed solution by the *Kennebec Coalition* was to deconstruct the Shawmut Dam in Fairfield, and at a minimum, to provide new and adequate fish passage facilities at the other three dams operated by Brookfield. Previously, other dam removals occurred at Winslow, Waterville, and farther upriver. Removal of the Shawmut Dam is still under adjudication. The Federal Court rejected Brookfield's appeal on July 8, 2024.[8] (The Shawmut Dam's location is depicted by Fig. 1.2). The Madison Electric Works on the Sandy River, a significant tributary above Waterville, was removed in 2006. The Fort Halifax Dam on the Sebasticook River, the largest tributary of the Kennebec, was removed in 2008. (Figure 1.2 depicts the locations of these dams.) These dam removals opened the upper river to fish runs. The run of herring on the Kennebec is now the largest in North America. Between 2009 and 2020, thirty-six million alewives have reached their spawning grounds above Benton Falls on the Sebasticook, and the largest concentrations of Bald Eagles on the US East Coast is now observed feeding on the herring runs on the Sebasticook River (as documented by NRCM, 2020). Nevertheless, the problems with the Atlantic salmon persisted.

[7] See www.fisheries.noaa.gov/species/atlantic-salmon-protected. See also *Natural Resources Council of Maine* "Restoring the Kennebec River above Waterville," www.NRCM.org/BLOG/.

[8] See www.eenews.net for 7/08/2024.

Environmentalist Steven Brooke praised the return of Bald Eagles (190 were counted in 2020), and the sight of them catching fish during the large fish runs on the Sebasticook. The half-pound alewives (a herring) spawn upstream in May and June. They are essential elements in the food chain. Gail Wippelhauser, a biologist for MDMR, says that after the Edwards and Fort Halifax Dams were removed, the alewife population exploded from 400,000 to more than five million. The alewives and blue-back herring are now commercially caught for Maine's lobster bait. The shad, striped bass, and eels are all back spawning on the Sebasticook.

Overall, the Kennebec is now closer to being a healthy ecosystem rather than an industrial-waste sewer. The water quality has dramatically improved with the increased flow and concomitant oxygenation. Aquatic insects, a key indicator of water quality, have returned. The striped bass, shad, and alewives quickly moved in large numbers to the upstream dam in Waterville, and then beyond when that dam was removed. Studies of the river above Augusta indicate significant increases in spawning and observed fish species.

The Penobscot River and Its Restoration[9]

Maine's Penobscot Bay is fifty miles north of Kennebec River's Merrymeeting Bay. The Penobscot River flows into Penobscot Bay after draining central Maine. (The River's location is depicted by Figs. 1.1 and 1.3.) Residing along that river, the *Penobscot Nation of Native People* have occupied central Maine for over 5,000 years. In their language, the name "Penobscot" is a mishearing of the Algonquin term that means "people of the extending white rocks." This river's main stem is 109 miles long, but together with its West and South Branches, its length is over 260 miles. Its watershed drains approximately 8,600 square miles. The *Penobscot Nation* provides the major restoration constituency for advocating dam removal and habitat restoration along this river. This *Nation* has made river remediation an issue of environmental justice, the journey of which has been long and difficult.[10]

[9] This section is partly based on Robinson (2023), Chapter 10, *Sect. 1.3.*

[10] Walls (2017, Chapter 10) explores Thoreau's mid-nineteenth exploration of this issue with respect to Maine's *Penobscot Nation.*

The Penobscot's banks were once heavily wooded. For more than two-and-a-half centuries, these riverbanks provided a substantial source of wood and pulp to Maine's economy. It is now being restored to once again providing a significant fishing area for Atlantic salmon, smallmouth bass, short-nose sturgeon, Atlantic sturgeon, shad, herring, smelts, and bottom fish that provide food for eagles, bears, and otters, and that also provide a significant source of sustenance for the *Penobscot Nation of People*.

The *Penobscot Nation* is part of the *Wabanaki Confederacy*.[11] This *Nation* includes 2,398 members as of their 2020 census, of whom 1,399 live in Maine, and 417 members live on the Nation's reservation at Indian Island near Old Town on the Penobscot River. This indigenous people have occupied Central Maine along this river for thousands of years. Burial sites in this area associated with an Algonquian speaking culture are 5,000 years old. The earliest records of land purchases from the indigenous people of the area date from 1639. In 1763, Massachusetts Bay took possession of all the islands and riverbank lands below the "head of the tide" of the Penobscot River, i.e. below Bangor. In 1794, Massachusetts negotiated a treaty with the Passamaquoddy People (an Algonquin tribe immediately north of the Penobscot along Maine's border with New Brunswick). This led to the Penobscot ceding 200,000 acres along the River. In 1818, the Penobscot ceded the remaining islands except for some small townships. Maine separated from Massachusetts and became a state in 1820 and assumed all of the obligations inherited from Massachusetts. In 1833, Maine purchased the townships and relegated the Penobscot People to Indian Island in the river just north of Old Town.

If the indigenous peoples of North America are, and were, separate nations, as has been the historic approach of the US federal government, then it should have been the federal government's task to negotiate treaties with these peoples. The individual states should not expropriate indigenous lands. But none of these land acquisitions occurred under the auspices of a federally ratified treaty. They occurred only under Maine's jurisdictional responsibility, i.e. not federally organized and recognized. In

[11] The Penobscot Nation is now a federally recognized indigenous tribal nation. The *Wabanaki Confederacy* also includes the Abenaki, the Passamaquoddy, the Maliseet, and the Mi'kmaq Nations, all of whom speak Algonquin languages.

1790, the US Congress passed the "Indian Trade and Intercourse Act"—commonly called the "Nonintercourse Act"—to eliminate exploitation of indigenous lands. But since the federal government took no actions against states for violating the Act, it was presumed that this Act did not protect Maine's tribes from abuse. By the mid-1830s, the Penobscot Nation had much of its territory expropriated. It retained only the Penobscot River and islands north of Indian Island. The state of Maine appointed Indian Agents to exercise total control over the reservation's food, clothing, shelter, and health care, all of which were purchased with funds from the sale of tribal lands. The Penobscot people thereafter lived bare subsistence lives. In the late nineteenth century, The Maine Supreme Court held that the Penobscot and Passamaquoddy peoples were not indigenous tribes and had no aboriginal rights.

In 1971, Tom Tureen and Francis O'Toole published an article in the *Maine Law Review* which argued that the area's indigenous tribes must fall under the federal and not state jurisdiction. Both of these tribes of Maine (the Passamaquoddy and Penobscot) had grievances about the state's management of tribal trust funds, the administration of hunting and fishing rights, and about tribal disenfranchisement. Tureen and the *Native American's Rights Fund* represented the Penobscot and Passamaquoddy in a suit filed in federal court in 1972 to reclaim tribal lands. The primary issue was that these two tribes were not recognized by the federal government. The federal First Circuit Court held that the plain meaning of the Act applied to any tribe, whether federally recognized or not. Furthermore, the Court found that the "Nonintercourse Act" requires the federal government to take a fiduciary role in protecting the lands of these tribes.

The US federal government agreed with the US District Court. The Passamaquoddy and Penobscot were federally recognized in 1976 thereby each becoming eligible for $5 million per year in housing, education, health care, and other services from the Bureau of Indian Affairs. Furthermore, the US Interior Department began to proceed under the Court's decision concerning these expropriated lands.

The 1790 "Nonintercourse Act" established that only the federal government could negotiate with independent nations, including indigenous nations. One-hundred and eighty-five years later, in January of 1975, the US District Court ruled that the "Nonintercourse Act" applied to the Penobscot and other Indigenous Nations. This opened the issue of whether it was legal for the states to appropriate indigenous lands. But finding that the State of Maine's actions violated the 1790 "Nonintercourse Act" was only a first step toward a just solution for Maine's indigenous peoples. Millions of acres in Maine were populated by hundreds of thousands who now needed to have their land titles adjudicated. The tribes were now federally recognized, and President Carter appointed a distinguished panel to organize negotiations between the indigenous tribes and the State of Maine. This led to Congress passing the "Maine Indian Claims Settlement Act" of 1980 (MICSA), which resulted from the negotiations.[12] In return for extinguishing all further indigenous land claims in Maine, MICSA allocated $81.5 million of federal funds to the indigenous tribes to purchase 300,000 acres in the state, of which the Penobscot purchased 143,685 acres for its Nation.[13]

The kernel of the culture of the Penobscot People was centered on the personification of the Penobscot River. The River provided the Nation with its food supplies (largely anadromous fish) and its spiritual context. Their ethic was one of mutual obligation to and with the river along which every location represented its own spiritual power. The Penobscot People's origin story centered on Klose-kur-beh, who provided the People with both spiritual and practical knowledge—for example, in how to build a birch canoe without killing the tree. Klose-kur-beh warned the Penobscot People about the white race whose lethal combination of greed and lust for power would "lead him to reach forth his hand to grasp all things for his comfort, and in the process, virtually destroy the world."[14] This spiritual position leads the Penobscot People to champion the protection of the River and its ecology. The MICSA decision therefore is highly relevant for the restoration of the Penobscot River, the body that lies at the heart of the Penobscot People's culture and existence.

By the late nineteenth century, the Penobscot River had primarily become a water transportation vehicle for logs from Maine's "Great North Woods," as the region was known. The sawmills were downstream in Old Town and Orono (as depicted by Fig. 1.3). The port city of Bangor is at the head of Penobscot Bay. In the early nineteenth century, Bangor shipped lumber to ports along America's Atlantic seaboard. In

[12] See Maine Indian Claims Settlement Act, Pub. L. No. 96–420, 94 Stat. 1786 (1980) (codified at 25 U.S.C. 1721–35).

[13] The State on Maine's civil and criminal jurisdictions remained over these indigenous lands.

[14] See Kolodny (2007, pp. 1–23).

the twentieth century, lumbering was replaced by the pulp and paper mills located between Millinocket and Bucksport. But the wood pulp and paper mills create dioxins as a byproduct of the chlorine bleaching process for paper. Dioxins—a cancer causing pathogen—were discharged from seven paper mills along the river (See the Appendix to Chapter 7 for a review of this pathogen). The *Penobscot Native People* were badly affected.

Following the lumber, pulp, and paper degradations, in the early twentieth century numerous hydropower dams were built along the Penobscot and its tributaries. Between the 1920s and World War II, inexpensive hydropower enabled textile and shoe manufacturing to become dominant in Maine, an industry that drew labor from French-Canadian immigrants. After the war, however, Maine became an important recreational destination. Restoration of the river then became essential for Maine's recreational destinations, and the resurgence of the fishing sector. The movement for dam removals became strong during the early years of the twenty-first century.

The Dams of the Penobscot

During the last century, the US led the world in dam construction. In the US, the Army Corps of Engineers has catalogued over 90,000 dams greater than six feet in height. These depleted fisheries, degraded ecosystem, and negatively altered our natural interactions. Many of these dams are now outdated, unsafe, and no longer serve their original purpose.[15] The particular disruptions caused by dams include:

- They prevent anadromous fish migrations to spawning grounds.
- They cause the river's silt to build up behind the blockage, and consequently not be deposited in the downstream delta where it might regenerate rich agricultural soils.[16]
- This upstream silting destroys the sandy and rocky spawning grounds of anadromous and non-anadromous fish.

[15] See www.americanrivers.org/threats-solutions/restoring-damaged-rivers/dam-removal-map/.

[16] See Robinson (2021, Chapter 11).

- The stagnant reservoir pools raise the water's temperature which alters the ecosystem and facilitates algae blooms with their consequent eutrophication and toxins.
- The peak demands on hydropower dams cause wide variation in water flows that further degrade the ecosystem.[17]

The *Penobscot Project* began in 1999 when the energy corporation PPL-Maine purchased a series of Maine dams. PPL approached the *Penobscot Nation* who engaged several conservation advocacy organizations in hopes of preventing the dams' renewal licensing. Their discussions led to the plan for removing the river's downstream dams while expanding power generation at other upstream dams. In 2004, an agreement concerning the dam removal process was reached by the *Penobscot River Restoration Trust* (PRRT)—the nonprofit organization that worked with state and federal agencies to implement the restoration of the river. *The Penobscot River Restoration Trust* (PRRT) has seven member organizations: *Penobscot Nation, American Rivers, Atlantic Salmon Federation, Maine Audubon, Natural Resource Council of Maine, The Nature Conservancy,* and *Trout Unlimited.* Note that these organizations also constitute the *Kennebec Coalition* with the exception of the *Penobscot Nation*. The PRRT provides dedicated leadership to the *Penobscot River Restoration Project* (PRRP) which largely consists of dam removals and fish bypass constructions aimed at restoring fish runs and the overall natural habitat of the river. The fish runs include twelve native species such as the anadromous short-nosed sturgeon, Atlantic sturgeon, Atlantic salmon, striped bass, steelhead trout, river herring and shad.

The PRRT purchased the Veazie, the Great Works and the Howland Dams in December of 2010. (Figure 1.3 depicts the locations of these dams.) With support from the *National Oceanographic and Atmospheric Administration* (NOAA) and funds from the "American Recovery and Reinvestment Act" of 2009, and also with half the necessary funding coming from private contributions, The Great Works Dam was removed in 2012, and the Veazie was removed in 2013. A channel that bypasses the Howland Dam was completed in 2015. Also, a new fish lift was installed at the Milford Dam in 2015. (Figure 1.3 depicts the location of this Dam.)

[17] Ibid.

All of this enabled the reconnection of thousands of miles of spawning habitat to the river's main channel for the first time in nearly 200 years.

The *Natural Resource Council of Maine* (NRCM) has been a significant leader in this dam elimination and habitat restoration movement. Also, NOAA has been the significant federal partner in this effort, and it designated the Penobscot as one of its ten nationwide *Habitat Focus Areas*. Since 2008, it has contributed $21 million toward the PRRP. This *Project* has already removed the obstructions of the two lowermost dams (Veazie and Great Works) in order to restore fish runs. It has also led to the decommissioning of Howland Dam and the construction of a fish bypass around it. The PRRP also improved fish passage at four other dam sites, and it improved hydropower production at six other dams so that the watershed's total hydroelectric generation has not decreased. Furthermore, the ongoing work in the watershed includes other dam removals, plus the construction of fish passage ways at the outlets of lakes, also the reconstructions of culverts that otherwise inhibit fish runs, and the surveying of the watershed to identify new habitat restoration initiatives.

The Penobscot and its tributaries flow from Mount Katahdin in central interior Maine. The river then flows southeasterly through the middle of Maine to the Penobscot Bay. It is the largest river system in Maine. Its watershed drains one-quarter of the state. This river provides the benefits of environmental interactions, cultural enrichment, and economic development. In particular, the PRRP will provide:

- improved access for 2,000 miles of river and stream habitat for anadromous fish,
- restored ecological systems that benefit native plants and animals in the river, its estuary, and the Gulf of Maine,
- support for the *Penobscot Nation's* culture and traditions,
- new opportunities for community development,
- new opportunities for natural interactions,
- enhanced hydropower generation.

The *Penobscot Project* has already restored links between the Gulf of Maine and the upper reaches of the watershed. Some native anadromous fish have already returned to their native spawning grounds. This has provided enhanced food sources for birds such as eagles and osprey, and also mammals such as porpoise and river otters. Previous to the dam removals, almost no shad passed Milford Dam. Now large amounts of shad are mechanically lifted passed Milford Dam to reach the upper river. Previous to the dam removals, almost no river herring passed the Milford Dam; but now over a million herring are lifted pass the Dam each year. Before the dam removals, sturgeon were not present at their ancient upriver spawning grounds, but they are now. Also, the *Penobscot Nation* has now hosted a series of national whitewater canoe races on the newly free-flowing river above Old Town. All of this has occurred while hydropower generation at six dams in the Penobscot watershed have increased their output of electric power sufficiently to more than offset the removal of the older facilities.[18]

For examples of the benefits derived from the above-mentioned projects, prior to the dam removals, river herring were previously counted by the hundreds. But in 2018 there were 2.8 million in the upper watershed after the dam removals. In 2016, Atlantic salmon returned to spawn in the upper river. John Banks, the Director of Natural Resources for the *Penobscot Nation* said, “For the tribe, the project represents much more than a fisheries and hydropower improvement effort. By ecologically reconnecting the watershed with the sea, this project repairs an age-old cycle that allowed the tribe to sustain itself for thousands of years.”[19] Kate Dempsey, Maine’s Director of *The Nature Conservancy* said, “Monitored populations of migratory freshwater fish have declined 76% in the past fifty years (worldwide), weakening connected ecosystems and fisheries that feed millions of people. That’s another reason the *Penobscot River Restoration Project* (PRRP) is so important; it has become an international beacon of hope and learning for how challenges from dams can be addressed, informing restorations all around the world.”[20]

As a particular example of restoration, the short-nose sturgeon, a species with a natural lifespan of 50 years, has now returned to the Penobscot. Along with the Atlantic sturgeon and Atlantic salmon, the short-nose had previously been listed as “endangered” by the EPA. But after the dam deconstructions, in 2015 the University of Maine researchers confirmed evidence that three female short-nose sturgeon were in waters between the remnants of the Veazie Dam and the town

[18] See www.nrcm.org/programs/waters/penobscot-river-restoration-project/.

[19] Ibid.

[20] Ibid, parentheses added.

of Orono. For tracing purposes, researchers previously implanted the sturgeon with small sound-emitting acoustic tags. Previous to the dam constructions, the short-nose and Atlantic sturgeon spawned in the river as far upstream as the Milford Dam, but the dams led to their decline and ultimately the EPA's listing as "endangered" in 1967. Sounding devices enable the tracking of the tagged fish. They now indicate that sturgeon have returned to their ancient winter habitat, and that they are now spawning in the Spring.

Habitat access is essential for the recovery of these species. The removal of the Veazie is a part of the PRRP. Along with the removal of the Great Works Dam in 2012, plus the Howland Dam fish bypass constructed in 2015, all of the habitat of the sturgeon is now restored along the Penobscot. This also restores the habitat for the nine other sea-run species native to the Penobscot which include Atlantic salmon and river herring. In addition, Molly Payne Wynne, the Monitoring Coordinator for the *Penobscot River Restoration Trust* (PRRT), states "The collaborative body of research on this project is among the most comprehensive when compared to the other river restoration projects across the country."[21] The project has become a model for other restoration projects in North America. It involves long-term monitoring that is funded by *NOAA Fisheries*, *The Penobscot River Restoration Trust*, and *The Nature Conservancy*.

Besides NOAA and *The Natural Resources Council*, the other partners in the *Penobscot River Restoration Project* include:

- the Penobscot Indian Nation,
- various agencies of the State of Maine,
- the Atlantic Salmon Federation,
- The Nature Conservancy,
- Maine Audubon Society,
- US Fish and Wildlife Service,
- Maine Sea Grant, and the
- Maine Heritage Trust.

Before the industrialization and dam constructions, the Atlantic salmon runs in the Penobscot numbered 50,000 or more. Shad and alewives

[21] Ibid.

migrated 100 miles upriver, and twenty-pound striped bass and sturgeon also spawned in the river. All of these species have now returned after the dam removals.

Penobscot Nation's Involvement in the Restoration

Nineteenth-century Maine provided the lumber for America's growth. The pine trees of Maine were cut and the Penobscot River was dammed to float the timber to sawmills. The bark and sawdust was dumped into the River which destroyed the benthos, i.e. the living organisms of the riverbed that supports the fish and waterfowl of the watershed. Thereafter, the paper mills and factories were established along the Penobscot. All dumped their wastes into the River. The established villages dumped their sewerage and other poisons so that the River became a dangerous body for humans, fish, and wildlife.

One of the more cogent articulations of the Penobscot River's degradation and subsequent restoration, and the importance of the River to the *Penobscot Nation*, was presented by Tribal Elder Butch Phillips in his polemic, "A River Runs Through Us."[22]

> The river became an open sewer. The resident fish, the pickerel, white perch, eels and others were no longer safe to eat because their flesh was contaminated and were a threat to The People and the creatures. The eagle and the osprey went away. The Sacred Circle of Life was broken.

Our laws for preventing the outrageous degradation of our rivers were ineffective until the late 1960s and early 1970s.[23] This particularly impacted American indigenous people since their culture was so entwined with the environments of rivers. Degradation of our rivers meant degradation of their cultures. The effects were explained in Phillips' polemic.

> The pollution continued and the sacred water that the ancestors had protected, cared for and blessed, had lost its spiritual power and had become a serious health risk to The People. The ancestors mourned the destruction and they wept, but the purity of their tears could not cleanse

[22] See www.nrcm-org/wp-content/uploads/2019/02/RiverRunsThroughUs_ButchPhillips.pdf.

[23] See Robinson (2023, Chapter 10) for a review of these developments.

their precious river. Native elders reminded The People of the ancient belief that if you harm any portion of the Sacred Circle of Life, you will eventually harm yourself. The People heeded the warning and went away from the river. That special connection to the river faded. The Sacred Circle of Life was broken. Our culture, which was based on a healthy environment and had evolved over hundreds of generations, was changed in just a few short generations.

The "Clean Water Act" of 1972 initiated the cleansing of America's rivers including the Penobscot. This Act established the effective regulations and regulatory bodies necessary to begin the cleaning of the Penobscot. The federal funding of sewerage treatment facilities provided the means for this cleaning. The "driving of logs" on the River was halted. The establishment of the EPA in 1970 halted the use of many dangerous and destructive pesticides and other chemicals. "Environmental Impact Statements" became required for construction projects in the Penobscot Watershed. The water became much cleaner. But dam removals and the establishment of effective fish passages remained to be accomplished. These improvements began in the early decades of the current century.

Now the Penobscot People's "Circle of Life" is being reestablished. The River is again an integral part of the Penobscot Nation. The pollution has substantially declined; the moose and eagles have returned. The River has returned to the Penobscot Nation, and the Nation to the River.

The Penobscot Is Now Restored?

After the efforts to restore the Penobscot, the river is now closer to being a pristine natural area as compared to its previous abandoned industrial character. The contemporary salmon runs are now considerably smaller than their pre-industrial levels, but after a century of building dams and alterations to the river, that must be expected. Nonetheless, approximately a thousand salmon are now returning to spawn each year which forms the largest Atlantic salmon run in the US. Their numbers will grow, and as the fisheries rebound, the wildlife that feeds on the migrating fish will also thrive.

The Penobscot restoration has political support. It has strong constituencies, particularly the *Penobscot Nation* whose lore recall how the river and its watershed once were, and who can therefore envision how they could be after restoration. The Penobscot therefore has strong advocacy constituents, strong funding from government and private sources, and the clear objectives of (i) dam removal, (ii) construction of fish byways, plus (iii) various habitat restorations.

The Beginning of the Dam Removal Movement

The dam removal movement began in Maine with deconstruction of its aged semi-derelict structures along the Kennebec and Penobscot Rivers. Why in Maine? In the US, the nineteenth and early twentieth centuries' industrialization began in the far northeast states. At its start, the degradation of America's natural assets appeared to be marginal incursions with most of the area's pristine natural assets being untouched. It was thought that the occasional dam constructions would not likely destroy the abundant fish runs, nor substantially affect eagles, ospreys, or offshore species. But the fish runs were destroyed. The sawmills did become pulp mills with dioxin poisons, and shoe factories did dump poisons into Maine's largest rivers. Species did decline at alarming rates. Maine's greatest assets—its healthy ecological water resources—were almost destroyed. This injustice to Maine's population, especially to its indigenous people, became a black mark of egregious sin in need of remediation. But the old polluting industries died, and the potential for environmental renewal emerged as the vision for a new Maine. The political forces changed. The clarion trumpets from a new environmental movement blew the Edwards Dam down. It also changed the federal licensing procedures (of FERC) for other hydro-power dams. This new dam removal movement merged with the existing environmental justice movement, and the door of opportunity for the *Penobscot Nation* opened to restore their traditional culture-based economy. This vision became politically dominant. The Penobscot and Kennebec Rivers have returned as natural assets. The Atlantic salmon, once severely endangered, is significantly returning along with other species.

After these successes, what else could be accomplished? What other pent-up political support for dam removals and restorations might now be unleashed? When Interior Secretary Babbitt appeared as the highest political guest at the Edwards Dam removal (July 1, 1999), the forces for more dam deconstructions were gathering. The environmental journalists were aggressive in their questioning of the Secretary. It was Maine's new Governor Angus King who optimistically answered the journalists' queries. The issue of dam removal was politically explosive nationwide because the forces of environmental justice and environmental restoration were congealing. The story they created and motivated is presented in the next chapters.

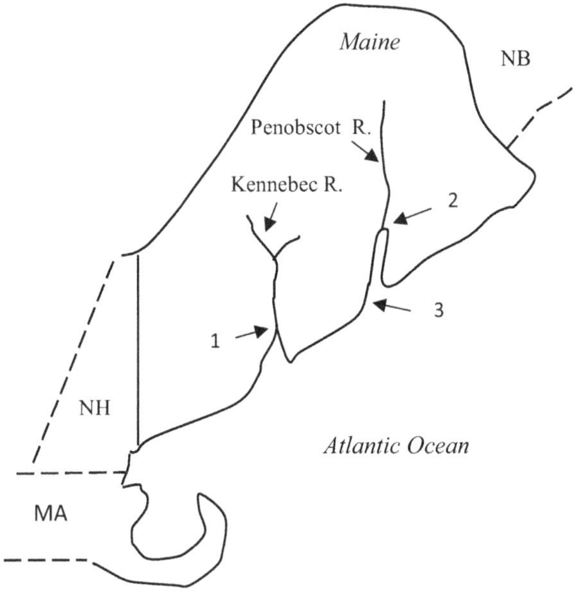

Fig. 1.1 Maine's Rivers

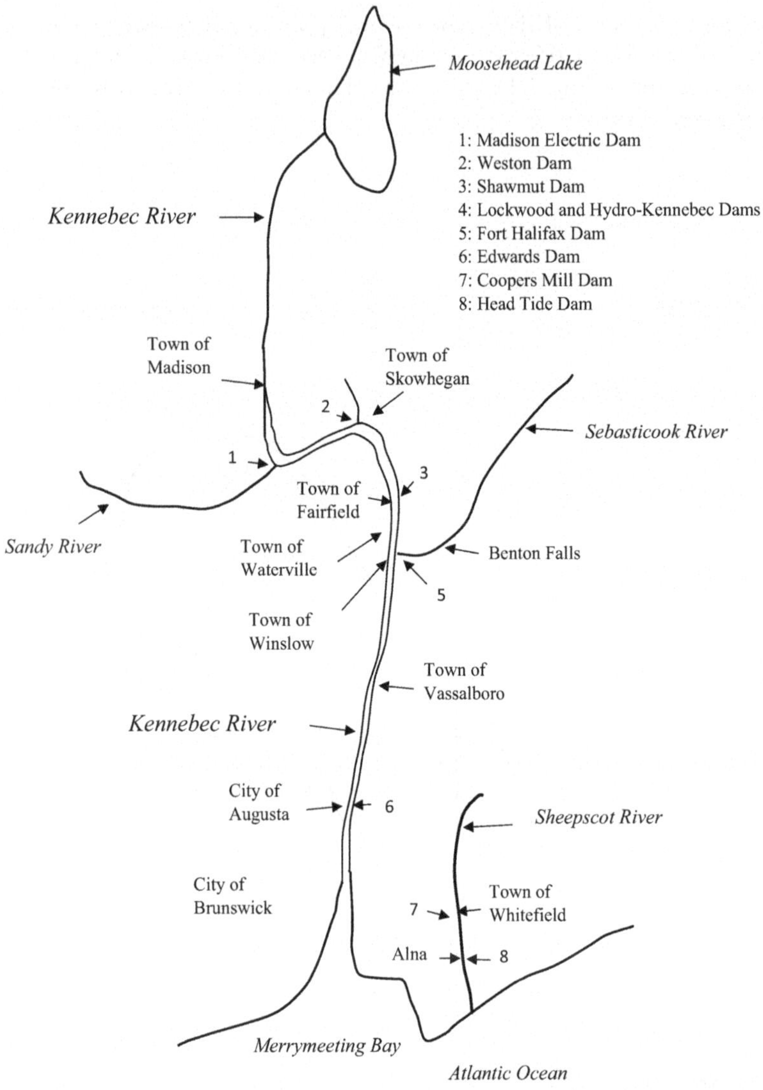

Fig. 1.2 Kennebec River Dams

1: Veazie Dam – removed in 2013
2: The Great Works Dam – removed in 2012
3: Milford Dam – fish lift installed in 2015
4: The Howland Dam – fish bypass constructed in 2015
5: Montgomery Dam

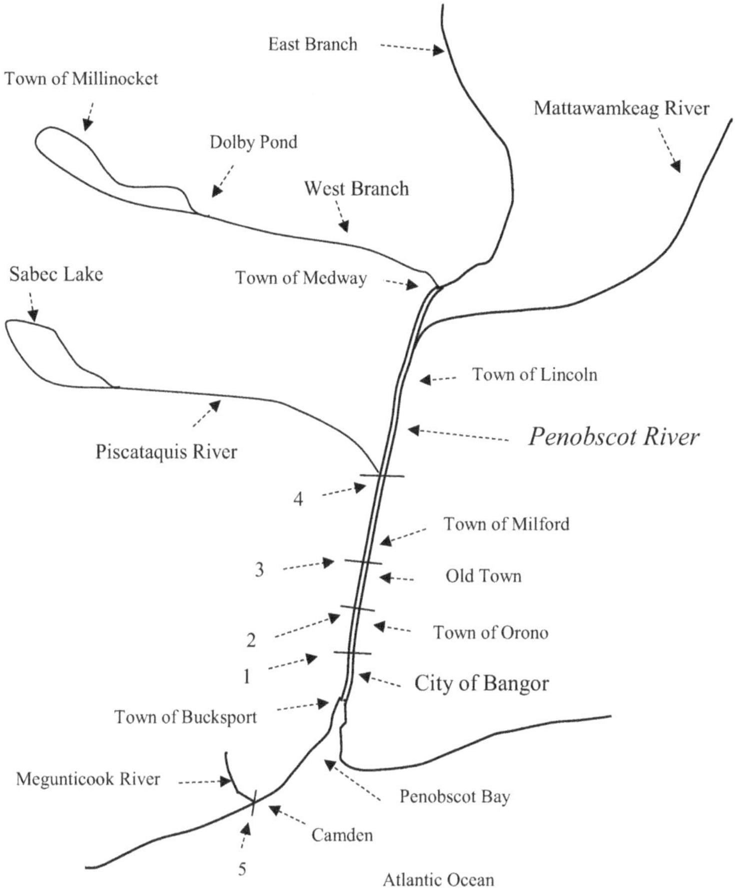

Fig. 1.3 Penobscot River

References

Cheever, D. (1997). Coalition Crows at Federal Decision. *Kennebec Journal*, 26 November, 1.

Grossman, E. (2002). *Watershed: The Undamming of America*. Counterpoint.

Kolodny, A. (2007). Rethinking the Ecological Indian: A Penobscot Precursor. *Interdisciplinary Studies in Literature and Environment'*, *14*(1), 1–23.

McPhee, J. (2002). *The Founding Fish*. Farrar.

NRCM (2021). Edwards Dam and Kennebec River Restoration. *Natural Resources Council of Maine*. www.NRCM.ORG/Blog/

Paulson, M. (1999, July 5). Potomac Watch: Babbitt Shows Abrasive Outlook by a Dam Site. *Seattle Post Intelligencer*. SeattleP-I.com

Robinson, R. (2023). *Environmental Advocacy and Local Restorations*. Palgrave Macmillan.

Robinson, R. (2021). *Environmental Organizations and Reasoned Public Discourse*. Palgrave Macmillan.

Walls, L. D. (2017). *Henry David Thoreau: A Life*. The University of Chicago Press.

CHAPTER 2

Big Concrete, the Tennessee River, and the New Deal

Developing the Tennessee Valley?

On November 29 in 1979, the flood gates of the Tellico Dam along the Little Tennessee River closed, and the Tellico Reservoir began to fill. This was to be the last dam constructed by the *Tennessee Valley Authority*, a federal agency created during the early days of Franklin Roosevelt's "New Deal." This new concept of a federal utility agency was created to tame the floods of the Tennessee River and bring prosperity to the severely depressed region. But this last dam was born and constructed in considerable controversy that encompassed "pork barrel politics," and environmental justice for the poor of the Little Tennessee River Valley and the long-suffering Cherokee Nation.[1]

The *Tennessee Valley Authority* (TVA) was created in 1933 by the Roosevelt Administration to (1) provide flood control to the Tennessee River, (2) provide hydroelectric power to the region, (3) facilitate navigation along one of America's more significant rivers, and (4) enhance economic development of the Tennessee Valley. The Tellico Dam was constructed for the purpose of economic development. It was to be the first critical element in the TVA's planned city of Timberlake with a projected population of 42,000. This is a "big concrete" dam of 129

[1] "Pork barrel politics" refers to expending federal funds on local projects so as to obtain political favor.

R. M. Robinson, *Restoring America's Rivers*,
https://doi.org/10.1007/978-3-031-81758-8_2

feet in height, and more than a half-mile in length. Its reservoir covers twenty-two square miles. More than 340 families were relocated by the TVA to enable the dam and reservoir's construction. This planned development was to serve the fourth of the above objectives. But the planned benefits did not materialize, and the TVA's analyses and purpose became questioned. As a result, the age of building any new big concrete dams in America ended with the Tellico Dam.

The Beginning of the Age of Electric Power

The Tennessee River is one of America's major arteries. It forms at the confluence of the Holston River and French Broad River near the City of Knoxville. (Figure 2.1 maps these rivers). It flows south-westerly through eastern Tennessee, and then turns south at Chattanooga. It then flows farther southward toward Huntsville in Alabama before turning northwesterly through Muscle Shoals and Florence in northwest Alabama. It then turns north to barely touch the State of Mississippi before flowing directly north to join the Ohio River at Paducah in Western Kentucky.

The Muscle Shoals area is in the far northwestern corner of Alabama. This section of the Tennessee River was once shallow and fast in variable currents with numerous rocky obstructions; it was therefore non-navigable. But the Wilson Dam—begun in 1918 and completed in 1924—created Lake Wilson which stretches from Florence-Muscle Shoals into north-central Alabama. The Lake now covers this once rough non-navigable section. With a single large lock, the Wilson Dam has allowed navigation into Lake Wilson (behind the dam) and then farther upstream. This dam-and-lock system allows navigation from the Ohio River far into Tennessee. The motivation for constructing the Wilson Dam, however, was not navigation; it was to produce military ordinance.

World War I began in 1914. The US entered this world war in April of 1917. For war preparation, President Woodrow Wilson sponsored "Preparedness Legislation" in 1916 that authorized the War Department to build a nitrate manufacturing plant to eliminate US dependence on foreign sources (mostly Chile). Nitrates are necessary for military munitions, but nitrate production plants required significant usage of electrical power. The large concrete hydropower Wilson Dam was constructed by the *US Army Corps of Engineers* to provide electricity to two nitrate plants in Florence-Muscle Shoals area.[2] (Figure 2.2 shows the location of the Wilson Dam.)

When World War I ended in 1918, the Wilson Dam was still incomplete. The Republican Presidents Warren Harding and Calvin Coolidge refused to complete the dam unless the Muscle Shoals hydro-facility was leased long-term to private interests. The problem concerned the then political debate of whether electrical power should be provided by public or private institutions. Should the Wilson Dam's federally generated power be sold directly to the public or sold to a privately owned utility which in turn would sell it to the public? Henry Ford offered to sign a 100-year lease for the Wilson Dam's hydropower and also offered to develop the nitrate fertilizer plants. He offered \$1.5 million per year lease payment, but the hydro-facility costs the federal government approximately \$150 million to build. The annual rate of return for this public investment would be a very small one percent if it were leased to Ford at such a low price.

Presidents Harding and Coolidge and Secretary of Commerce Herbert Hoover (the elected president following Coolidge) were in favor of Ford's lease, and the legislation passed the House of Representatives quickly. In the Senate, however, Nebraska's George Norris—a consistent champion of public power as opposed to private utilities—thought Ford's offer to be paltry given the public's large investment. Norris believed that the facility should and would generate billions of dollars in revenue from the sale of the electric power to the Tennessee Valley's region. Norris managed to block approval of Ford's lease. By doing this, he maintained the possibility of municipal power and rural electrification at low costs to the general public.

Construction of the Wilson Dam was completed in 1924, and its electrical generation began serving the area in September of 1925. Norris passed his "Muscle Shoals" bill through Congress, but it was vetoed in 1931 by President Herbert Hoover. But in 1933, almost immediately after taking office, President Franklin Roosevelt created the *Tennessee Valley Authority* (TVA) and began TVA's "rural electrification program" as one of Roosevelt's initial "New Deal" programs. In addition, the nitrate plants were maintained to service the US military's demands

[2] On *YouTube*, see the old newsreels "Construction of Wilson Dam," and/or "History of the Wilson Dam on the Tennessee River," and/or "Video of the Building of Wilson Dam Florence and Alabama 1918."

throughout World War II. Also, over 3,700 vessels passed through the Wilson Dam's locks each year, and the TVA ultimately constructed 29 new hydroelectric facilities in the Tennessee River's watershed.

The classical architecture of the Wilson Dam is currently still standing in Muscle Shoals. The Dam is one-hundred years old, and almost a mile long and 137 feet high, with a production capacity of 663 megawatts of electric power. It was the first construction of a big concrete hydropower dam. It was born from the fear of the World War I emergency, and it spawned the controversies of public versus private utility power and associated federal subsidization for rural electrification.

The New Deal and Electric Power

The 1920s was the decade that initially electrified America. Whereas the incandescent light bulb, and home and street lighting began in the 1880s, the real advent of home lighting, radio, electric motor-driven projectors for movie theaters, electric starters for automobiles, and even electric stove cooking, all substantially began and proliferated in the 1920s. Most American urban homes were connected to central electric grids during the 1920s, but only ten percent of rural farms were connected.[3]

At the end of World War I, small electric generating companies competed to provide local service, primarily in urban areas. Government regulation of this new industry was nonexistent. Early in the 1920s, "holding companies" began purchasing controlling interests in these local electric generating companies. By the 1930s, nine holding companies controlled seventy-five percent of the industry.[4] Since these "holding companies" controlled utilities in multiple broad geographic areas (i.e. groupings of states), state regulation was difficult to form. These electric grids were too extensive to easily allow state regulation, and the holding companies had state-level political influence that prevented any attempts at regulation. As a result, pyramids of holding companies were formed to operate across state jurisdictions in order to funnel investment capital into this rapidly expanding industry. By the early 1930s, these pyramids were siphoning large dividend amounts to the few owners of these holding companies.

[3] See Hession and Sardy (1969, pp. 618–621).

[4] See Riggs (2020).

In the pyramid structure, the first level (lowest level) of companies were generating and distributing the electrical power. A second level of companies would purchase enough shares in several local generating companies to have control of an area's electric distributions. But a tertiary-level company would purchase enough shares of two or three of these secondary-level holding companies to gain control of them. The tertiary companies were owned by only a few individuals. In this way, power was concentrated in a sort of monopoly system. These electric utility holding companies would, of course, not compete with each other for local business since they were controlled by the same owners. In this system, the profits to the owners were large. According to the Federal Trade Commission, the 1924 annual rate of return to these pyramid owners ranged from 19 to 55 percent.[5]

With this system, expansion of electric service into rural areas completely stalled. The economies of scale in this industry were considerable in servicing the dense populations of urban areas, but these economies were not present for rural communities. The high fixed costs of building a generating facility, and building the distribution grid, could not be covered for servicing only a few farms. But in 1933, President Franklin Roosevelt created the *Tennessee Valley Authority* (TVA), the first federally owned public utility company. Its purpose was to provide electricity to the rural areas of the Tennessee Valley region, and also to provide flood control and help finance economic development. When Roosevelt signed the authorizing legislation, malaria still affected thirty percent of the population of the Tennessee Valley, and the average household income was merely $640. The Valley's agricultural soils were also largely depleted by the 1930s.[6]

In his 1932 campaign for president, Franklin Roosevelt gave a speech in Portland, Oregon, which initially explained his plan for electrifying rural America and also for controlling holding companies. This speech was to a nationwide radio audience and indicated Roosevelt's plan for massive dam building and rural electrification. His creation of the TVA was the first step in this direction. In the following "off year" elections of 1934, democratic majorities increased in both houses of Congress. In his "state of the union address" of January 4, 1935, Roosevelt asked for

[5] Ibid.

[6] See Eschner (2017).

the elimination of "the evil of holding companies," the highly profitable organizational structures that appeared to have stymied rural electrification. Since the 1929 stock market crash, utility holding companies were not popular. The Congressional hearings concerning the financial debacle that occurred between October, 1929, and March, 1933, focused on the problems with these holding institutions, particularly with Samuel Insull's utility empire—Associated Gas & Electric—which consisted of sixty-three subordinated firms across twenty-three states.[7]

Once FDR became president in March of 1933, he focused on employment programs like the TVA's dam constructions.[8] Similar construction programs began on the Columbia River, the Snake River, the Ohio River, and the Colorado River. Roosevelt focused on hydropower and flood control, especially the former as a source of both industrial electric power and rural electrification. This placed Roosevelt in direct confrontation with utility holding companies, especially Commonwealth & Southern, the Tennessee Valley's major utility with its charismatic president Wendell Willkie. Wilkie became a considerable political foe leading the opposition to holding company regulation. He rose to receive the Republican nomination for president in 1940, but he lost to Roosevelt.

Roosevelt insisted that the existing electric generating companies should remain private, but he clearly favored an extensive federal hydropower program. This threatened utility barons like Willkie. The *Public Utilities Holding Company Act* passed Congress and was signed by Roosevelt in 1935. It limited the ability of holding companies to own utilities in multiple states. The market for electric power consequently became more competitive. Between 1933 and 1942, the previously high prices that electric companies charged fell forty percent in real terms. But as President Roosevelt promised, the existing electric power generating utilities remained private companies.[9]

In May of 1935, President Roosevelt created the *Rural Electrification Administration* (REA). It was to "initiate, formulate, administer, and supervise a program of approved projects with respect to the generation, transmission, and distribution of electric energy in rural areas." In 1936,

[7] See Wasik (2006), and Hughes (1979).

[8] Roosevelt's inauguration of March 4, 1933, was the last presidential inauguration of that date. The 20th amendment to the US Constitution changed the inauguration date thereafter to January 20.

[9] See Sablik (2020).

Congress authorized $100 million to enable the agency to finance the REA's activities, and to make loans to electronically wire homes and to finance lighting and electric appliances. By then, private utilities became suspicious of the REA and resisted its intrusion into rural areas. As a result, the REA turned to a familiar rural area institution, the cooperative organization (or co-op). Every rural county had co-ops that served various rural purposes. A co-op is collectively owned by its members. In rural areas, farmers organized co-ops for the purchase of seed, fertilizer, equipment, storage of grains and the like, and also for marketing the area's production. There were even a few examples of rural co-ops for electricity generation. But many states did not have laws for the governance and organization of co-ops so that in 1937, the REA created a model, "The Electric Cooperative Corporation Act" for state consideration. This model indicated that the co-ops would be nonprofits governed by elected boards with members having one vote each.

Through the co-ops, farmers could borrow at low interest rates of 2 to 3 percent to finance construction of transmission lines and to pay for wiring their farms and the purchase of home and farm appliances. REA engineers would design the grid infrastructure, and the REA purchased the materials for these constructions. As a result, during the 1930s the REA reduced the costs of rural power lines from $2,000 per mile to $825 per mile. All of this stimulated electrification expansion and allowed economies of scale to be exploited.

Rural co-ops usually did not generate the power for rural electrification. They purchased the power wholesale from private electric generating utilities or from the TVA of other federally created entities such as Bonneville Power Authority. If this was not possible, then the REA would assist in establishing a power generating facility such as a local hydropower dam.

The REA created competition for rural territories that were not previously serviced by private utilities, but that perhaps planned to address the rural needs at a later time. In the 1920s, when the urban areas were being electrified, the market was more competitive and utilities had an incentive to expand quickly to established customers. But in the late 1920s and early 1930s, the urban utilities had the incentives to consolidate under holding companies and faced little pressure to develop rural areas as customers. But by the end of World War II, approximately half of America's farms were electrified primarily through the REA's efforts.

The Massive Concrete of Hoover Dam

Wilson Dam was the first big concrete dam in the US. It stands 137 feet high, and 4,541 feet long, with a maximum base width of 160 feet. Its large length spans the width of the relatively flat Muscle Shoals section of the Tennessee River. It was completed in 1924. The Hoover Dam on the Colorado River, however, was completed a decade after Wilson Dam. The Hoover Dam is a very different shape. It stands 720 feet high, and 1,200 feet long with a base of 650 feet wide. Its architecture, therefore, differs substantially from the Wilson Dam. Hoover Dam spans a hard-rock canyon that is very high and narrow. It was constructed in Black Canyon on the Colorado River as it flows between Nevada and Arizona. (Figure 4.1 in the next chapter maps these rivers.) Planning this giant structure began in 1928 prior to Franklin Roosevelts Administration and its New Deal. It was built, however, between 1931 and 1936 which is largely during the New Deal.

Hoover Dam was constructed by the *Federal Bureau of Reclamation*. It is located in a high canyon of hard rock near Boulder City, Nevada. This type of canyon is particularly ideal for a location. It differs from most of the broader dam locations of the Tennessee Valley reviewed below. Dams such as the Wilson or Wheeler were constructed on flatter geography. In essence, the harder impervious rock of Black Canyon was designed into the Hoover Dam's obstruction of the Colorado River. Hoover Dam is an enormous concrete arch that bows toward the upstream current. This curving arch acts as a wedge that channels the river's force onto the hard-rock walls of the canyon. The dam is 660 feet thick at the bottom foundation and narrows to 45 feet at the top which forms a highway bridge between Arizona and Nevada. Compared to the Wilson and most other dams, the construction of Hoover Dam was difficult and dangerous for workers, but applications for construction jobs were abundant due to the unemployment crisis of the 1930s.

In 1931, the town of Boulder City was built adjacent to the site of the construction. Boulder City was built in 15 months and was government owned and operated as the dwelling for construction workers and other employees. It was connected by rail to Las Vegas in Nevada. In 1932, more than 10,000 unemployed workers arrived in Las Vegas seeking employment; only 3,000 were hired. Chinese workers were discriminated against and consequently not hired. This discrimination was actually specified in the construction's contract.[10] Only 30 blacks were employed for this construction. Union activity was severely discouraged even amid the dangerous and difficult working conditions.[11] The high heat of the Arizona desert and dangerous heights of the canyon walls created the hazardous conditions. Arranging explosives along the high canyon walls motivated workers to create their own "hard hats" from dried and rigid tar to protect them from rockfall. Observing this, modern hard hats were ordered and issued by the government's contractors.

Two temporary diversion dams and four lengthy diversion tunnels (16,000 feet in total) were constructed to allow concrete to be placed. To reduce the foundation site to hard rock, 1.5 million cubic yards of soil and lose rock was removed to prevent seepage. The first concrete was poured into the foundation in June of 1933, eighteen months ahead of the planned schedule. But concrete both heats and contracts in its hardening process, and the potential for uneven contraction and consequent cracking would weaken the final structure. A cableway of wire structures was erected that allowed large buckets (7 feet in diameter) of wet concrete to move to the foundation and pour the concrete into rectangular molds. Approximately 600 miles of cooling pipes were inserted into the structure to cool the hardening concrete. After cooling, grout was injected into the very narrow spaces between the blocked pourings. A total of 3.25 million cubic yards of concrete were ultimately poured over the two years of construction that concluded on May 29, 1933.[12] The dam was completed in 1936, two years earlier than planned. At that time, it was the largest concrete structure ever built. Ninety-six deaths occurred due to construction accidents within the very high heights of the canyon walls.[13]

Lake Mead, created behind the dam, began filling on February 1, 1935. In the latter half of 1936, water levels were sufficiently high to allow power generation with three turbine generators. Eventually, eight

[10] See Hiltzik (2010, p. 315).

[11] Ibid, pp. 70–73.

[12] This was enough concrete to build a two-lane highway between San Francisco and New York City. See Billington et al. (2005).

[13] See the four informative *YouTube* videos of (1) "Building Hoover Dam," (2) "Hoover Dam Construction (Part 1) 1931," (3) "Construction of Hoover Dam, One of the Largest Dams of the World," and (4) "Blood Concrete and Dynamite: Building the Hoover Dam."

generators were installed. The dam was dedicated by President Roosevelt on September 30, 1935. By September of 1939, the dam's hydropower facility became the largest electric generating plant in the world. It was called Boulder Dam until 1947 when the name was changed to Hoover Dam.

Hoover's hydro-plant generates a maximum of 2,080 megawatts of peak power. But note that this power generation has been decreasing due to the lower water levels associated with the prolonged drought that began in 2000. Lake Mead and its downstream flows continue to provide irrigation and municipal water use for Southern California, plus it services the area around Yuma, and the Central Arizona Water Project. (Chapter 4 explains these projects.) Hoover's hydropower is distributed to Nevada, Arizona, and California with more than one-half going to Southern California's entities.[14] Almost twenty-nine percent of Hoover's hydropower is used for pumping water for irrigation districts. A significant portion is also used to power the City of Las Vegas in Nevada and its various establishments.

The enormous concrete structure of Hoover Dam was facilitated by the economics of the Great Depression. The initial financial crisis of October 1929 evolved into three waves of bank closures in 1930, 1931, and 1932, which together generated unemployment levels of at least one-quarter of the US workforce.[15] The big government employment programs of Franklin Roosevelt's "New Deal" became the response. The Hoover Dam project set the initial standard for large employment projects with its politically popular promise of large-scale hydroelectric generation to power the irrigation and municipalities of California and Arizona. The Hoover Dam project then set the direction for other big concrete dams to be constructed on the Colorado, Columbia, and Tennessee Rivers, all of these being in response to the emergency of the Great Depression.

[14] Bureau of Reclamation (2017), "Frequently Asked Questions: Hydropower," at www.usbr.gov/lc/hooverdam/faqs/powerfaq.html.

[15] See Hession and Sardy (1969, pp. 629–634).

Tennessee Valley Authority's Dams

Table 2.1 lists TVA's significant dams and their locations; Fig. 2.1 (presented at the end of this chapter) maps these locations. There are nine TVA dams on the main body of the Tennessee River. They are operated by the TVA to (1) provide hydroelectric power to the Valley, (2) to control flooding, and (3) to provide a minimum channel of nine feet for navigation over the entire length of the River, from Paducah up to Knoxville. Eleven of TVA's significant dams are on the tributaries of the Tennessee River: the Toccoa, Clinch, Hiawassee, Holston, Nottley, Little Tennessee, Watuga and French Broad Rivers. All but two of TVA's significant dams are entirely concrete; two of these have earthen abutments (mostly clay and rock) on their ends, and two are combinations of rock, aggregate, and clay. All of these dams produce hydropower. With the exception of the Wilson Dam, which is reviewed above, the purpose for the constructions of each of these dams, and the specifics of each, is presented below.

The main dams on the Tennessee River are briefly described in the appendix to this chapter. The significant dams on the Tennessee's tributaries are briefly described after. Note that the Tellico Dam is reviewed in a separate section.

Flood Control in the Tennessee Valley

Controlling the Tennessee River's floods requires the TVA to coordinate the operations of its dams. The TVA currently operates 49 dams in the Tennessee Valley, 29 of these are hydroelectric, but all play a role in flood control. The drainage in the Tennessee Valley flows from northeastern Tennessee and the surrounding areas of Virginia, Georgia, and North Carolina toward Chattanooga which is located in south-central Tennessee. The Tennessee River causes Chattanooga to be the focal point of flooding and flood damage in the Valley. The severe flood of 1867 set a type of disaster standard for the history of the Tennessee Valley, but recent climate change-related effects on rainfall in the Valley appear to be particularly threatening in that new flood levels have exceeded historic levels.

Prior to the creation of the TVA, Chattanooga was almost completely underwater during the major floods of 1867 and 1917. The flood of 1867 remains the most significant flood recorded in the eastern Tennessee Valley. The Upper Tennessee Valley has been especially susceptible to flooding because of its location between the Smokey Mountains on its east side, and the Cumberland Plateau on its west side. The Tennessee Valley gently slopes from southwest Virginia toward the Tennessee River and then through Chattanooga.

During the first seven days of March in 1867, more than twelve inches of rain fell across eastern Tennessee, and at least six inches fell in the rest of the Valley. Snowmelt was also heavy across the higher elevations. The

Table 2.1 TVA Operated Dams Built During the New Deal

Dam	*Dates of construction*	*Type*	*Location*
Blue Ridge	1925–31	Earthen	Toccoa River, Georgia
Wilson	1931–36	Concrete	Tennessee River, Alabama
Norris	1933–36	Concrete	Clinch River, Tennessee
Wheeler	1933–36	Concrete	Tennessee River, Alabama
Pickwick	1935–38	Concrete	Tennessee River, Tennessee
Guntersville	1935–39	Concrete	Tennessee River, Alabama
Chickamauga	1936–40	Concrete	Tennessee River, Tennessee
Hiawassee	1936–40	Concrete	Hiwassee River, N. Carolina
Kentucky	1938–44	Concrete	Tennessee River, Kentucky
Watts Bar	1939–42	Concrete	Tennessee River, Tennessee
Cherokee	1940–41	Concrete and earth	Holston River, Tennessee
Fort Loudoun	1940–43	Concrete	Tennessee River, Tennessee
Chatuge	1941–42	Earthen	Hiwassee River, N. Carolina
Apalachia	1941–43	Concrete	Hiwassee River, N. Carolina
Nottley	1941–42	Mostly concrete	Nottley River, Georgia
Fontana	1941–44	Concrete	Little Tennessee River, N. Carolina
Douglas	1942–43	Concrete	French Broad River, Tennessee

course of the flood through the Valley was described by the *Report of Chief of Engineers, 1875–1876.*

> The flood of 1867 far exceeded all precedents for the past 90 years. It consisted of one great rise due to furious rain storms which covered its entire valley, particularly the mountain region. At Kingsport, on the Holston, rain fell nearly continuously from February 28 to March 7. At noon on March 7 the river attained its highest point, being 30 feet above low water and 4 feet above any other flood. .... At Chattanooga the rise began on March 4, overflowed the banks on March 8, and attained height on March 11, being 53 feet above low water and 15.5 feet above the high water of 1847, the highest on record. .... The destruction of property and life by this flood was beyond parallel in the history of the Tennessee Valley (Taken from "The Chattanooga Flood Control Problem," 76th Congress, 1st Session, House Document No. 91, 1939, page 71)

In March of 1973, the Tennessee Valley suffered one of its worst and costliest floods. Heavy rains began in the evening of Wednesday, March 14, in South Middle Tennessee and in Northern Alabama. The heavy rains lasted through the morning of Friday, March 16. Tributary creeks of the Tennessee River such as Paint Rock, Limestone, Indian, Big Nance Creeks suffered heavy flooding. In Florence, Alabama, the Tennessee River set records for maximum stream flow. Roads and bridges were washed away, and there was one road death near Big Nance Creek.

Chattanooga was the hardest hit of the large cities where massive flooding occurred along South Chickamauga Creek, flooding that totally covered the airport. The Tennessee River crested at 10 feet above flood level. Damage was estimated at more than $35 million with hundreds of homes and businesses flooded. More than 300 families were evacuated with one-fifth of the city inundated. Flooding also forced evacuations from Knoxville, and travel was blocked into the Smoky Mountains because of landslides.

In 1973, in locales within the Tennessee Valley, between Wednesday, March 14 and Friday, March 16, the smallest local recorded rainfall was 2.05 inches, and the largest was 10.54 inches. In the northern portion of the watershed, most of the recordings were above 7 inches, and most were in the 7 to 9-inch range. Besides Chattanooga, the other particularly severely affected areas were Huntsville in Alabama, Knoxville in Tennessee, and the Elk Creek area. Water levels reached twice what was previously considered the 100-year flood crest. Water rose along the entire

Table 2.2 Annual Chattanooga rainfall totals since 2013 (Average over the last 130 years is 51 inches)

Year	*Inches*	*Year*	*Inches*
2013	64	2017	56
2014	53	2018	67
2015	59	2019	66
2016	43	2020	70
		2021	61

Tennessee River to its highest historical levels. For example, in Florence, Alabama, the road over the Wilson Dam was closed for the first time. Many businesses and homes were evacuated in Muscle Shoals. Measured peak flows were double the previous records, and because of high levels behind the dams, the TVA was forced to release water from the Chickamauga Dam, Nickajack Dam, and Guntersville Dam. Flood records were also broken in the Elk River Basin.

Since it was created in 1933, the TVA claims to have prevented an average of $260 million per year in flood damage. In February of 2019, record high rains again fell across the Tennessee Valley, and the TVA used its dams and reservoirs to help avert nearly $1.6 billion of flood damage in Chattanooga.[16] Table 2.2 presents the annual rainfall for the years 2013 through 2021. Note that the long-term average annual rainfall over the last 130 years is 51 inches. All of the nine years of 2013–2021, however, had rainfalls greater than 51 inches. If the measure of annual rainfall is randomly distributed about the long-term average of 51 inches (if the probability is one-half for any single year's rainfall being above the long-term average of 51 inches), then the probability that for all nine years the rainfall would exceed 51 inches is only two-tenths of one percent. We must conclude that recently the average has increased and now exceeds 51 inches. The problem of Tennessee River flooding appears to be increasing in severity.

The TVA's Last Dam Construction

The TVA's last dam completion was the Tellico. Planning for what became the Tellico Dam began in 1936 when the TVA suggested a dam across the mouth of the Little Tennessee at Fort Loudoun. World

[16] See the TVA's Webpage.

War II, however, interrupted the finance for this construction. But in 1959, the TVA again began planning for this dam. It was then called the Tellico Project and the City of Timberlake Plan. Its purpose was to develop a reservoir for recreation—not hydropower or for flood control. Timberlake was the TVA's ambitious plan for a city of 42,000 that would develop the Little Tennessee Valley as a recreational destination and would also serve as a transportation hub for industries that needed access to the Tennessee River's navigation. Twenty-three-thousand acres were purchased and cleared by the TVA for this project. But then strong opposition to the project developed, and this surprised TVA.

In 1963, groups of landowners in the affected area, and also small-business owners, formed to voice strong opposition to what they described as TVA's intrusion. The TVA then attempted to organize local interests in support of the project, but citing the loss of prime farmland, the Tennessee Farm Bureau voiced its opposition to the dam in 1964. Then in 1965, the Cherokee Nation sued the TVA for planning to flood their sacred ancient homeland along the Little Tennessee River. Also, in 1971, Economics Professor Keith Phillips of the University of Tennessee criticized the TVA's cost–benefit analysis used to justify the dam.[17] Philips criticism was supported by Tennessee's Governor Winfield Dunn who said in a letter to the TVA Administrator that, "The Little Tennessee as it exists now is a waterway too valuable for the State of Tennessee to sacrifice."[18]

In 1972, the Little Tennessee Valley's farmers and environmentalists formed the Environmental Defense Fund (EDF). The EDF sued the TVA under the *National Environmental Policy Act* (NEPA) claiming that no environmental impact study was conducted as required by NEPA. The TVA provided the analysis, however, and the dam's construction was allowed to continue.

In August of 1973, University of Tennessee Professor David Etnier and his students conducted a search for possible endangered species on the Little Tennessee. Eitner identified several snail darters that were previously unknown.[19] Four months later, President Nixon signed the *Endangered Species Act* (ESA), which provided protection for the habitat of identified endangered species. But at this time, the Tellico Dam was almost

[17] See Plater (2013a, 2013b, May 9).

[18] Ibid. Also see Appendix C to this volume.

[19] See Plater (2013b, June 18).

complete with sunken expenditures of $53 million. In November of 1975, the snail darter was placed on the *Endangered Species List* by the US *Fish and Wildlife Service* (FWS). This appeared to stop Tellico's completion. To continue the story of the Tellico, however, we need to diverge into the history of the ESA.

A Brief History of the *Endangered Species Act* of 1973[20]

In 1972, President Nixon declared that the current species conservation effort was inadequate. Congress responded, and the "Endangered Species Act" (ESA) was signed at the end of December, 1973. As part of this Act, Congress authorized the *Secretary of the Interior* to acquire habitat lands necessary for the conservation of threatened or endangered species. The purpose of the Act is to not only protect species, but also the ecosystem habitat upon which they depend. The ESA is administered by the *Fish and Wildlife Service* and the *National Marine Fisheries Service*. The significant provisions of the Act include:

- The federal government must determine whether a species is endangered or threatened. If so, it must list the species for protection as either "threatened or endangered."
- The critical habitat of an endangered or threatened species must be identified and designated for protection.
- Federal agencies must use their authority to conserve all threatened and endangered species so listed, and this includes preserving their habitats.

For a species to be listed as either threatened or endangered, it usually means that its habitat is itself being endangered, i.e. the preservation of habitat is necessary. Therefore the habitat also poses the point of confrontation with so-called economic interests. In 1973, the ESA was written and worded so as to not take so-called economic opportunity costs associated with habitat preservation into consideration, but only to consider the existence of the species. The ESA, therefore, poses a

[20] This brief history section is taken from Robinson (2023, pp. 13–16).

program that supports a general and natural preservation of a species and its habitat.

In a 1978 amendment, however, Congress added the words "… taking into consideration the economic impact …" in the Act's provision concerning critical habitat designation. This 1978 amendment changed the listing procedure to consider the economic impacts of preserving the critical habitat. This change almost completely halted new listings, with almost 2,000 species being withdrawn from consideration following the amendment's passage. The habitat in question might have had other valuable commercial value, but the reservation demand for preservation was not considered under this 1978 amendment.[21] Preserving habitat, however, not only means preservation for the particular species, but also preserving areas in a wild state for other species and also for the human visitors who do not disturb but only observe the habitat. This sort of preservation likely stimulates reservation demands and the benefits from ecotourism.

A 1982 amendment, however, further changed the wording of the ESA by adding the word "solely" to specify that only the biological status of the species should be considered, not the economic impact that the habitat preservation might pose. This placed the impetus back on habitat preservation. But the ESA and its amendments also require that for any listing, the EPA must solicit comments from the public, and that one or more public hearings must be held for this purpose. This opened the process to public evaluation and expression concerning preservation of any specific habitat locale.

Since being placed on the endangered species list, the following have increased in population size as of January 2019:

- The Bald Eagle increased from 417 to 11,040 pairs between 1963 and 2007. It was removed from the endangered list in 2007.
- The Whooping Crane increased from 54 to 436 birds between 1971 and 2003.
- Kirtland's Warbler increased from 210 to 1,415 pairs between 1971 and 2005.

[21] For an examination of reservation demand, see Robinson (2021, Chapter 6). In environmental matters, "reservation demand" means that individuals are willing to pay for preservations or restorations of habitat even though the individuals have no explicit plans to visit these habitats.

- The Peregrine Falcon increased from324 to 1,700 pairs between 1975 and 2000 and was removed from the list in 1999.
- The Gray Wolf populations increased dramatically in the Northern Rockies and Western Great Lakes States.
- The Mexican Wolf increased to its minimum sustainable population of 109 wolves in 2014.
- The Red Wolf increased from 17 in 1980 to 257 in 2003.
- The Gray Whale increased from 13,095 to 26,635 between 1968 and 1998, and it was removed from the "endangered" list.
- The Grizzly Bear increased from 271 to 580 in the Yellowstone area between 1975 and 2005.
- California's Southern Sea Otter increased from 1,789 to 2,735 between 1976 and 2005.
- The San Clemente Indian Paint Brush increased from 500 plants to 3,500 between 1979 and 1997.
- Florida's Key Deer increased from 200 to 750 between 1971 and 2001.
- Texas' Big Bend Gambusia increased from 24 to over 50,000.
- The Hawaiian Goose increased from 400 to 1,275 between 1980 and 2003.
- Virginia's Big-Eared Bat increased from 3,500 to 18,442 between 1979 and 2004.
- The Black-Footed Ferret increased from 18 to 600 between 1986 and 2006.

Unfortunately, eleven other species have become extinct since the 1973 Act was passed. In general, however, the ESA has been somewhat effective for the preservations of both species and natural habitats.

The current problem of species proliferation poses a new phase for the environmental movement, i.e. the environmental restoration at various specific locales. This current problem of restoration concerns species proliferations in suitably regenerated areas and appropriate human interactions within these locales.

The Snail Darter, the Cherokee, and the Tellico

A law student at the University of Tennessee, Hiram Hill, together with Professor Etnier sued the TVA to protect the endangered snail darter. In *Tennessee Valley Authority versus Hill* (1976), the US Supreme

Court decided in favor of the ESA's protection of the snail darter and construction of Tellico was stopped.

The economics of completing the Tellico Dam was not favorable in 1977. Charles Schultze, Chairman of the Council of Economic Advisors stated, "... if one takes just the cost of finishing it against the benefits and does it properly, it doesn't pay, which says something about the original design."[22] The Tellico Project was, however, favored by Tennessee's powerful Senator Howard Baker who referred to the snail darter as his "nemesis."[23] He added a rider to an unrelated public works bill that exempted the Tellico Dam Project from the ESA. President Carter, under Congressional pressure concerning other important actions, signed the public works bill in September of 1979. But also in that year, Cherokee tribes sued the TVA to prevent the Tellico Reservoir's flooding of their sacred homeland sites (Sequoyah versus TVA). The reservoir area contained numerous *18th Century Overhill Cherokee Villages* (Tanasi, Chota, Toqua, Tomotley, Citico, Mialoquo, and Tuskegee) that were occupied for as much as 1,000 years prior to the "Trail of Tears" genocidal debacle.[24] These sites had earthwork platform mounds for the "council houses" of the Appalachian Cherokee tribes. The Cherokee, however, did not succeed in their lawsuit.

The Cherokee are bitter toward the TVA. They speak of prophecies from the old and past generations as voiced by Goliath George, a wrinkled 78-year old who speaks only Cherokee. In an affidavit he recalled listening as a small boy to a medicine man who "would talk to my people from a hickory stump that was notched so he could climb on top and look out over the valley." (Rawls. November 14, 1979) "He talked of what would happen in four or five generations. He said the Valley would be covered with water—our forefathers would be on the bottom of the Valley looking up through a wall of glass. Tears rolled down his cheeks when he said that one day the people would once again be put to the test of holding on to that which is sacred or giving up forever part of their lives."[25] At Chota and other sites including the Village of Tenasi—from which the State of

[22] See Plater (2008, April).

[23] Matthiessen (1980, February 7).

[24] The "Trail of Tears" refers to President Jackson's forcing of the Cheroke to leave their native East Tennessee and surrounding area's homeland and walk to the newly formed reservations in Oklahoma Territory in the 1830s.

[25] See Rawls (1979, November 11).

Tennessee derives its name—archeologists dug feverishly to save whatever Cherokee artifacts they could find. They found evidence of habitation from 7500 years BC. These ancestral grounds were inundated when the dam's gate was closed on November 29, 1979.

A Just Deal for Nellie McCall

The Cherokee and the snail darter were affected by the Tellico Dam, and so were the more than 340 other residents who were relocated. On the day the Tellico's gates closed and the Reservoir began to fill, Nellie McCall was an eighty-four-year-old widow and resister to TVA's use of eminent domain to take her family's 90-acre farm. "How would you like it if somebody was trying to steal your land?"[26] Her confrontation was representative of forty-five years of TVA's conflict with the Valley's residents, including with the Cherokee indigenous population. The TVA offered to pay Nellie $76,000 for her farm and dwelling, but she refused to sell. "It's a land grab. The lake would take about an acre and a half, and I offered to give that to them for nothing, but they said everybody had to go."[27]

Nellie McCall and her husband purchased their ninety-acre farm almost fifty years previously with a loan that required payments of $250 per year for forty years. On the day when TVA forced her to leave her farm, she flipped through her photo album pointing to her deceased husband and a favorite cow. She claims they paid off their loan in nine years. Her husband, two farming brothers, and a sister all "grieved to death" in the previous five years over the loss of their farms to TVA's eminent domain. About the TVA, she claims, "You can't trust them. You can't trust somebody that'll steal your land and then tell you its for your own good. They're like a yaller dog."[28]

With respect to Nellie's offer to give TVA the inundated acreage if she could retain her other acreage, David Freeman, the TVA's Administrator explains that the land along the lake will increase in value. "TVA made it more valuable. Why shouldn't TVA get the benefit from selling the land for a plant that will bring more jobs to the area?" Such is the

[26] Ibid.

[27] Ibid.

[28] Ibid.

expressed arrogant logic of the TVA, a government agency meant to serve the public. Such is the treatment of Nellie McCall. Freeman's industrial "plant" was never developed.

On November 29, 1979, the flood gates closed on the Tellico Dam, and the TVA began its sales efforts of lakefront acreage of Tellico Reservoir.[29] Analysis of TVA's acquisition methods for the Tellico Project, however, raise concerns of abuse.[30] The land purchased under eminent domain has suffered from failed development efforts. These efforts include a failed toxic waste dump at Tellico Peninsula. This site is now a yet-to-be developed mixed-use real estate project. The failed Timberlake Project was also relaunched in 1984 as a retirement golf-course community. The developer, however, was federally charged with mail fraud and money laundering. Through all of this, the 42,000-person village of Timberlake was not established, nor the industrial commercial business developed, nor its associated employment expansion.

As of 2022, the Tellico Dam was the last dam constructed by the TVA. The Tellico Project initiated the age of questioning the ethical and economic justifications for these "big concrete" structures. One factor in the decline of dam construction is the scarcity of appropriate dam sites within America's significant watersheds. Another significant factor, however, is the reevaluation of America's need for natural spaces, especially along its rivers and streams. But a third factor is the decrease in willingness of the American society to disrupt the lives of property owners in the proposed dam's areas (This is a question examined repeatedly in latter chapters).

TVA's Last Attempt on the Duck River

In the 1980s, the TVA attempted to construct an $83 million dam similar to the Tellico Project, but on the Duck River (Figure 2.1 shows the location of this River.) This effort failed. The Duck River flows 209 miles through Middle Tennessee. It is one of Tennessee's most scenic waterways. According to *The Nature Conservancy*, it is North America's most biologically diverse freshwater river. According to the US Geological Survey, its ecology is one of the world's most diverse. But this

[29] See Madden (1981, July 2).

[30] See Venable (2022, October 13).

river is the center of an outdoor recreational area with 150,000 anglers, kayakers, and canoers. It also serves as the sole water resource for Middle Tennessee's 250,000 population with its explosive growth according to the 2020 Census. Consequentially, there is great pressure to increase water withdrawals from the Duck River.

Since 1999, *The Naturae Conservancy* and other environmental organizations are working to ensure that the water withdrawals from the Duck River will not disrupt the river's incredible wildlife, especially its endangered or threatened species. These environmental strategies include:

- working with farmers to implement conservation measures,
- protecting riparian areas with conservation easements,
- reducing sedimentation in the watershed, and
- promoting development of greenways along the river, and increased access to the river.

The Duck River has more diversity in species of aquatic life than any other North American river. This includes 60 species of mussels and 22 species of aquatic snails. Many of these species are threatened or endangered. They all play a role in promoting water quality in that they filter large amounts of water every day, and they stabilize stream channels.

The TVA's attempt at dam building on the Duck River failed because its cost–benefit analysis was faulty. Upon public review, the TVA was found to have overestimated the benefits of dam construction and underestimated its costs. Recreation had shifted to small boaters (kayakers and canoers) using the river, and away from boaters on reservoirs. The TVA was slow to recognize this. In 1999, the partly constructed Duck River Dam was demolished due to environmental concerns and the escalating costs of completing the construction.

The Case of the Tennessee River

Latter chapters will explore the construction of "big concrete dams" on the Colorado River, the Columbia River, the Missouri River, and some of North America's smaller rivers. The problems these dams engender for the environment, and for the populations that rely on these rivers, are significant and explored. The geography of the Tennessee River Valley, however, requires that its floods be controlled. The big concrete structures with controlling spillways enable coordinated taming of this artery. This currently appears necessary for the prosperity and civilization of the Valley.

For example, the TVA's flood control network of dams averted an estimated $137 million in flood damages in Chattanooga following ten inches of rain in March of 2021. Also, according to TVA's estimate, $772 million was averted in the winter of 2020 when Chattanooga was deluged in February. These dams, the TVA argues, are strongly justified by this flood damage aversion.

TVA's hydropower generation is now only ten percent of all the electric power the Authority generates. Its dam's existence, however, will be less justified by its power generation in the future as alternative energy sources come online. Also, with modern highways and rail transport, the Tennessee River's dams and locks will be of lower relative importance than in the 1920–1950 period. But flood controls now appear to be increasingly necessary as climate change is increasingly generating increased precipitation and stronger storms. Eliminating the big concrete and returning the Tennessee to a more freely flowing river appears to be unlikely and perhaps unjustified. Latter chapters, however, will pose some very different stories for other river systems.

Appendix

Fort Loudoun Dam is located 50 miles downstream from Knoxville on the Tennessee River. It is a "big concrete" dam of 122 feet high and 4,190 feet long.[31] Fort Loudoun Lake is 14,600 acres in area. Its locks facilitate navigation up to Knoxville. The dam generates 155.6 megawatts of peak electric power. It was constructed during the World War II production emergency of 1940 to 1943, and its power was used for military production. For its construction, TVA purchased 16,200 acres and displaced 317 residents.

Norris Dam was the TVA's initial construction as authorized in 1933. It is a big concrete dam (1,860 feet long and 265 feet high) constructed on the Clinch River in eastern Tennessee. The Clinch River flows 300 miles from Virginia through the relatively unpopulated hill country of

[31] A mile is 5,280 feet.

northeastern Tennessee. The dam was named after Nebraska's Senator George Norris, the great champion of public power who was considerably responsible for both the building of the Wilson Dam (see above) and the establishment of the TVA.

This dam was considered necessary for control of the rampant floods of the Tennessee River. Private utilities—such as Knoxville Power and Light—sought to construct a dam on the Clinch River for hydroelectric generation, but Senator Norris argued that a dam was needed for both power generation and flood control in coordination with other TVA dams. A private utility would not be interested in such coordinated public interests.

Construction of Norris Dam required the controversial purchase of more than 152,000 acres of family farms and small towns. Approximately 3,000 families were relocated to accommodate the 53 square miles of Norris Lake behind the dam. This dam generates 126 megawatts of hydropower.

Watts Bar Dam and Lake are located approximately half-way between Chattanooga and Knoxville. This dam is a "big concrete" structure of 112 feet high and stretches 2,960 feet across the Tennessee River. It generates 175 megawatts of electric power. It facilitates navigation up to Knoxville, and also into the Clinch River. For this navigation, Watts Bar Lake is long thin extending 72.4 miles to Fort Loudoun Dam. To construct this dam and lake, the TVA purchased 54,600 acres and displaced 832 families. It was constructed between 1939 and 1942 when nearby military airplane manufacturing needed electric power for aluminum smelters.

Chickamauga Dam was primarily constructed to protect the City of Chattanooga—located just to the west of the dam—from its severe flooding problem (see the section below) and also to protect navigation into Central Tennessee. It is a "big concrete" dam of more than a mile in length, and 129 feet in height. It does, however, generate 119 megawatts of electric power. Its reservoir extends 59 miles to the base of Watts Bar Dam. This dam's construction required the purchase of 61,350 acres and displaced 903 families. It has a single lock that needed substantial federal funding to be restored in 2013.

Guntersville Dam is named after the nearby Alabama town of Guntersville. It is a "big concrete" hydroelectric dam that generates 140,400 kilowatts at peak capacity. It is 94 feet high and three-quarters of a mile long as it stretches across the Tennessee River. Although it generates hydropower it was constructed between 1935 and 1939 for flood control and navigation with a single lock that enables transport beyond Wheeler Dam. The dam's reservoir is 73 miles long up to the base of Nickajack Dam in Alabama. Its construction required the purchase of 110,145 acres and the relocation of 1,182 families. It was completed in January of 1939.

Wheeler Dam is fifteen miles upstream from Wilson Dam and the City of Muscle Shoals in Alabama. It was constructed by the TVA between 1933 and 1936. It is a "big concrete" dam of 1.2 miles in length, and 72 feet in height. It generates 411,800 kilowatts of electrical power at its peak capacity. Its reservoir is 67,070 acres, and stretches to the upstream Guntersville Dam to the east. Wheeler Dam has two locks to facilitate navigation. To construct this dam and reservoir, 103,400 acres were purchased by TVA which displaced 840 families.

Pickwick Landing Dam was constructed by the TVA as necessary for allowing navigation between the lower section of the Tennessee River and the Wilson Dam's reservoir at Muscle Shoals. The dam is in Tennessee. It was constructed between 1935 and 1938, and is "big concrete" with a height of 113 feet and length of 7,715 feet. It generates 240,200 kilowatts at its peak. Pickwick Lake, the dam's reservoir, is 43,100 acres and 53 miles long, reaching to the base of Wilson Dam. It has two locks to enable navigation. Building the dam and reservoir required the purchase of only 63,700 acres and displaced 506 families.

Kentucky Dam is the lowermost of TVA's nine dams along the Tennessee River; it is also the largest. It is 206 feet high and 1.6 miles long with an electric generating capacity of 233 megawatts. It is a "big concrete" dam with the TVA's biggest storage reservoir—Kentucky Lake—which two-hundred square miles of storage stretching 184 miles to the base of Pickwick Landing Dam near the Tennessee—Alabama border. It was constructed between 1938 and 1944, and is TVA's most expensive project at $118 million in 1944 dollars. To construct this dam, TVA purchased 320,244 acres and displaced 2,609 families.

Kentucky Lake is located in far Western Kentucky near Paducah, where the Tennessee River flows into the Ohio River. This dam and lake were constructed to reduce flooding on the lower Ohio River and the Lower Mississippi River. Four percent of the water flow of the Lower Mississippi originates in the Tennessee River Watershed. As such, Kentucky Lake is the largest artificial reservoir in the Eastern US.

Since Kentucky Dam is located close to the New Madrid Seismic Zone which experiences large earthquakes, it was constructed to withstand major shocks. In addition, the locks of the Kentucky Dam enable large barges to navigate into Kentucky Lake. This has enabled a significant complex of chemical plants to locate below the dam.

Blue Ridge Dam is upriver from three other dams on the Toccoa/ Ocoee River operated by the TVA. It is constructed close to the point where the states of Georgia, Tennessee, and North Carolina meet. The dam impounds the 3,300-acre Blue Ridge Lake on the border of the Blue Ridge Mountains. The Chattahoochee National Forest surrounds Blue Ridge Dam and its lake. The City of Blue Ridge is immediately west of the dam.

Toccoa Electric Power Company (a subsidiary of Tennessee Electric Power Company) built and operated Blue Ridge Dam in 1931. The passage of the TVA Act in 1933 gave oversight of the Tennessee River Watershed to the TVA. When the Supreme Court upheld the Act in 1939, the dam and its facilities was sold to the TVA. This hydro dam produces 22 megawatts of electric power, but construction/maintenance problems have been present.

Hiawassee Dam is a hydroelectric and flood control dam constructed on the Hiawassee River of North Carolina, a tributary of the Tennessee River. It is a high (307 feet) and long (1,376 feet) "big concrete" structure that generates 185,000 kilowatts of peak electric power. The Hiawassee River was particularly flood prone and added substantially to the flooding problems of the Tennessee River and therefore controlling it during high water stages was considered necessary.

The Hiawassee River flows from northeast Georgia northward through Western North Carolina. The dam is ten miles upstream from Apalachia Dam—close to the Tennessee and North Carolina border—and forty-five miles downstream from Chatuge Dam. Hiawassee Dam and reservoir are surrounded by Nantahala National Forest. The Hiawassee Lake stretches twenty-two miles upstream. Constructing this dam required the purchase of 24,102 acres and displaced 261 families.

In the 1950s, TVA began pumping storage water into Hiawassee Lake. It used turbine pumps during low-demand times to pump water into the reservoir, and used the released water flow to facilitate high power generation during peak demand times. This was the first US hydro system to operate this way.

Cherokee Dam and Lake are on the Holston River which initiates near Kingsport in Tennessee. It then flows140 miles southwest across northeastern Tennessee to join French Broad River in Knoxville and form the Tennessee River. Cherokee Dam is located 52 miles upstream from the Tennessee River. Cherokee Lake stretches 59 miles upstream from the dam. This dam has no locks that facilitate navigation. It does provide some flood control, and it does generate 135,200 kilowatts of power. Its construction required TVA to purchase 45,158 acres and displaced 875 families. It was constructed in 1940–1941.

Cherokee Dam's primary purpose was to generate hydroelectric power for the nearby aluminum production for the World War II military airplane industry.

Chatuge Dam is an earth and rock structure built for flood control on the Hiwassee River in North Carolina. It's the uppermost of the three dams on this river. The TVA constructed this dam in the early 1940s (1941–1942). This dam is located 121 miles upstream from the mouth of the Hiwassee River as it flows into the Tennessee River near Cleveland, Tennessee. Lake Chatuge covers eleven square miles on the border of Georgia and North Carolina. Chatuge Dam was constructed to regulate flow into the Tennessee River to prevent flooding of nearby Chattanooga which is only 30 miles from Cleveland and fifty miles from the Hiwassee River. To construct this dam, the TVA needed to purchase only 11,641 acres and displace 278 families.

Because Chatuge Dam was constructed quickly to contain the winter rains of 1941–42. It was constructed of rock and rip-rap (smaller rock fill) and clay with only a concrete spillway. A ten megawatt hydrogenator was added later to provide power to East Tennessee's World War II aluminum industry. The dam is 144 feet high and more than half-mile long.

Apalachia Dam is a purely hydroelectric structure on the Hiwassee River in North Carolina, and is the lowest of the three dams constructed by the TVA on that river. It was constructed to provide electrical power to East Tennessee's World War II aluminum industry. An 8.3-mile underground conduit transfers water to a hydrogenator downstream from the dam. The dam is named for the nearby Apalachia community and its L&N railroad terminal. It is located 66 miles upstream from the Hiwassee's joining with the Tennessee River.

The Apalachia Dam is of "big concrete" construction of 150 feet in height and 1,308 feet in length. Apalachia Lake extends 9.8 miles to the base of Hiwassee Dam and is 1,070 acres in surface. TVA purchased

4,500 acres and relocated 22 families to build this dam. It was constructed between 1941 and 1943.

Nottely Dam is of earth and rock construction. It is 184 feet in height and 2,300 feet in length. It was built in 1941–42. It is primarily a flood storage structure on the Nottely River in Georgia close to the Tennessee border. It purpose is to control flood flows into the Hiwassee River and therefore into the Tennessee River and possible floods of Chattanooga. Lake Nottely has a surface area of 4,180 acres and is 20 miles in length.

Nottely Dam is located 21 miles upstream from its junction with the Hiwassee River. This dam and lake are entirely surrounded by the Chattahoochee National Forest. To construct this dam, TVA purchased 7,984 acres and relocated 91 families. A hydrogenator was added in the 1950s that generates 15 megawatts of power.

Fontana Dam is a hydroelectric facility on the Little Tennessee River in North Carolina. It was constructed to satisfy the high demands of World War II's aluminum industry in the eastern Tennessee Valley. It was constructed in the 1942–44 period. Fontana Dam is a "big concrete" structure of 480 feet in height (the highest dam East of the Mississippi) and 2,365 feet in length. Its three hydro-generators produce a peak capacity of 293.6 megawatts. Fontana Lake covers 10,230 acres of surface, and 29 miles of the Little Tennessee River plus 11 miles of its tributary the Tuckasegee River.

The Little Tennessee River flows 135 miles from its origins in North Georgia to its junction with the Tennessee River at Lenoir City. Fontana Dam is 61 miles upstream. The small Fontana Village is just south of the dam.

The Aluminum Company of America (Alcoa) purchased land along the Little Tennessee in the early 1930s, and did so with anticipation of constructing and using hydropower dams for aluminum smelters. TVA negotiated for these areas and emergency World War II measures allowed it to finance the purchase of 68,292 acres. TVA relocated 1,311 families for the Fontana Dam's construction. The hydroelectric power from the Fontana Dam did power Alcoa's aluminum production.

Douglas Dam is a hydroelectric facility on the French Broad River in Tennessee. It was constructed in the very quick time of approximately one year at the height of World War II—1942 to 1943—in order to satisfy the electric power demands of the Manhattan Project's nearby Oak Ridge Plant for separating uranium isotopes. Douglas Dam is a "big concrete" structure of 202 feet in height and 1,705 feet in length. Douglas Lake has a surface of 28,420 acres.

French Broad River flows westward from the Appalachian Mountains. It absorbs tributary flows from the Pigeon River and the Nolichucky River, and then joins the Holston River at Knoxville to form the Tennessee River. Douglas Dam is 32 miles upstream from the Holston River.

During 1941, President Roosevelt requested that Congress allocate funds for Douglas Dam's construction because of the need for war production. But this dam would flood significant farming areas and Congress hesitated to allocate the funding. After the Pearl Harbor attack, however, and the organization of the Office of War Production Management, the need for a radical increase in TVA's power production became a priority. The Douglas Dam Project was completed in February of 1943, and began electric generation in March of 1943. Figure 2.1 indicates the location of the Holston River, and the Douglas Dam is 32 miles upstream. The legend in Fig. 2.2 explains the locations listed in Figs. 2.1, 2.2 and 2.3.

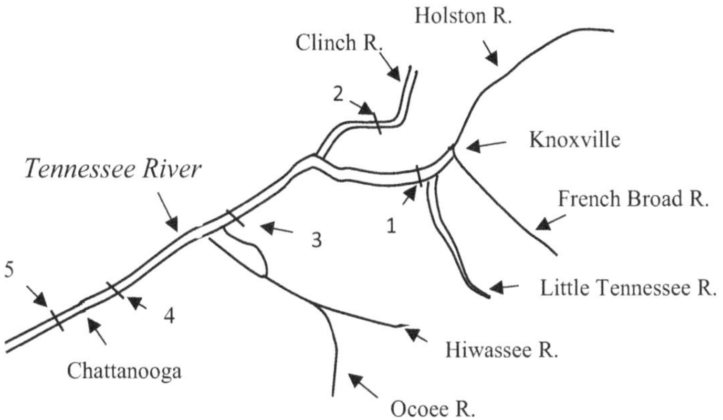

Fig. 2.1 Tennessee River from Knoxville to Chattanooga

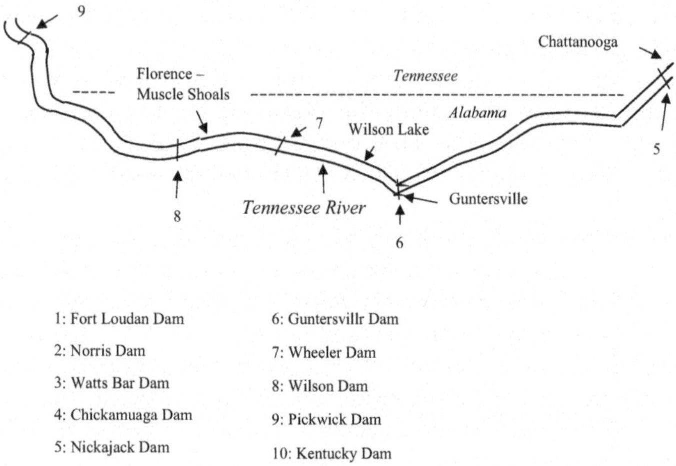

Fig. 2.2 Tennessee River from Chattanooga to Florence—Muscle Shoals

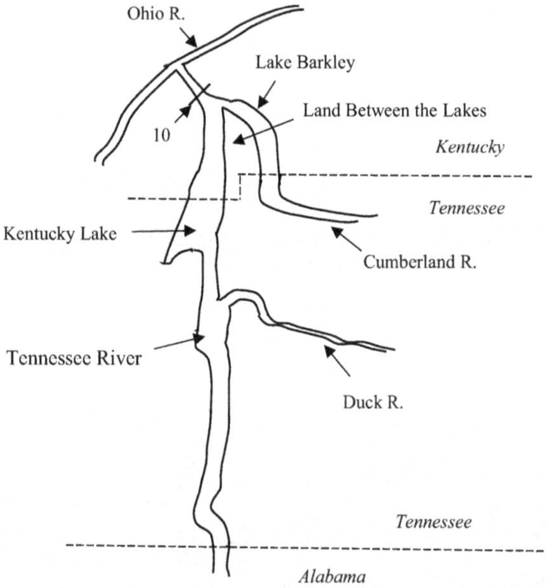

Fig. 2.3 Tennessee River from Alabama to Ohio River

References

Billington, D., Donald, J., & Martin, M. (2005). *The History of Large Federal Dams: Planning. Design, and Construction in the Era of Big Dams,* US Department of the Interior, Bureau of Reclamation, Denver, CO.

Eschner, K. (2017, May 18). Here's How FDR Explained Making Electricity Public. *Smithsonian Magazine*. www.smithsonianmag.com/smart-news-heres-how-fdr-explained-making-electricity-public-180963286/

Hession, C. H., & Sardy, H. (1969). *Ascent to Affluence: A History of American Economic Development*. Allyn and Bacon, INC.

Hiltzik, M. A. (2010). *Colossus: Hoover Dam and the Making of the American Century*. Free Press.

Hughes, T. (1979). The Electrification of America: The System Builders. *Technology and Culture, 20*(1), 124–161.

Madden, T. (1981, July 2). Private Land TVA Claimed for Lake to be Given Away to Developers. *UPI* archived.

Matthiessen, P. (1980, February 7). How to Kill a Valley. *The New York Review*. www.nybooks.com/srticles/1980/02/07/how-to-kill-a-valley

Plater, Z. (2008, April). Tiny Fish/Big Battle. *Tennessee Bar Journal, 44*(4), at the Wayback Machine.

Plater, Z. (2013a, May 9). *Annotated Timeline of TVA v. Hill and Related Events*. https://core.ac.uk/download/pdf/71464142.pdf

Plater, Z. (2013b, June 18). *The Snail Darter and the Dam: How Pork-Barrel Politics Endangered a Little Fish and Killed a Reservoir*, Yale University Press, New Haven, CT.

Rawls, W. (1979, November 11). Forgotten People of the Tellico Dam Fight. *New York Times*.

Riggs, J. A. (2020, November 9). Roosevelt's Drive to Break Up Utilities Brought Power to the People. *Historynet*. www.historynet.com/roosevelts-drive-to-break-up-utilities-brought-power-to-the-people/

Robinson, R. M. (2023). *Environmental Advocacy and Local Restorations*. Palgrave-Macmillan.

Robinson, R. M. (2021). *Environmental Organizations and Reasoned Discourse*. Palgrave-Macmillan.

Sablik, T. (2020). Electrifying Rural America. *Econ Focus*, First Quarter, Federal Reserve Bank of Richmond.

Venable, S. (2022, October 13). Many Decades Later, Ramifications of the Tellico Project Endure. *Knoxville News Sentinel*. Knoxville, TN.

Wasik, J. F. (2006). *The Merchant of Power*. Palgrave Macmillan.

CHAPTER 3

Environmental Justice in the Olympics and on the Ottaway

Introduction

This chapter explores two of the more significant river restoration movements initiated through dam removals. Both were organized by American indigenous tribes: the removals of dams along the Elwha River in the Olympic Peninsula of Washington State, and the dam removals along the Ottaway River of Michigan. Skillful approaches to bringing these deconstructions to actuality have been exhibited by the Klallam tribe of the Olympics, and by the Ottawa and Ojibwe Tribes of Michigan's Great Lakes Region. In particular, the new experimental *FishPass* technology being applied to the last remaining lower dam on the Ottaway River is of very significant importance for controlling invasive destructive species in the Great Lakes tributaries (and also for globally important river systems such as the Amazon, Nile, and Mekong). These important developments are reviewed here.

Justice for the Klallam

The Olympic Peninsula is on the northwest corner of Washington State. It lies across the Puget Sound from Seattle and contains the Olympic National Park. The Peninsula is bordered on the east by the Puget Sound, on the west by the Pacific Ocean, and on the north by the Strait of Juan de Fuca. This strait separates Washington State from Vancouver Island in

R. M. Robinson, *Restoring America's Rivers*,
https://doi.org/10.1007/978-3-031-81758-8_3

Canada. Port Angeles is in the middle of the north coast of the Peninsula (Fig. 3.1 maps this area.) In size, the Peninsula is approximately 3,600 square miles. The Olympic National Park occupies the center of the Peninsula. At 1,370 square miles, it covers more than a third of the Peninsula. The Olympic Mountains occupy this National Park with a series of six-to-eight-thousand foot glaciated peaks, the highest being Mt. Olympus at 7,965 feet which is located slightly northwest of the Olympic National Park's center.

The Elwha River flows northerly from the glaciated center of the Olympic National Park toward Port Angeles on the Strait of Juan de Fuca. Prior to its damming in the early twentieth century, the Elwha River was one of the few rivers in the contiguous US that supported all of the Pacific Northwest's native anadromous salmonid species: Chinook, chum, coho, silver, and sockeye salmon, plus steelhead, bull, and cutthroat trout. The river provided a prolific environment, especially for very large Chinook that were frequently over forty pounds. Prior to the early twentieth-century's construction of two dams on the Elwha, almost 400,000 fish returned annually to spawn. By the late twentieth century, only approximately 3,000 returned in any year. Silver salmon—often called pink salmon—were once estimated at more than 250,000 returning adults per year. But by the 1980s, these returns were almost nonexistent.

Eighty percent of the Elwha River's watershed lies within the Olympic National Park where it is consequently protected. But the Elwha Dam was privately constructed between 1910 and 1912, before the National Park was created. This blocked fish passage and limited salmon and steelhead migrations to only the lower 4.9 miles of the approximately sixty miles of the River. In 1927, Glines Canyon Dam was also constructed seven miles upriver from the Elwha Dam. When the Olympic National Park was created in 1938, it did not include the Glines Canyon Dam. But in 1940, the Park's boundaries were expanded to include this structure although dams were incompatible with the Park's ecological preservation purpose. It would eventually be deconstructed after a lengthy contentious fight which began, as reviewed in Chapter 1, with the events of the 1999 Edward's Dam removal in Maine.

Before reviewing the events related to the damming of the Elwha River, however, it is necessary to review the indigenous people who lived along the Elwha River. The Coastal Salish are indigenous people who have resided for thousands of years along the Pacific Northwest Coast: British Columbia, the Olympic Peninsula adjacent to the Puget Sound, plus other Washington coastal areas, and the Oregon Coast as far south as Tillamook (about half-way down the Oregon Coast toward California). In 2013, there were 56,590 members of the Coastal Salish: 28,406 with *Indigenous Status* in Canada's British Columbia, and 28,284 in federally recognized tribes in Washington State. The S'Klallam consists of three local tribes who are Coastal Salish people by language and customs. The three S'Klallam Tribes include (1) the Jamestown S'Klallam who established the town on the Olympic Peninsula that bears their name, (2) the Port Gamble S'Klallam who reside largely on Vancouver Island in British Columbia, and (3) the Lower Elwha River S'Klallam who reside along that River. These three indigenous tribes were largely sustained from the marine bounty of the area, and were therefore seasonally mobile. But they wintered in traditional villages of rambling wooden longhouses. It is important to note, however, that after the European settlers invaded the S'Klallam's traditional areas during the nineteenth century, the indigenous tribes did react by working at the new sawmill in Jamestown, and also by selling their harvests to the white settlers, and thereby investing their income to repurchase lands along the Peninsula to renew their own settlements. Note that they had to previously cede their lands according to the *Point No Point Treaty of 1855.*

The indigenous Klallam Tribe has lived along the Lower Elwha River for thousands of years (Although the Klallam include the Jamestown, Port Gamble, and the Lower Elwha Klallam, the latter tribe is referred to here as just the "Klallam" which means "strong people"). The damming of the Elwha River, therefore, badly and significantly interrupted their culture (This damming is reviewed below). The loss of salmon due to the Elwha River's dams devastated the culture and livelihood of the Klallam. The Elwha Dam was tragically constructed on the Klallam's traditional "site of its creation." This was an unconscionable imposition on the Klallam's culture. Before this story is told, however, the many court fights concerning salmon and the indigenous of the Northwest should be reviewed. These battles center on the famous *Boldt Decision* and the *Treaties of 1855*.

Salmon, the Indigenous Tribes, and the Courts

In 1974, the US Ninth District Court under Judge Boldt decided that the State of Washington infringed on the Indigenous' Treaty Fishing Rights established by the collection of agreements referred to as the *Treaties of 1855*. In particular, the decision found that:

- The Indigenous have the right of access through private property to its traditional fishing locations.
- The state cannot charge the Indigenous any fees to fish, nor discriminate against the tribe's fishing methods.
- The Indigenous must have an equitable opportunity to share in the total harvest.

In 1975, the Ninth Circuit Court of Appeals upheld Judge Boldt's decision, and the US Supreme Court refused to further hear an appeal. But the State of Washington refused to enforce the *Decision*, so that Judge Boldt had to order the US Coast Guard and other federal agencies to enforce the Court's decree.

These *1855 Treaties* consisted of a series of agreements with numerous Northwest Tribes such as the Quinault, Muchleshoot, Squaxin, Sauk-Suiattle, Stillaguamish, Queets, Makah, Lummi, Hoh, Yakama, Skagit, Quileute, Nez Perce, Umattila, Klallam, and others. In particular, these agreements generally established the following:

> The right of taking fish at all usual and accustomed grounds and stations is secured to said Indians in common with all citizens of the Territory, and of erecting temporary houses for the purpose of curing the same; together with the privilege of hunting, gathering roots and berries, and pasturing their horses. (Blum & Swift, 1998, p. 421)

These various Treaties included the Treaty of Medicine Creek, Treaty of Point Elliot, Treaty of Neah Bay, and the Treaty of Point-No-Point among others. The latter treaty was signed on January 26, 1855, and included the Jamestown, Port Gamble, and Lower Elwha Klallam, plus the Skokomish who also resided in the Puget Sound area. All of these *1855 Treaties* used similar language that assured traditional indigenous fishing rights off of the reservations. In each treaty, indigenous tribes were coerced into parting with their land and moving onto reservations, all supposedly with the object of protecting their traditional fishing practices at customary locations.

By 1883, however, settlers had established more than forty salmon canneries in the Pacific Northwest. By 1905, there were more than twenty-four canneries just on the Puget Sound alone. Salmon harvests were being affected by industrialized fishing. When Washington became a state in 1889, the legislature passed regulations supposedly designed to curtail the indigenous tribes' fishing for conservation reasons. But these regulations were actually designed to protect the nonindigenous fishing harvests. For example, the legislature banned the use of "weirs" for fishing, a customary method used by the indigenous.[1] This was blatant discrimination in apparent violation of the *Treaties of 1855*. In response, the tribes sought legal recourse.

In 1887, the Yakama sought to fish on traditional locations including privately owned non-reservation lands. For example, this included Celilo Falls at the Dalles on the Columbia River for which the US Supreme Court upheld indigenous fishing rights of access. In 1914, the US Supreme Court reaffirmed this indigenous access right.

In *Tulee v. Washington*, the US Supreme Court also prevented the State of Washington from imposing fees on indigenous fishing. Also, in a series of decisions in 1968, 1969, and 1973, in *Puyallup Tribe v. Department of Game of Washington State*, the US Supreme Court upheld the tribes' rights to use fishing nets. It found that the State of Washington was discriminating against the indigenous tribes by prohibiting them from using nets while simultaneously allowing nonindigenous commercial fishermen to do so. Also, the Court held that any fishing regulation must only be for the purpose of conservation and not be merely an excuse to discriminate against tribal rights. In this Puyallup case, the Court also held that the use of the word "equitable" in this treaty indicated that the indigenous tribes have the right to pursue half the harvest of salmon every year, and the state cannot take actions aimed at otherwise limiting this attempt relative to what it allows for nonindigenous fishermen. The State's regulatory attempts at fish-run preservation must also be fully enforced for nonindigenous fishing if it is imposed on the tribes.

The Puget Sound Gillnetters Association and the Washington State Commercial Fishing Vessel Association filed lawsuits in state court to

[1] A "weir" is a barrier trap that channels the fish into particular directions that enable them to be more easily caught.

block the new federal court-imposed regulations that supported the indigenous rights. These Associations won in state courts, but in 1976 the US Supreme Court overruled the state courts. The federally imposed regulations would stand.

In 2001, twenty-one Northwest Tribes asked the US District Court to recognized that the State has a treaty-based obligation to preserve the salmon and steelhead runs and their spawning habitat, and that this applies particularly to the problem of the many State-installed poorly designed street culverts that impeded fish migrations. In 2007, the Court found against the State of Washington and ordered it to significantly increase its efforts to change these culverts and otherwise improve spawning habitat so as to protect the salmon and steelhead runs.

The public's reaction to this series of Court decisions was antagonistic. Nonindigenous commercial fishermen ignored the court's decrees, and the state was reluctant to enforce them. In 1984, Washington State voters even passed a referendum-initiative that declared an end to the "special rights" of the indigenous, but the State also refused to enforce this voter initiative. By 1984, tribal members were engaged in commercial fishing, and had forty-nine percent of the year's catch. By this time, the tribes had become co-managers of the salmon fishery. They hired biologists and staff to advise and aid in their fisheries' management. The legal struggles to resolve the fishing rights of the Northwest Tribes were somewhat completed after seventy years of court battles, but this was certainly no more than a pyric victory if the salmon and steelhead runs were nonexistent. Deconstructing the two Elwha River dams were the key next step for the Klallam Tribe, a step aimed at preserving their sustenance and culture.

The Two Dams

By signing the *1855 Treaty of Point No Point*, the Lower Elwha Klallam were forced to relinquish their former lands and be restricted to their reservation on the Elwha's delta along the Strait of Juan de Fuca, a short distance west of Port Angeles.[2] (See Fig. 3.1.) In return, they were guaranteed the right to forever harvest salmon and steelhead from all of their

[2] See "Treaty of Point No Point, 1855" at HistoryLink.org, Essay 5617.

traditional sites along the Elwha River. This treaty explicitly promised "... the right of taking fish at usual and accustomed grounds and stations ..."

The Elwha and Glines Canyon Dams, however, were constructed in 1913 and 1926 respectively for the purpose of generating electric power for Port Angeles on the Washington side of Juan de Fuca Strait. The electric power was mostly used by the Nippon Paper Mill in Port Angeles. But the power provided was only slightly more than a third of that required for this single facility. Overall, the power provided by the Elwha River's hydroelectric facilities was only a small fraction of the total power used by Port Angeles.

Given that the dams would obstruct the salmon and steelhead runs, and given that these runs were at the heart of the Elwha's culture and sustenance, the construction and existence of these dams were always opposed by the Klallam. The Elwha Dam was actually constructed outside the legal process of either federal or Washington State regulations. Since 1890, the State regulations required fish passage facilities for any dam, but this was not provided at the Elwha Dam. The Dam's construction never met any of the State's licensing requirements. The Glines Canyon Dam, however, was granted a fifty-year license under the Federal Power Act in 1926. But this dam was also constructed without a fish passage facility.

When the Glines Canyon Dam's fifty-year license was expiring in the 1970s, the owner of the dams (Crown Zellerback at the time) applied to FERC for a license renewal that would include both dams. But the Klallam obtained federal recognition as a tribe in 1968, and this "recognition" allowed their complaints to be officially considered by FERC in its licensing process. (See Chapter 1 for a review of the Federal Energy Regulatory Commission—FERC—and its licensing requirements.) In 1978, however, the Elwha Dam failed a safety inspection, and consequently the Klallam perceived the opportunity to have the structure decommissioned. It hired an engineering firm to analyze the probability of failure and provided this analysis to FERC. The Elwha Dam had some flaws of weakness at its foundation. The Klallam called for the Dam's removal.

In 1979, FERC declared it would treat the Elwha and Glines Canyon Dams together as a single licensing consideration. It refused a multi-year license for either of the combined entities, and issued only a series of single-year permits while the longer-term licensing was under consideration. This series of single-year permits lasted until the mid-1980s. But the fact that Glines Canyon Dam was within the confines of the Olympic

National Park, and FERC could not issue a license to a privately operated hydroelectric dam within the confines of a National Park, meant that the relicensing of the combination of dams was unlikely.

In addition, the 1974 *Boldt Decision* favored habitat preservation. Relicensing would likely not serve the interests of preservation and therefore would conflict with *Boldt*. Also, in 1986 the "Electric Consumers Protection Act" required FERC "to give equal consideration to the purposes of energy conservation; ...; enhancement of fish and wildlife; the protection of recreational opportunities; and the preservation of other aspects of environmental quality while issuing licenses."

In 1987, Crown Zellerback was purchased by the James River Corporation which sought to sell the dams to the federal government. The Klallam once again called for the dams' removal. But this time, the Tribe provided studies that showed that the dams' deconstruction was the best environmental option, and was also more cost effective than all the other alternatives.[3] Before FERC could announce its decision, however, Congress acted in 1992 and passed the "Elwha River Ecosystem and Fisheries Restoration Act." This Act called for "... full restoration of the Elwha River ecosystem and the native anadromous fisheries." The James River Corporation finally sold the dams to the federal government in February of 2000. The two dams could now be removed, but given the problems associated with the removal of such large dams, they could not be removed quickly. The deconstructions were not completed until September of 2014. Note that these dams were the largest dam deconstructions up to that date.

The total cost of the Elwha River's restoration was estimated by NOAA at $351.4 million. This figure included the purchase of the two dams for deconstruction and related facilities, plus construction of two water treatment plants to remove the new sediment from the public water intakes, plus construction of flood control facilities, and a greenhouse for raising native plants for revegetation of the riverbanks.

This "full ecosystem approach" to restoration called for by the 1992 Act reflects the holistic philosophy of the Klallam as expressed by an elder, "Its not just about taking the dams out, or even just putting the fish back. Its about the whole picture, the human population, marine predators,

[3] See Guarino (2013).

overfishing, the works. If the system is addressed, then maybe restoration will work."[4]

The Sediment of the Elwha

As with the other dam deconstructions considered in this book, sediment disposal was a major consideration for the Elwha dams. This sediment was substantial, and once the dams were constructed, it accumulated behind the Glines Canyon Dam for nearly 100 years. This was sediment that the fast-flowing currents of the undammed river previously transported to the bluffs of Port Angeles and Ediz Hook (see Fig. 3.1). The sediment previously established the River's once substantial delta at the Strait of Juan de Fuca, and also established the Ediz Hook. But when the sediment flow stopped at the Glines Canyon Dam, then without this replenishment the wave action and currents through the Strait eroded the sandy beaches until they became rocky and pebbly slopes. The delta shrunk by several acres which eliminated the clam beds. After all this erosion, Ediz Hook's maintenance required that the Army Corps of Engineers replace the barrier with rip-rap to protect the Port Angeles Harbor.[5] The Corps spent $100,000 annually to offset the erosion of Ediz Hook.

Previously, the undammed Elwha River was fast-moving and carried considerable sediment. Prior to dam removal, Lake Mills, the reservoir behind Glines Canyon Dam, contained thirty-three million tons of this sediment. The US Geological Survey (USGS) estimates that five years after undamming, eight-million tons of silt accumulated along the River's banks, and this included the formation of a new and currently expanding delta. Fourteen million tons of this old sediment has also dispersed into the Strait of Juan de Fuca and the Pacific Ocean.

An important area of investigation is the nearshore environment and estuary system, and also the delta, and the shallow water seagrass beds. Salmon and other fish require a healthy nearshore environment to survive, and for the Elwha River, these areas tend to be outside the National

[4] See quotes from tribal members presented at http://0318b58.netsolhost.com/effectsontheelwhaklallampeople.html.

[5] "Rip-rap" consists of the collection of smaller and larger rock used as a breakwater.

Park's protection.[6] A river system is a primary source of sediment nutrients which accumulate in nearshore environments. With the Elwha dams' constructions, the nearshore areas were starved for nutrients for one-hundred years. But within weeks of the dams' removals, the Elwha began reestablishing nearshore habitat. Two years after the dams' removals, the riverbed at the mouth of the Elwha was thirty-three feet higher than pre-removal levels. The Elwha's delta was beginning to be reestablished.[7] All the salmon species and steelhead were returning, as well as the Dungeness Crab, clams, sea birds, and other species. The shoreline is also transforming from rough cobblestone to sandy beach.

Every year since the dam removals, snorkeling teams have surveyed the returns of steelhead to the Elwha above the former dams' positions. The increase in summer-run steelhead has been substantial from only a few for the first year to 340 after five years. Of particular importance is the observation that very few were hatchery bred; only six of the 340 observed in 2019 were produced by hatcheries.[8]

A wide array of studies by NOAA Fisheries examined the results of the dam removals after a five-year interval. They concluded that by all the measures observed, the River is now more of a healthy entity than it was before the undamming.

Michigan's Ottaway River, Fish Passage Technology, and the Grand Traverse Band of Ottawa

The *Grand Traverse Band of Ottawa and Chippewa* (GTB) have resided for thousands of years in the watershed of the Ottaway River (The Chippewa are also known as the "Ojibwe," and the Ottawa as the "Odawa.") In 1855, these tribal nations signed a treaty that relegated them to a federal reservation that is now located in the Ottaway River's watershed, close to Traverse City. (Figure 3.2 maps this area.) Under the "Indian Reorganization Act of 1934," the GTB, however, were not federally recognized as a functioning tribe until 1980. The recognized

[6] See Howard (2016, June 2).

[7] Ibid.

[8] The hatchery-bred trout are easily identified by their notched fin.

federal reservation of the GTB is the "Grand Traverse Indian Reservation," a somewhat disconnected grouping of federally granted and tribally purchased parcels with approximately 1,100 combined acres to the south and east of Traverse Bay. There is slightly more than 1,600 tribal members living on the reservation, and approximately 4,000 tribal members in total. They were the first tribe to create a casino, an institution that generates income for tribal members.

The Ottawa (or "Odawa") is derived from the Algonquian word for "trader." They are an Algonquian speaking people who settled around the Great Lakes Area of North America. Both the Ottawa and Chippewa were recognized by the treaty of 1855 for the purpose of creating the above-mentioned reservation. The Chippewa (or "Ojibwe") are the largest indigenous people by population size north of the Rio Grande. Like the Ottawa, the Ojibwe are also located in the Canadian and US Great Lakes region. There are nineteen separate federally recognized tribal bands of Odawa and Ojibwe in this area. Note that the GTB have been involved in dam deconstructions and have an active "Natural Resources Department" of fish and wildlife biologists, game wardens, and other scientists dedicated to preservation and restoration of the area's waters and wildlife resources. This Department is highly relevant for operating the technology reviewed below.

The Ottaway River (the River's original name was "Ottaway," but it is also known as the Boardman River) flows into Grand Traverse Bay in the northwest corner of Michigan's Lower Peninsula. (Figure 3.2 maps this River.) The main stem of the Ottaway flows approximately twenty-eight miles, but combined with its tributaries they flow one-hundred and thirty miles to drain three-hundred square miles of watershed. From its headwaters, the river's course is westerly and then northerly into Boardman Lake about one-mile upstream from Grand Traverse Bay. This Lake is the reservoir formed by Union Street Dam in Traverse City which is adjacent to the Lake.

Traverse City is a small town of 8.5 square miles. Its population is only 15,678 according to the 2020 census. Its surrounding townships, however, exhibit considerable suburban sprawl with the bordering four counties having a combined population of 153,448. The City is located at the head of the east and west arms of Grand Traverse Bay, a thirty-two-mile inlet of Lake Michigan. The Bay is divided into two seventeen-mile arms by the "Grand Traverse Peninsula," now called the "Old Mission Peninsula." The "Grand Traverse" was the name originally applied to the area by Reverend Peter Doughtery who established the original settlement, a mission on the Grand Traverse. The Reverend also planted the first cherry orchard on the Grand Traverse, the fruit for which the area is now known for.

In 1872, the railroad connected to Traverse City primarily to serve the logging industry of Michigan. Logs from northwest Michigan were transported to sawmills in Traverse City, and lumber was shipped through Lake Michigan. Also, Reverend Doughtery's cherry orchard was expanded into an industry by the early 1900s, and the first "National Cherry Festival" began in Traverse City in 1925. It currently brings 500,000 visitors to Traverse City.

The climate of the area exhibits mild summers and severe winters with lake-effect snows from Lake Michigan. Upstream from the sawmills, the river has no industry so the river is clean and cold, and is designated as a "Blue Ribbon Trout Stream" by the State of Michigan (Michigan Department of Natural Resources).

At the end of the nineteenth and beginning of the twentieth centuries, Traverse City Light and Power (TCLP) constructed five dams on the Ottaway River. The Sabin Dam is upstream from the Union Street Dam, and the Boardman Dam is farther upstream, and then the Keystone Dam, and then the Brown Bridge Dam, all constructed on the Ottaway's twenty-eight miles. They were erected for hydropower generation and flood control. In 1961, the Keystone Dam was washed away by floodwaters. In October of 2012, the Brown Bridge Dam—the highest and largest by displacement—was washed away during preparations to demolish it. Its remains were removed in January in 2013. In 2017, the Boardman Bridge was demolished. The Sabin Dam was demolished in 2018. Only the Union Street Dam remains. Its deconstruction has been proposed by the local government of Traverse City.

The GTB have been instrumental in removal of three of the Ottaway River Dams, and of rebuilding a fourth dam to facilitate a controlled fish passage. Tribal leadership has been the essential ingredient for the Ottaway's restoration. They secured funds for a feasibility study and organized community support. The result is "the most comprehensive dam removal and watershed restoration in Michigan's history." (Fox et al., 2022, p. 38)

Brown Bridge Dam's license was expiring in 2003. It generated at most one megawatt of power, and FERC requirements for upgrades (especially for a rebuilt spillway to accommodate flood flows) would be more that $1 million. TCLP decided that continued operation would not be justified. Given the prospect of river restoration through dam removals, GTB arranged a settlement in 2005 that organized a community implementation team of citizens and environmental groups—the Boardman River Dam Committee (BRDC) for raising funds and arranging contractors. GTB raised $750,000 from the US Fish and Wildlife Service for a feasibility study of dam removal and restoration of the Ottaway River's ecosystem. It also raised $1.35 million from the Bureau of Indian Affairs for removal. This led to plans for removing the three upriver dams and modification of the downriver dam. The Brown Bridge Dam was removed in 2012, the Boardman in 2016, and Sabin in 2018. Construction designed to radically change the Union Street Dam (as substantially reviewed below) is currently underway.

One problem with the last remaining obstruction on the Ottaway River—the Union Street Dam—is that it is old and crumbling. In 2021, "Michigan's Department of Environmental Quality's" (MDEQ's) "Dam Safety Unit" downgraded this earthen dam from "Good Condition" to "Fair/Poor Condition." Twenty-seven trees had grown on the earthen dam which weakened its structure, and caused leaking. Frank Dituri, the Director of Public Works for Traverse City states, "If you walk along the base of the dam come springtime, you can see water actually seeping, weeping through it in places. It needs to be replaced." (LaFond, January 17, 2020)

There is, however, a second problem with the Union Street Dam; it is the only remaining obstruction that prevents invasive species from invading the upper Ottaway River. These invasions pose a massive problem throughout the Great Lakes region, and also in other globally significant river systems such as the Amazon, Nile, and Mekong. The invading species include the Sea Lamprey, the Asian Carp, the Round Goby, the Ruffe (see the Chapter Appendix), and also many other species such as America's steelhead trout (both Brown and Rainbow Trout) and Pacific Salmon (Coho, Sockeye, Silver, and Chinook). Salmon and steelhead would devour the native Grayling's eggs and young. They also harm the Lake Sturgeon, the walleye, and the smallmouth bass, all native to the river. Although the Grayling is actually in the salmon family, it is a highly desirable brook-trout type of fish that only reside in cold, clean water. They are native to the Ottaway River. (See the Appendix to this Chapter for a comprehensive review of the invasive species threatening the Great Lakes.)

Given the two significant problems cited above, the Union Street Dam cannot just be allowed to be washed away by flood waters, or even to be deconstructed by an orderly well-engineered process. It must be replaced while preventing invasive fish passage. But the cost of the latter option would be in excess of $12 million, a sizeable burden for Traverse City's small budget.

The solution to both of the above problems would be to install the new experimental dam technology *FishPass* which would be funded by a $20 million federal grant from the "Great Lakes Restoration Initiative" (GLRI.) (See Robinson, 2023, Chapter 3 for a detailed review of the GLRI). *FishPass* is a newly proposed untried experimental technology looking for a dam in need of replacement. It would ultimately lead to a new sort of dam. Its purpose is to allow some desirable anadromous fish—such as Ottaway's Lake Sturgeon—to pass upstream, but not other undesirable species, especially the Sea Lamprey which is a parasite that quickly kills larger fish. The GTB along with the MDNR would decide which of the species encountered would be appropriate to pass. The Ottaway was one of twelve candidates for this experimental technology for a GLRI subsidy. It was judged the most suitable.

FishPass is the planned new barrier that will replace the Union Street Dam. It involves a collaboration between the Great lakes Fishery Commission, the GTB, Traverse City, the Michigan Department of Natural Resources, the MDNR, and the US Army Corps of Engineers.[9] It is a modification of the "weir technology," the ancient indigenous method of channeling a river's current to a narrower opening that forces the anadromous fish into the passage for harvesting. This proposed weir construction will span slightly more than half the Ottaway River, and it will block all fish passage, allow flood currents to flow through, but maintain the level of the Boardman Lake. The constructed weir will have a semi-circle that resembles gear teeth. Two 400-foot concrete channels will be constructed on the side of the weir, and the migrating fish will be forced into the channels. There they will be separated, and the desired species allowed to pass; the undesired will be obstructed. The structure will be straddled by an overhead gantry crane which facilitates species identification and selection. The 400-foot channels will apply new digital-electronic fish identification methods, and even variation in light at the

[9] The Great Lakes Fishery Commission was established in 1955 by the *Canadian/US Convention on Great Lakes*. It coordinates fishing research, and controls invasive species.

openings for fish passages, plus variation in the stream's current speeds that would discourage passage for some species. Even the sound effects on the migrating fish will be part of the experimentation. This experimentation ultimately allows for the identification of what encourages passage of desirous species and discourages passage of those not desired. Perhaps, for example, certain flow rates encourage salmon to pass, but not other fish. The experimentation to discover what works best is estimated to require ten years. Upstream from the fish passages there will be an obstruction barrier for all fish so that only the appropriate species will ultimately be allowed to enter the upstream river via a "trap and transfer method." With *FishPass*, the GTB and MDNR will ultimately decide which fish species can pass.

Dan Zielinski, an engineer with the Great Lakes Fishery Commission and the *FishPass* lead engineer, points out, "So this is really the first opportunity where we can look at combining all these technologies that have been under research over the last fifty years by the Great Lakes Fishery Commission and, you know, over 100 years of research on fish passage in general." (LaFond, 2020, January 17) This experimental technology will allow discovery of the appropriate strength of current, and perhaps the light appropriate for the desired species' passage. This experimentation has great promise for appropriate barrier designed and constructions throughout the Great Lakes area, and also various river systems. In this application, the Sea Lamprey is the primary target invader.

The construction of the *FishPass* technology was formerly planned for 2021. The political context of this planning included the Edenville Dam failure in May of 2020. This was an old hydroelectric and flood control dam close to Midland in central Michigan. The Federal Energy Regulatory Commission (FERC) terminated its hydropower license in 2018 because of the dam's flaws. Michigan's Department of Environment, Great Lakes, and Energy (EGLE) assumed supervision of the dam and claimed it was structurally sound. Disputes over the appropriate water levels for the dam ensued, but when heavy rains occurred, the dam failed. The towns of Edenville, Sanford, and sections of Midland needed evacuations. Permanent damage occurred. Subsequent forensic evaluations indicated that the dam had original construction flaws that caused its ultimate failure. All of these disputes delayed funding for *FishPass*, but construction of *FishPass* is now planned for April of 2024.

It is possible that because of the *FishPass* technology, the devastating impact of dams on fisheries will be considerably modified, and significant control of invasive and destructive species will begin. Because of the political pressure to avoid failure of the Union Street Dam, and the experience of a catastrophe similar to Edenville's Dam failure, the *FishPass* alternative with its GLRI subsidy was readily accepted as a replacement for the last dam on the Ottaway. But at this time, a suit was filed by a Traverse City resident. The suit noted that the dam and adjacent park were part of Traverse City's parks, and an experimental facility such as *FishPass* was inconsistent with the park's stated purpose. It would "eviscerate the park's calm vibe." This legal dispute was ultimately settled by Michigan's Supreme Court in August of 2023. The *FishPass* facility will be constructed and is planned for Spring of 2024.

For the GTB, the Ottaway River is a living and sacred entity. It has a voice that speaks to the Tribe, and is then manifested through the Tribe's songs and ceremonies. (For a review of this culture, see Cook, 2017). Dam removals were considered as only the initial step in healing the Ottaway River's ecosystem, and ultimately in healing the Tribe itself. The Tribe believes that the "fish nation," the "deer nation," and all the living entities of the ecosystem have a voice through the Tribe. (Bailey, 2017 reviews this tribal philosophy.) This ecological voice led to the GTB organizing 200 public meetings with over 1,000 different community members participating in exploring the options for the dams' removals and/or modifications. The removals were controversial, especially among those with property adjacent to the River, and among others who used the reservoir for recreation. But the GTB persevered in attempts to coordinate the community's interests along with the efforts of government agencies. A member of the Army Corps of Engineers commented,

> The leadership role that the Grand Traverse Band has taken on this project has been something that I have never seen before. Not only that, understanding and working with the Tribe and really beginning for the first time in my own life to realize the issues that they face, the determination that they have to restore this environment has really been something that I will take away and that is going to last with me for a long time. (Kalish, 2017)

A member of the MDNR reinforced this notion stating, "I do know that the respect and admiration of the Tribe was immense through this project as a result of their deep, deep commitment not only to the watershed but to this project as a whole and to the community." (Kalish, 2017) A GTB member also commented, "We came together as a people, Native and non-Native people, and thought what was best for the water." (Shomin, 2017)

After the upstream dams were removed, the debate switched to what to do with the Union Street Dam. This was resolved largely by an overwhelming consensus for the *FishPass* technology. The Union Street Dam was selected as the remedy largely because the GTB patiently and skillfully built this community consensus from their generally recognized knowledge and community "good will."

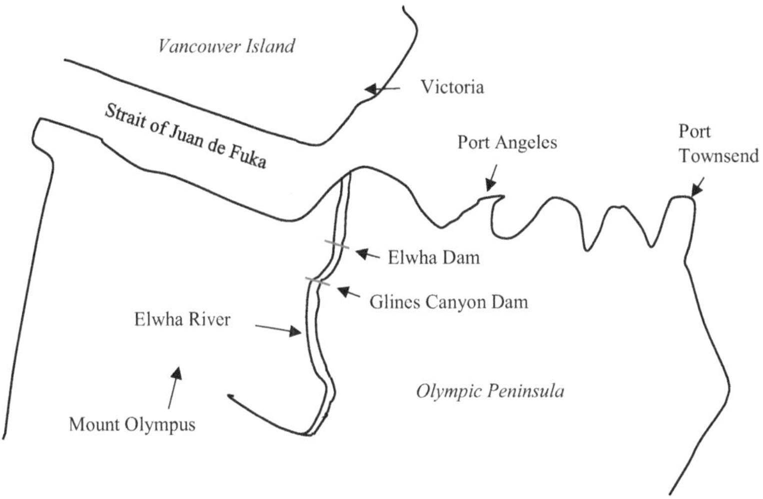

Fig. 3.1 Elwha River and Dams

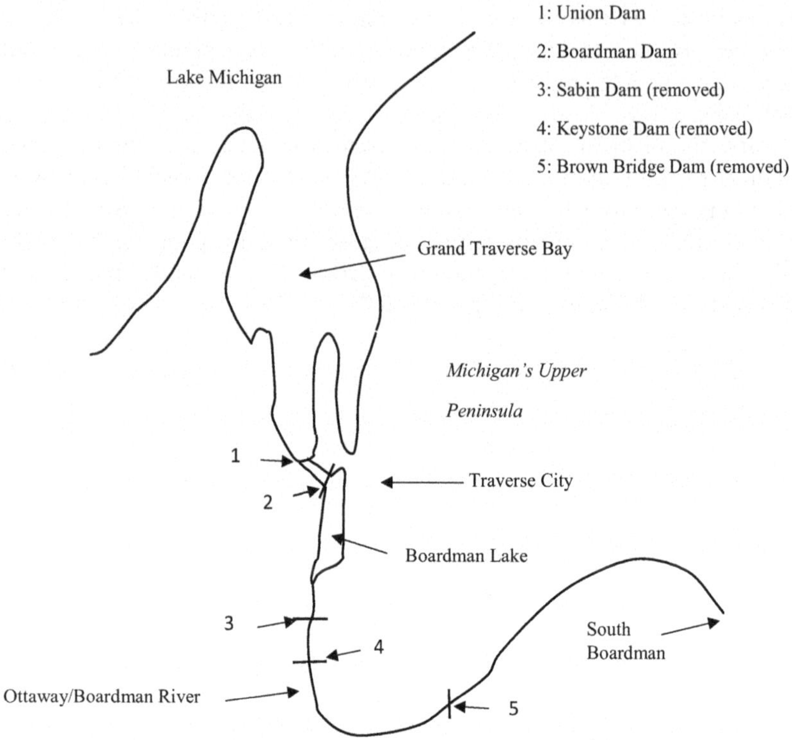

Fig. 3.2 Ottaway/Boardman River

Appendix: Invasive Aquatic Species[10]

Introduction to Invaders[11]

Trade, be it centuries ago or more recently during the massive increase in global trade since the 1970s, largely occurs by marine freighters. These large vessels are capable of spreading the invasive species that once introduced into an ecosystem, have no predatory controls. These are most often spread first in our river-ports. Ship water-ballast and anchors have a tendency to facilitate these invasions. It is not, therefore, just humanity that upsets or destroys the natural environment; other species can be rather destructive as well. Some of these recent destructive invasions are documented in this appendix.

In this appendix, eight invasive aquatic species (animals) are reviewed. They are invasive and destructive of the marine environments in the Great Lakes and our significant rivers on both coasts. In general, they are non-native to North America. Their destructive damage and methods of remediation are also reviewed. These eight species include:

- the Zebra Mussel,
- the Quagga Mussel,
- the Round Goby,
- the Asian Carp,
- the Ruffe,
- the Alewife,
- the Sea Lamprey, and
- the Spiny Water Flea.

These invasive animals are examined here in turn. Others could have been included, but these eight appear to be the most destructive up to now. We should have no doubt, however, that others will follow.

[10] This material is based on Robinson (2023).

[11] See "Least Wanted" AIS List, Great Lakes, St. Lawrence Governors & Premiers at www.gsgp.org/projects/aquatic-invasive-species/. The information for this section comes from numerous sources that include *National Geographic, Wikipedia, US EPA,* the *Natural Resources Research Institute, The National Park Service,* and the *Minnesota Aquatic Invasive Species Research Center (MAISRC),* and others as cited.

Zebra Mussels[12]

The zebra mussel is a small freshwater mollusk. It was originally native to the lakes of southern Russia and the Ukraine, but it has been accidentally introduced elsewhere as a globally aggressive invasive species. Since the 1980s, it has aggressively invaded the Great Lakes and the Hudson River with considerable resulting damage. This mollusk was first identified in 1769 in the Ural, Volga, and Dnieper Rivers. This mollusk gets its name from the striped pattern commonly found on its shell. It can grow to 2 inches in length. It is easily identified by the flat-bottomed "D" shape to its shells, which allows it to lie flat against a solid surface thereby enabling its strong attachment.

Zebra mussels are filter feeders that can process up to one liter of water per day. They deposit feces on the floor of the waterbody. Also, nonfood particles are combined with mucus and deposited as pseudo-feces. Since the zebra mussel became established in Lake Erie, the water clarity increased from six inches to up to three feet in the areas where this mussel predominates. This increased water clarity allows sunlight to penetrate deeper which enables submerged aquatic plants to grow more than they otherwise would. When these plants die and decay, they wash up on shorelines and foul beaches as in the case of Saginaw Bay reviewed previously in Robinson (2023, Chapter 7).

Lake floor food supplies are enriched by zebra mussels as they filter pollution from the water. This biomass becomes available to bottom-feeding fish and the fish that feed on them. The catch of yellow perch, for example, increased by five-fold since the invasion of zebra mussels into Lake St. Clair.

Zebra mussels attach to most surfaces below the water line. This includes sand, silt, and harder surfaces, but juveniles prefer harder rockier surfaces. Other mussel species are frequently the most stable objects in silty water beds, and zebras often attach to them which kills the native mussels. Zebras also build colonies on native clams, reducing their ability to move, feed, and breed. This eventually leads to their extinction. This led to the near extinction of native clams in Lake St. Clair and the western basin of Lake Erie. This pattern is being repeated in other Great

[12] See Hoddle, M.S. "Quagga & Zebra Mussels," *Center for Invasive Species Research, UC Riverside,* at http://cisr.ucr.edu/quagga_zebra_mussels.html. See also, "Species Profile—Zebra Mussel, *National Invasive Species Information Center,* United States National Agricultural Library.

Lakes' waterbodies. Zebra mussels, for example, are ca[able of fouling the *FishPass* facility reviewed in Chapter 3.

The lifespan of a zebra mussel is four to five years. An adult female zebra can produce over one- million eggs per year. Free swimming microscopic larvae, called "veligers," can drift in the water for several weeks before settling on a hard surface (any hard submerged surface). These mussels can tolerate a wide range of environmental conditions, and adults can survive out of water for about seven days. Research on natural enemies has focused on predators, particularly birds (36 species) and fish (15 species that eat veligers and 38 that eat attached mussels). Wintering waterbirds at Lake Constance have been found to decrease zebra mussel biomass in shallow waters by 90%. But almost all of zebra's natural enemies are not present in North America.

Crayfish could have a significant impact on zebra mussels. An adult crayfish can consume more than 100 zebra mussels per day. These predation rates, however, are significantly reduced at lower water temperatures. But some fish, such as the smallmouth bass, are predators of zebra mussels in the Great Lakes. Also, Canadian conservation authorities have successfully used a liquid fertilizer that kills zebras. Other tests of biopesticides have proved effective in Illinois, Minnesota, and Michigan.

Zebra mussels were first detected in Lake St. Clair. They were thought to have been inadvertently introduced into the Great Lakes by the ballast water of ocean-going ships, or perhaps ship anchors. They enter the Great Lakes through the St. Lawrence Seaway. From their first appearance in American waters in 1988, zebra mussels have spread to a large number of waterways including Lake Simcoe, the Mississippi, Hudson, Ohio, Cumberland, Missouri, Colorado, Arkansas, and Tennessee Rivers, plus various lakes. Scientists predict that this mussel will continue to spread to other rivers and lakes through recreational boats. Congressional researchers have estimated that the zebra mussel has cost communities over $5 billion and power companies $3 billion.[13]

In 1993, a major decrease in water dissolved oxygen was observed in the Seneca River in central New York. This was caused by a high concentration of zebra mussels. The Seneca also experienced significant reduction in water dissolved chlorophyll because of the zebra mussels.

[13] See Connelly, NA, and C.R. O'Neill et al. (2007), "Economic impacts of zebra mussels on drinking water treatment and electric power generation facilities," *Environmental Management*, 40 (1): pp. 105–112.

Also, zebra mussels are the source of a deadly avian botulism that killed tens of thousands of birds in the Great Lakes since the late 1990s. Zebras are edible and accumulate biotoxins.

The Quagga Mussel

Like the zebra mussel, the quagga (Dreissena rostriformis) is a freshwater mussel—an aquatic bivalve mollusk. It is indigenous to the Dnieper River of the Ukraine. Because of the pattern of its stripes, it is named after an extinct subspecies of African zebra. It is a damaging invasive species that is currently of major concern in the Great Lakes.[14]

The shell of the quagga mussel is usually black or yellow, but a distinct variation in Lake Erie is pale or white. The quagga is paler than the zebra mussel and smaller—about the size of a thumbnail. As a mussel, the quagga is a filter feeder; it uses its cilia to draw water into its siphon to feed on phytoplankton, zooplankton, and algae. It is a prolific breeder and spreads rapidly in the Great Lakes. Like the zebra mussel, it is capable of fouling the *FishPass* facility.

The quagga mussel was first observed in North America in 1989 when it was discovered in Lake Erie along the Ontario coast. It was identified as a distinct species in 1991. By 2021, the biomass of quagga in the lower four Great Lakes was estimated to be so great that it was the primary regulator of phosphorous, remetabolizing and recirculating it instead of allowing it to sedimentize as it normally would.

The introduction of quagga into the Great Lakes appears to be the result of ballast water discharged from transoceanic ships that were carrying juveniles or adult mussels. The quagga is highly prolific with high potential for rapid adaption and expansive colonization. Quagga's ability to rapidly colonize hard surfaces causes serious economic costs. This biofouling organism clogs water-intake structures such as pipes and screens, which reduces pumping capabilities for power and water treatment plants. Docks, break-walls, buoys, boats, and beaches have all been heavily colonized. Many of the potential impacts of quagga are still unknown, but its rapid spread across Europe's differing regions indicate its high potential for rapid adaptation to varied climates. It therefore might have high capability to adapt to climate change. It causes the same

[14] See Reintelen, T. and D. Van Damme (2011), "Quagga Mussel Dreissena" at www.iucnredlist.org/soecies/188911/8661357, IUCN Red List of Threatened Species.

problems as zebra mussels. It destroys life supporting algae, and damages boats, power plants, harbors, and native mussel populations.

Quaggas filter large amounts of water to feed on substantial amounts of phytoplankton and suspended particulates. This decreases chlorophyll concentrations and accumulates pseudo-feces. By removing phytoplankton, quagga mussels decrease the food available for zooplankton thereby altering the food chain. This filtration also increases the water's transparency which causes a proliferation of aquatic plants. This alters the entire ecosystem. The pseudo-feces accumulate on the bottom. As the waste particles decompose, however, oxygen is depleted, water acidity increases, and toxic byproducts are produced. The quagga mussels also accumulate pollutants in their tissues which are then passed up the food chain thereby increasing wildlife exposure to these pollutants. Native freshwater mussels are eliminated along with the plankton that previously supported the native fish. The yellow perch and the sunfish of the Great Lakes feed on the quagga, but the poisons do bioaccumulate thus contaminating these fish stocks.

The Round Goby[15]

The *Round Goby* is a small bottom dwelling fish that is native to Central Eurasia and the Black and Caspian Seas. They have established large non-native populations in several Eurasian rivers, and the North American Great Lakes. The round goby is a small soft-bodied fish characterized be a distinctive black spot on the first dorsal fin. The eyes are large and protrude slightly from the top of the head. They range in length from 4 to 10 inches. When they mature, round gobies become mottled with gray, black, brown, and olive gray markings. Since 1990, round gobies have rapidly expanded into the Great Lakes and its tributaries, and recently were discovered in Cayuga Lake—one of New York State's "Finger Lakes."[16]

Since it was first discovered in the Great Lakes in 1990, the round goby has been considered an invasive species with significant ecological consequences. But these consequences are complex because the fish both competes with native species and provides an abundant source of food

[15] See https://en.wikipedia.org/wiki/Roundgoby.

[16] See Jude, D.J. and R.H. Reider and G.R. Smith (1992), "Establishment of Gobiidae in the Great Lakes Basin," *Canadian Journal of Fish and Aquatic Science*, 49 (2): pp. 416–421.

for them while consuming other invasive species. In this latter sense, the round goby acts as an invasive control. It is aggressive and outcompetes native species such as the sculpin and logperch for food (largely snails and mussels) thus substantially reducing their numbers. Round gobies are voracious predators of the eggs of native fish. They are also very adaptable to degraded environmental conditions.

Many native predatory fish such as smallmouth bass, largemouth bass, walleye, salmon, and trout prey on the round goby. But the goby preys on the zebra and quagga mussels which concentrates the toxins they accumulate into the flesh of the fish that eat the goby. In this process, gobies inhibit the spread of the invasive zebra and quagga mussels but contaminate the native predatory fish. In addition, the round goby has become food for the Lake Erie Watersnake, once listed as a threatened species. The goby now accounts for 90% of the Watersnake's diet.

As a result of all this, while the round goby is invasive and preys on the eggs of native fish, it is also helps to limit other invasive species, and has become an important food source for some native fish.

Asian Carp

There are four species of what is commonly called Asian carp. These are all invasive and destructive of native fish populations. These four include:

- the *bighead carp* (Hypophtyalmichthys nobilis) is a filter feeder of the plankton that feeds the young native fish and native mussels.
- The *silver carp* (Hypophtyalmichthys molitrix) is also a filter feeder of the plankton that is the food for young native fish and native mussels.
- The *black carp* (Hypophtyalmichthys piceus), is carnivorous and feeds on native mussels and snails, some of which are already endangered.
- The *grass carp* (Hypophtyalmichthys idella) is herbivores and feeds on aquatic plants to such an extent that the food web of native fish and invertebrates is impacted.

These four species proliferate to great numbers in the Mississippi, and potentially can dominate the Great Lakes and its tributaries thereby destroying the native fish populations. The extremely high abundance of bighead and silver carp in the Mississippi Basin can potentially limit the plankton food and habitat space of native fish populations. The grass and black carp are bottom feeders that with their great abundance, also destroys the habitat and food sources of native fish.

It is the proliferation of the carp throughout the upper mid-west of the US that poses the urgent problem. When they enter a waterbody, they proliferate quickly and the native species decline. For example, between 2010 and 2019, Illinois contract fishermen harvested 7.5 million pounds of carp from the Illinois River.[17] These carp uproot vegetation and create muddy water through their rooting in the mud for food. This causes the decline in the native fish.[18]

A report issued in 2012 by the Great Lakes Commission concludes that physical separation of the Great Lakes and the Mississippi River watersheds is the best long-term solution to prevent Asian carp and other invasive species from migrating between the waterbodies.[19] Stopping these invasive species from spreading into Lake Erie is a particularly significant concern because Lake Erie provides the ideal habitat for carp. In October of 2013, scientists documented that carp had reproduced in Ohio's Sandusky River, a large tributary of Lake Erie.[20] A 2013 study detailed the devastating potential of a possible carp invasion on the native fish of Lake Erie.[21]

Silver carp can grow to 100 pounds, and are notorious jumpers (up to 8 feet above the water). They are scared easily by boats, and numerous boaters have been severely injured by collisions with airborne fish. According to an EPA report, injuries include cuts caused by fish fins, black eyes, broken bones, back injuries, and concussions.[22]

Eurasian Ruffe

The *Eurasian ruffe* is a small freshwater fish native to temperate regions of Eurasia. It is invasive in the Great Lakes and is reproducing faster than

[17] See Roman, Joe "Asian Carp" at http://eattheinvaders.org/Asian-carp/.

[18] See Fuller, Pam (July 6, 2005), "Species Fact Sheet: Cyprinus carpio," *US Geological Survey*, at https://nas.er.usgs.gov/queries/FactSheet.asp?speciesID=509.

[19] See Lundgren, Timothy and LLP Varnum (February 6, 2012), Report Recommends Great Lakes Separation to Address Asian Carp," *The National Law Review*, at www.natlawreviewe.com/article/report-recommends-great-lakes-separation-to-address-asian-carp.

[20] See USA Today, October 28, 2013.

[21] See Christian Science Monitor, January 5, 2016.

[22] See www.epa.gov/EPA-IMPACT/2007/July/Day-10/i13371.htm.

other species. The ruffe's color and markings are similar to the walleye—an olive-brown to golden-brown color on its back, paler on the sides with a yellowish white underside. It grows to almost ten inches in length. It has a large spiny dorsal fin that is spiked to keep away predators. It a very aggressive fish. The ruffe's diet is primarily zoobenthos: small water bugs, eggs, and larvae. It has the capacity to reproduce at a very high rate.

The accidental introduction of the ruffe to Lake Superior at Duluth is causing significant damage to native fish. It effectively competes for habitat space and food. It has similar eating habits to perch but it also has accelerated reproduction rate compared to other fish. Having more ruffe, therefore, leads to less food for other fish. The ruffe is unique in its ability to adapt to many habitats and temperatures, resulting in its success in spreading despite factors such as climate change. The ruffe also has an exceptional ability to detect water vibrations through organs called neuromasts. This trait both aids the ruffe's ability to find food and to avoid predators. These organs advance to being more sensitive as the ruffe matures, while the perch's neuromasts weaken as it matures.

The ruffe has the ability to overpopulate and crowd out many other fish species, and consequently to damage the Great Lakes ecosystem. According to the US Geological Survey, without concerted intervention, the ruffe might ruin Lake Superior.[23]

The ruffe is the first invasive species classified as a nuisance by the Non-indigenous Aquatic Nuisance Prevention and Control Program. Along with it being the most populous fish in the St. Louis River basin (in Duluth, Minnesota), it has disrupted ecosystems all across the Great Lakes. The invasion was first noticed in the 1980s by Minnesota's Department of Natural Resources. They suggest that the fish was introduced to the lake by ballast water dumped into Duluth's harbor by freighter ships. Ever since the ruffe were detected, studies show that the ruffe and the yellow perch are closely related and are rivals for food and habitat. The ruffe is expanding in this area, and the perch are contracting.

Ever since the ruffe was detected in the Great Lakes, the authorities have been trying to control them. The first method was to increase the Walleye and Northern Pike because they are natural predators. If a large school of ruffe is found, they can be poisoned, but if some survive,

[23] See "Ruffe—FactSheet," https://nas.er.usgs.gov/queries/factsheet.aspx?SpeciesID=7.

they will reproduce rapidly. The chemical *Lampricide TFM* kills ruffe but leaves other fish unharmed.

Pheromones are also being investigated as a control. After extensive study, it has been discovered that ruffe are repelled by their own pheromones. When injured, a ruffe will release pheromone into the water to warn other ruffe. Scientists have concluded that:

- The pheromone is species specific so it only repels ruffe, not other fish.
- The pheromone is resilient to freezing so it could be used during Minnesota's long winter season.

Through this method, scientists can block ruffe from their natural mating spots and produce a population decline. Their goal, of course, is to eliminate the species in the Great Lakes.

Alewife[24]

The *alewife* (Alosha pseudoharengus) is an anadromous species of herring found in North America. It is a North American shad. As an adult, it is a marine species of the northwest Atlantic. It moves into estuaries before swimming upstream to breed in freshwater habitats, but some populations never leave freshwater. It invaded the Great Lakes by using the Welland Canal to bypass Niagara Falls. In the Great Lakes, its population rose, peaking between 1950 and 1980, mostly in Lake Huron and Lake Michigan, to the detriment of many native species. To control them, Pacific salmon (coho and Chinook) were imported with partial success. The alewife is a National Marine Fisheries Service "Species of Concern."

The average length of an alewife is 10 inches, but it can grow to 16 inches. The front of the body is deep and larger than other fish found in the same waters. Both anadromous and landlocked forms occur (The landlock form is called a "sawbelly.") Adult alewife are caught during their spring spawning migration upstream by being scooped by nets. They are the preferred bait for the spring lobster fishery in Maine. Alewife are eaten by people, mostly after being smoked.

[24] See Fisheries, NOAA (2021). Also see "Mystic River Herring Monitoring Project," at https://mysticriver.org/herring-monitoring.

In spite of the introduction of Pacific salmon, alewife contributed to the decline of many native Great Lakes species. Alewife is also a common predator of numerous native and non-native zooplankton.

Several factors limit the spread of alewife. These factors include loss of the spawning habitat due to dam constructions, and recently because of the recovering of striped bass populations. Recently, because of the declines in the alewife, the states of Massachusetts, Rhode Island, Connecticut, Virginia, and North Carolina have established moratoria on harvesting them. The alewife is a US National Marine Fisheries Service Species of Concern.

In the spring, alewife migrate upstream in north Atlantic rivers to spawning grounds in ponds and lakes. After spawning, they return back to the ocean. After the eggs hatch, the young fish live in the lake eating the zooplankton that drift with the currents. After some growth, they feed on bottom dwelling crustaceans. After growing above one inch, they journey to the ocean. And grow to adults. They have numerous predators such as bass, trout, and birds.

In Cambridge Massachusetts, the Alewife Brook flows into the Mystic River watershed. Alewife Station, at the end of the Boston MTA's Red Line, is named after this once numerous fish. But today, there no longer are alewife in the Alewife Brook. According to the watershed scientist Andy Hrycyna of the *Mystic River Watershed Association*, this is because the brook is overly polluted with very low dissolved oxygen. In other sections of the Mystic, alewife is still abundant. In 2012, the *Mystic River Watershed Association* has established a fish ladder and elevator that enable the fish to reach its spawning ground in the Upper Mystic Lake.[25]

Sea Lamprey[26]

The *sea lamprey* (Petromyzon marinus) is a parasitic native to the Northern Hemisphere. It has an eel-like body without paired fins. Its mouth is jawless, round and sucker-like, and wider than the head with sharp teeth arranged in concentric circular rows. Sea lampreys are olive or brown-yellow on the dorsal and lateral part of the body with some

[25] Farley, Jeanine (April 9, 2022), "Everything there is to know about the alewife," *Cambridge Day*.

[26] See US Geological Survey, "Nonindigenous Aquatic Species Factsheet: Petromyzon marinus," USGS, https://nas.er.usgs.gov/queries/FactSheet.asp?speciesID=836. Also see "Great Lakes Fishery Commission—Sea Lamprey," at www.glfc.org/sea-lamprey.php.

black and lighter coloration on the belly. Adults can grow to 47 inches in length and a weight of over 5 pounds.

This species is found in the northwest Atlantic and is invasive in the Great Lakes. They have been found at depths of 4,000 meters and can tolerate a wide temperature range. Sea lampreys are anadromous. From their lake or sea habitats they migrate up rivers to spawn. After spawning, the lampreys die. Larvae filter feed on plankton. After several years, the larvae undergo a metamorphous that allows them to migrate to the sea or lake, and to start adult bloodsucking for feeding. Some start feeding prior to reaching the sea where they prey on a wide variety of fish.

The lamprey uses its suction cup-like mouth to attach itself to its prey, and uses its sharp tongue to rasp away flesh and to start bloodsucking. Victims die from excessive blood loss. After a year and a half of feeding, the lampreys return to the river to spawn and die.

Sea lampreys are a nuisance in the Great Lakes. They are native to the inland Finger Lakes, Lake Champlain, and Lake Ontario. Improvements in the Welland Canal at Niagara Falls in 1919 are thought to have allowed its spread from Lake Ontario to Lake Erie, and then to Lake Michigan where it decimated the native fish populations (in particular lake trout, but also fish like alewife) in the 1930s and 1940s. The lake trout should be an apex predator in Lake Superior. But the sea lamprey has become the actual apex predator since it preys on the lake trout, and has decimated its population. The relationship between predators and prey are now unbalanced. The control program includes both electric current and chemical lampricides, but they have had only partial success. The control program is under the Great Lakes Fishery Commission, a joint Canadian-US body. Michigan State University researchers are leading experiments in a newly synthesized pheromone that interrupts lamprey spawning. But as of 2017, the most effective control is a selective pesticide. The current lamprey control program relies on lampricides being applied when the lampreys are congregated in the Great Lakes tributaries in either the adult or larval stages. For example, above the Union Street Dam, lampricide has been applied seventeen times to the Ottaway River.

Previously, Niagara Falls served as a natural barrier that prevented lamprey from entering the Great Lakes. By 1938, however, sea lampreys had spread throughout the entire Great Lakes system. These invasive fish were able to thrive once they invaded the Great Lakes because of the availability of excellent spawning and larval habitat, an abundance of host fish, a lack of predators, and their high reproductive ability.

Sea Lampreys have had an enormous negative impact on the Great Lakes fishery. Before their invasion, Canada and the US harvested about 15 million pounds of lake trout in the upper Great Lakes each year. By the late 1940s, lamprey populations had exploded. They fed on large numbers of lake trout and other fish that were once the mainstay of a thriving fishery. By the early 1960s, the catch had dropped to 300,000 pounds, about two percent of the previous average.

Spiny Water Flea[27]

The *spiny water flea* (Bythrotrephes cederstroemi) is a planktonic crustacean of approximately one-half inch in length. It is native to freshwaters of Eurasia, but in 1980 it was accidentally introduced to the Great lakes by ship ballast water. It has a long abdominal spine with several barbs which protect it from predators. This spine is about seventy percent of its body length. The adult has three to four barbs along the spine. This crustacean has one large eye that is usually black or red. It has four pairs of legs, and it also has mandibles for the consumption of its prey.

Planktonic crustaceans are usually only a few millimeters long. They are the primary food source for all the juvenile fish in the Great Lakes. The most important of these planktons is the *Daphnia*—a genus that comprises some 100 freshwater species. These tiny animals are critical to lake health. Besides providing food for fish, they graze on floating algae. By beating their legs constantly to create tiny currents, they pull algae toward them. *Daphnia* keep algae in balance, but they are declining in Lake Superior and other Great Lakes. This is due to the invasive spiny water flea, a predator several times larger than *Daphnia*.

The spiny water flea is now established in dozens of lakes across the upper Midwest and Ontario. It is voracious in feeding on *Daphnia* and other zooplankton. Recent studies found that since the arrival of the spiny water flea, native plankton has declined by sixty percent in some of Minnesota's lakes. This impacts the populations of game fish such as yellow perch and walleye. The juvenile stage of these game fish cannot eat the spiny water flea because of the spines. But the water flea is decimating the zooplankton that juvenile fish do feed on. In addition, the water

[27] The material of this section is largely taken from Folger, Tim (December 2, 2020), "This ferocious water flea is mauling the Great Lakes," *National Geographic,* at www.nationalgeographic.com/environment/article/invasive-water-fleas-decimating-plankton-in-great-lakes.

flea fouls fishing lines, nets, and other equipment. There are currently no controls for this invasive predator.

Invasive Species and the Future of the Great Lakes

The plankton shredding spiny water flea is one of many invaders of the Great Lakes—there are more than 180 other invaders. As reviewed above, lampreys invaded in the mid-nineteenth century. Many of the worst invaders arrived in the ballast water of freighter ships. Also as reviewed above, quagga and zebra mussels have radically changed the ecology of Lake Michigan and Lake Huron. These invaders now number in the trillions. In Lake Michigan, they now filter about half the Lake's water every few days. In commenting on the clarity of the water, Hugh MacIssac, an invasive species biologist from the University of Windsor in Ontario, said "Lake Michigan now almost looks like open Caribbean water."[28] But only a few decades ago, Lake Michigan was full of plankton and more murky as a result. Now these waters are clear because they are lifeless, devoid of the plankton necessary for the Lake's food chain.

To oppose the problems of invasive species, we now have new water-ballast laws. Since 2008, all ships entering the St. Lawrence are required to flush their ballast tanks with salt water, which kills any freshwater species. Also, in 2017 a United Nations regulation now mandates that all new ships be equipped with ballast-water treatment systems. These new regulations have had effects. "We went from almost two invasive species per year (in ballast water) to zero in the last decade," observed Rochelle Sturtevant, and ecologist with the NOAA.[29]

Unfortunately, there are no currently effective ways to control invasive mussels or spiny water fleas. Fish don't eat them because the barbed tail gets caught in their throats, so they spit them out. Spiny water flea populations now become so dense that they clog fishing lines with gooey blobs that prevent the line from passing through the eyelets of fishing rods. While it is unlikely that *Daphnia* and other plankton in the Great Lakes will be entirely eliminated by the water flea, this invader has substantially altered the food web upon which the fish survive. The future of the Great Lakes, therefore, remains uncertain.

[28] Ibid.

[29] Ibid.

References

Bailey, J. (2017). The Ottaway, A River Reborn. *Into the Outdoors*. https://vimeo.com/222235168

Blumm, M. C., & Brett, M. S. (1998). The Indian Treaty Piscary Proft and Habitat Protection in the Pacific Northwest: A Property Rights Approach. *69 University of Colorado Law Review.*

Cook, J. (2017). The Ottaway, A River Reborn. *Into the Outdoors*. https://vimeo.com/222235168

Fox, C., Reo, N., Fessell, B., & Dituri, F. (2022). Native American tribes and dam removal: Restoring the Ottaway, Penobscot and Elwha Rivers. *Water Alternatives, 15*(1), 31–55.

Guarino, J. (2013). Tribal advocacy and the art of dam removal: The Lower Elwha Klallam and the Elwha Dams. *American Indian Law Journal, 2*(1), 114–145.

Howard, B. C. (2016, June 2). River Revives After Largest Dam Removal in US History. *National Geographic.*

Kalish, T. (2017). The Ottaway, A River Reborn. *Into the Outdoors*. https://vimeo.com/222235168

LaFond, K. (2020, January 17). Traverse City will be the First in the World to Sort Fish for Passage Upriver. *Interlochen Public Radio*. Published by Michigan Environment.

Robinson, R. (2023). *Environmental Advocacy and Local Restorations*. Palgrave Macmillan.

Shomin, T. (2017). The Ottaway, A River Reborn. *Into the Outdoors*. https://vimeo.com/222235168

CHAPTER 4

The Glen Canyon Mistake

Introduction

Marc Reisner's influential book *Cadillac Desert* was published in 1987. In that book, Reisner documented the Federal *Bureau of Reclamation's* considerable folly in attempting to convert America's Southwest Desert into a prosperous paradise by damming every river and minor tributary possible under the guise of flood control, irrigation for agriculture, hydro power, and service to the lawns and gardens of growing Southwest cities and suburbs. In 1997, as an indicator of the book's considerable political and cultural importance, PBS created a three-part *Frontline* documentary from Reisner's seminal work. The documentary is also titled *Cadillac Desert*.[1] In the third episode, two of the more significant protagonists and antagonists of this effort to re-engineer nature were interviewed by Reisner: former Senator Barry Goldwater from Arizona, and David Brower, former President of the environmental advocacy organization, *The Sierra Club*. Senator Goldwater had always supported water projects of any sort in his desert state of Arizona. Brower had strongly opposed environmental destruction, especially in the form of dam construction which he viewed as irrational attempts at real estate and agricultural development for the sake of short-term vested interests. As a result of Senator

[1] The PBS documentary had the same title as the book. It can be viewed currently on YouTube.

R. M. Robinson, *Restoring America's Rivers*,
https://doi.org/10.1007/978-3-031-81758-8_4

Goldwater's support for the *Bureau of Reclamation,* the Glen Canyon Dam was constructed on the Colorado River within the confines of the Grand Canyon. To establish this dam, Brower and *The Sierra Club* had agreed to not oppose its construction in return for the *Bureau's* cancelation of another proposed dam along the Green River tributary of the Colorado River, i.e. the Echo Park Dam located in the *Dinosaur National Monument*. Brower called this capitulation agreement, "the biggest sin I ever committed." (Adler, 2007, p. 15)

At the time of the interviews, both Goldwater and Brower were elderly and close to the end of their lives. During his interview, Goldwater looked out over the sprawling City of Phoenix and uttered, "What have we done to this beautiful desert, our wild rivers? All this dam building on the Colorado, across the West, was a big mistake. What in the world were we thinking?" When Brower was asked how he reacted to the Senator's admission, Brower responded, "I reached for the phone and called and I said, Barry let's do the right thing, help me take out Glen Canyon Dam. He said he would! Then he died a few months later." David Brower died a few months after the interview.

This chapter examines the phenomenon of Glen Canyon Dam and its creation of Lake Powell, a very large storage lake on the Colorado River that is currently threatened by the natural trends of drought and human overdevelopment. This chapter examines the personalities involved in this drama of development of the American West: Senator Goldwater, the priest of the High Sierras John Muir, the leader of the *Sierra Club* David Brower, and the Commissioner of the *Bureau of Reclamation* Floyd Dominy. The Glen Canyon Dam's construction is a tragic event of environmental destruction, but it is also an ongoing phenomenon resulting from the four individuals mentioned above.

This volume explores this human "phenomenon" of, "If you can build it, then build it large!" There is a social frailty suggested by this assertion, a frailty illustrated by the examples presented here and in the chapters of this book. Our current environmental movement is a reaction to ignorance, a human weakness that unfortunately continues. We have built dams and other large projects out of ignorance of their environmental effects, and also of the more feasible and less destructive alternatives. Strong personalities and the struggle to emerge from this ignorance and toward knowledge compose the sagas presented below.

The Colorado River Storage Project and Echo Park

In 1956, the Federal Bureau of Reclamation conceived and planned the *Colorado River Storage Project*. This significant chain of dams provides hydropower, flood control, and water storage for the *Upper Colorado River-Basin States* of Utah, New Mexico, Colorado, and Wyoming. (Fig. 4.1 maps this area and the "Storage Project.") The impetus for such a project actually began in 1922 when the *Colorado River Compact* was signed by these Upper Basin States plus the Lower-Basin States of California, Arizona, and Nevada. The Compact assured a flow of at least 7.5 million acre-feet per year to the lower states. The problem with this commitment is imbedded in the variability of the natural flow of the river. In drought years the annual natural flow is below 4 million acre-feet, but in flood years the natural annual flow can reach 22 million acre-feet. How would the drought years be managed? How would the surplus years be managed? To control this highly variable natural flow, the "Storage Project" was initially proposed in 1950. It was thereafter designed and constructed to smooth this flow.

The original storage plan included a dam in *Dinosaur National Monument* on the Green River at Echo Park. This proposed dam was strongly opposed by preservationists who viewed the Monument as protected lands permanently set aside from development. The *Bureau of Reclamation*, however, had no sentiment for preservation. It did not perceive a National Monument or National Park as society's promise of permanent preservation, but only a temporary "set aside" designed to serve the interests of a small number of enthusiastic tourists, academics, and environmentalists. The Bureau generally believed that the "preservation" could itself be "set aside" when circumstances warranted, and the Bureau generally judged that water management posed justifiable circumstances.

The Echo Park Dam was actually first proposed in 1941 to be within the small *Dinosaur National Monument*. Dinosaur fossils were first found along the Green River in 1909, and an 80-acre park encompassing these fossil beds—the *Dinosaur National Monument*—was established in 1915. The Monument was expanded in 1938 to 200,000 acres which then included the canyons upstream from the fossil beds. These canyons had the potential for supporting a large dam and lake. But in July of 1950, the *Saturday Evening Post* published an article by Bernard DeVoto, "Shall We Allow Them to Ruin Our National Parks?" This article compared the potential flooding of Echo Park to the politically notorious 1923 damming and flooding of Hetch Hetchy Valley in Yosemite, a damming that provided San Francisco with a convenient water supply, but deprived the Nation of a spectacular Yosemite Valley. As explained below, this comparison of Echo Park with Hetch Hetchy led directly to the Sierra Club's trade of saving Echo Park at the expense of damming Glen Canyon.

The Sierra Club's reaction to the threat of damming Echo Park has its roots in the Club's historical and strong connection to its support and defense of Yosemite Valley against the City of San Francisco's notorious construction of the Hetch Hetchy dam. This early fierce environmental-preservationist battle involved the iconic John Muir and the origins and purpose of *The Sierra Club*. It also involved the environmentalists' perception of society's "lying promise of preservation." The moral turpitude of this "lying promise" stimulates strong outrage among those who pursue the environmental cause. If "lying promises" are viewed as acceptable in our society, then ignorance will generally prevail, and environmental preservation and restoration will always lose.

The original leader of those who passionately pursued the environmental cause during the twentieth century was John Muir, the guru and high priest of the Sierras. His presence and efforts are reviewed first.

John Muir and the Sierra Club as Defender of Nature

Yosemite is akin only to *Yellowstone National Park* in its perceived environmental sanctity. In 1864, the magnificence of Yosemite Valley was protected by the State of California as a state park. The great natural philosopher and explorer of Yosemite was John Muir (1838–1914), a native of Scotland. His family emigrated to Wisconsin in 1849. With his brother, Muir further emigrated to the Western US in 1862–64. He settled in San Francisco from where he quickly traveled to Yosemite for an initial overwhelming week of exploration. In 1868, Muir obtained a shepherd's position tending sheep in the Sierra's summer meadows. He built a small cabin along Yosemite Creek where he lived for two years. He chronicled his experience in *First Summer in the Sierra*, published in 1911.

Muir became an avid reader of Henry David Thoreau and his fellow Concord transcendental-naturalist Ralph Waldo Emerson. On his explorations of Yosemite, Muir traveled alone carrying "only a tin cup, a handful of tea, a loaf of bread, and a copy of Emerson." (Tallmadge, 1997, pp. 52–53) He became the expert explorer of the Sierras. He was respected for his knowledge of natural history, as well as his skills as a guide leading other explorers, and his narratives of nature rooted in his excursions and observations. Intellectual visitors, artists, and celebrities sought out Muir. In 1871, his visitors included Emerson and his academic friends. After a day's lengthy conversation, Emerson offered John Muir a position at Harvard (according to Tallmadge, 1997, p. 53). Muir declined.

In 1892, Professor Henry Senger of the University of California, Berkeley, suggested to John Muir the necessity of forming *The Sierra Club* as an environmental organization dedicated to exploration and preservation.[2] The first meeting of the Club was in May, 1892. Muir was elected President. The Club's first efforts were to defend Yosemite by bringing it under federal control.

John Muir considered the Sierra and Yosemite as wilderness. He considered the grazing of domestic livestock as the biggest threat to the high meadows, especially the grazing of sheep in Yosemite's meadows. In June of 1889, the influential associate editor of *The Century* magazine—Robert Underwood Johnson—camped with Muir in Tuolumne Meadows and witnessed the damage that this grazing caused. Muir wrote two *Century* articles for Johnson to publish in 1890: "The Treasures of the Yosemite," and "Features of the Proposed National Park." They argued for exclusion of grazing in the Sierra's high meadows, and for the Yosemite National Park to be created with a status similar to *Yellowstone National Park*. Johnson agreed to lobby Congress to make Yosemite into a national park. Yosemite, however, remained under California State control.

Muir tried to persuade Gifford Pinchot, the first head of the US Forest Service and the leading advocate for the sustainable but commercial use of natural resources, that areas such as Yosemite should be preserved for their spiritual and transcendental influence, as places for "rest, inspiration, and prayers." But "forestry as tree farming" was Pinchot's approach. The

[2] *The Sierra Club* was modeled on the *Appalachian Mountain Club* formed in 1876, and dedicated to exploration and preservation.

two men debated their respective views in various magazines: *Outlook, Harper's Weekly, Atlantic Monthly, World's Work,* and *The Century.* In the debate as to whether to dam Hetch Hetchy—a debate described below—Muir and Pinchot considerably diverged. In 1903, John Muir accompanied President Theodore Roosevelt to Yosemite. Muir told the president about the state's mismanagement of Yosemite, and was able to convince Roosevelt to extend federal control to the area. Roosevelt later exclaimed, "Lying out at night under those giant Sequoias was like lying in a temple built by no hand of man, a temple grander than any human architect could by any possibility build." (Meyer, Winter 1997) In 1906, Yosemite became a National Park.

With population growth in the San Francisco Bay area, political pressure mounted to dam the Hetch Hetchy Valley of Yosemite for use as a reservoir. Muir, *The Sierra Club* and Robert Underwood Johnson strongly opposed this damming. Hetch Hetchy Valley was adjacent to the iconic Yosemite Valley and was very similar in appearance and substance. Muir asserted, "Dam Hetch Hetchy! As well dam for water tanks the people's cathedrals and churches, for no holier temple has ever been consecrated by the hearts of man." (Meyer, 1997, Winter) Woodrow Wilson, however, signed the dam bill in December of 1913. Muir lost his last great battle, and the Sierra Club lost its first great battle. Muir died in December of 1914 at age 76.

The loss of Hetch Hetchy stung *The Sierra Club*, and the sting lasted. Yosemite, including Hetch Hethchy, was preserved after a considerable lengthy campaign, but the preservation did not last. The damming of Hetch Hetchy Valley was considered immoral theft. After all the effort of Muir and the Sierra Club, Yosemite's preservation was merely a classic "lying promise" propagated by society, where nature can be discarded cavalierly for convenience. Echo Canyon was considered a similar immoral theft of all the effort to preserve, and that is why the reaction from *The Sierra Club* was so strong and swift. But the problem was more extensive than immediately apparent in that Glen Canyon had hardly been explored. Environmentalists were ignorant of its considerable valuable natural assets: its canyons, grottos, and glens. In environmental concerns, ignorance is both ugly and destructive!

David Brower, the Environmental Campaigner

America's environmental movement is founded on strong personalities—"crusaders all" who campaigned against the destruction of nature by modernity. These leaders include outsized personalities such as Thoreau, Teddy Roosevelt, John Muir, and certainly David Brower (1912 – 2000), the first "Executive Director of the Sierra Club," 1952–1969. Brower began as a mountaineer with 70 first ascents along the High Sierra's more difficult peaks. He was a native and long-term resident of Berkeley, California, where he was an editor of the *University of California Press*. In 1942, he published *The Manual of Ski Mountaineering*, and subsequently served as a lieutenant in the "10th Mountain Division" during World War II when he earned a "Bronze Star for Gallantry."

After World War II, Brower returned to the *University of California Press*, editing the *Sierra Club Bulletin* in 1946. He also managed the Sierra Club's "annual high mountain trips" between 1947 and 1954. He was both a persuasive speaker and publisher of persuasive books on wilderness and other environmental causes, volumes such as *This is the American Earth* (1962), and *In Wilderness Is the Preservation of the World* (1962). Under Brower's leadership, between 1952 and 1969, the Sierra Club's membership expanded from 7,000 to 70,000. Brower provided the "fire in the belly" for the Sierra Club, leading it to become the strongest environmental advocacy organization in the world, especially for wilderness causes (along with the *Wilderness Society*).

Brower also founded the environmental advocacy organizations *Friends of the Earth* (1969), the *League of Conservation Voters* (1970), the *Environmental Policy Center* (1971), and *Earth Island Institute* (1982). Through all these activities, David Brower not only branded the "environmental movement" with his adversarial leadership style, but he also effectively focused the movement's attention on federal government agencies and policies.

Two significant and influential books focused on the considerable efforts of David Brower: Marc Reisner's *Cadillac Desert* (1987) and John McPhee's *Encounters with the Archdruid* (1971). McPhee's tag of "Archdruid" for David Brower was embraced by him.

The Bureau and the Colorado River

The 1902 "Reclamation Act" created the federal "Reclamation Service" for managing the non-navigable rivers in the Western US. This "Service" was renamed the *Bureau of Reclamation* in 1923, a federal agency within the *US Department of the Interior.* The operational mission of the *Bureau* is to oversee and operate the hydro systems of dams and water delivery projects that the agency builds throughout the Western US. These systems are in support of flood control, municipal water supplies, electric power, and irrigation to 10 million acres of farmland. It is the second-largest producer of hydropower in the US. It currently operates 180 projects throughout the 17 Western US States.

The *Bureau's* initial investigation of damming the Colorado River was in 1922. In general, the 1920s was the great decade of America's electrification for lighting, radios, telephones, industry, and household appliances. Prior to 1922, hydropower transmission was limited to 80 miles, and all potential dam sites along the Colorado were not within the 80-mile distance from significantly large populations. But the engineering of power transmission advanced sufficiently between 1890 and the early 1920s to make long-distance transmissions feasible for hundreds of miles. Large hydro-power dams therefore became a possible economic method of electrical generation for distant population centers. In 1922, the *Bureau* (called the "Reclamation Service" then) was led by Arthur Powell Davis, and the Secretary of the Interior was Albert Fall. In that year, the Bureau issued the "Fall-Davis Report" which called for damming the Lower Colorado for the purpose of hydropower and flood control. Boulder Canyon was one of the more feasible potential sites investigated because of its hard rock on both sides of the River. The proposed project was thereafter referred to as the "Boulder Canyon Project."

America's Southwest Desert extends through seven states—California, Arizona, Nevada, Utah, New Mexico, Colorado, and Wyoming. The only major river flowing through these seven states is the Colorado (It drains a portion of Southwestern Wyoming). Its potential for supplying water to the desert sections of these states—especially for Southern California, Arizona, Nevada, Utah, and northern sections of New Mexico—is obvious. The inevitability of competition among the seven states for the limited water resources of the Colorado is also obvious.

In 1917, the "League of the Southwest" was formed by the seven states along the Colorado for the purpose of developing and managing the River's assets. But in 1922, in *Wyoming v. Colorado* the US Supreme Court undermined the claims of the upriver states with a threat to allocate Colorado River's water according to precedent use, i.e. the Court's so-called rule of prior appropriation. This meant that upriver states might not get to use the Colorado River's water that flowed through its territory because downriver states already had claims to it. Since the upriver states where the Colorado River originates was relatively under-developed as compared to California with its already considerable water usage, the upriver states perceived the threat of being under allocated if this "precedent usage" was the deciding factor. In addition, the explorations of dam construction on the lower Colorado—as explored in the June, 1922, "Fall-Davis Report"—added to the anxiety of the upriver states. The dams would be close to the California border where diversions to the Imperial Valley had already begun (See below for a review of the Imperial Valley and its use of the Colorado River water). Under this political anxiety and pressure, a "Compact" agreement was signed by the seven states in November of 1922. The agreement sought to allocate the water equally between the Upper Basin States (Wyoming, Colorado, New Mexico, and Utah) and the Lower Basin States (Arizona, Nevada, and California). By supposedly assuring water for the Upper Basin, the "Compact" removed much of the political opposition to the proposed Boulder Dam on the Lower Colorado.

The role of the *Bureau of Reclamation* in constructing dams along the Colorado is a paramount component of the story of the American West. Building the Hoover Dam was one of the Bureau's initial tasks. The Boulder Dam, later renamed the Hoover Dam, began construction in 1928, and was completed in 1935. Glen Canyon Dam, constructed between 1956 and 1966, is generally considered one of the last of the big dam constructions. The term "big dams" refers to the very large concrete constructions that provide hydropower, flood control, agricultural irrigation, municipal water supplies to households and business, and also recreation on the lakes formed behind these constructions. "Big dams" were built on the Columbia, the Snake, the Tennessee, and the Colorado Rivers. But smaller dams were constructed on the tributaries within these river basins, and also on many other smaller but significant Western rivers such as the Sacramento and the Rogue Rivers. Most of these were built by the *Bureau of Reclamation*.

Floyd Dominy served as Commissioner of the *Bureau* from May 1, 1959, to December 1, 1969. He was the Bureau's most aggressive defender and proponent of Western river damming. He was born in 1909 into a poor Hastings, Nebraska, homestead of 160 acres—a subsistence family farm. He graduated from the University of Wyoming with a degree in Agricultural Economics in 1933, the worst year of the Great Depression. Dominy's childhood experience left him less of a proponent of the small family farm, and more of a proponent of large-scale agricultural development through exploitation of economies of scale.[3] He viewed rivers as resources to be engineered and not left in their natural flowing state. Irrigation, power generation, municipal use, and recreation by boating on the lakes behind the dams were Dominy's acceptable uses, not ecotourism.[4]

In the 1930s, Dominy served in Campbell County, Wyoming, as a "county agricultural agent"—a US Department of Agriculture position. With respect to the grazing ranches of Wyoming, in an interview Dominy recalled, "I saw there the fruits of the 640-acre Homestead Act … that most of those 640 acres could not sustain a family under any reasonable economic conditions that have prevailed then and now." He viewed the small farm as an institution of the past given the standard of living of most Americans of the 1950s and 1960s. In a speech given upon leaving office in 1969, Dominy observed, "The general trend now is … the abandonment of family-sized farms and the deterioration of small communities into ghost towns. In our modern mechanized and high-speed civilization, I see no major changes likely in this trend." The rural standard of living had changed dramatically over the first half of the twentieth century and providing irrigation for subsistence farming was not of interest to Floyd Dominy.

Floyd Dominy joined the Bureau of Reclamation in 1946, and rose to Associate Commissioner in 1956 and then Commissioner in 1959. During the 1960s, he supervised the completion of the *Colorado River Storage Project* which included construction of the Glen Canyon, Flaming Gorge, and Navajo Dams. The *Colorado River Storage Project* was not designed to serve the small family farm, but to meet the requirements of the *Colorado River Compact*. In particular, this series of upriver dams would provide flows to meet the considerable demands of the downriver interests. The largest component of the Storage Project is the Glen

[3] See Pisani (2002). Also see Reisner (1987).

[4] See Dominy's interview in the documentary series *Cadillac Desert*.

Canyon Dam with its Lake Powell. But the Storage Project also includes the Flaming Gorge Dam, the Navajo Dam, and the Aspinall Complex of three dams, the Blue Messa Dam, the Morrow Point Dam, and the Crystal Dam. (Fig. 4.1 indicates the locations.)

Lake Powell lies behind Glen Canyon Dam, near Page in northern Arizona. But most of Lake Powell is in southern Utah. This lake was designed to provide 64% of the storage capacity, and 75% of the hydroelectric generating capacity of the Project. (The environmental benefits of deconstructing this dam are reviewed below.)

The Decline of Large Dam Constructions

By the end of the 1960s, Americans considered the age of big dam construction to be over. The constructions of the 1930s were in part due to the need for "employment projects" to combat the Great Depression. But in the post-World War II era, that need lessened considerably when various factors made big dams less attractive even to Western residents. These factors include the massive loss of water in transit due to leakage and evaporation, a loss that is often greater than the amount of water delivered (The current search for solutions to these problems of leakage and evaporation is reviewed below). It is likely to be more efficient for people to locate close to the water sources, and this relocation potentially negates the need for large-scale dams with large lakes behind them. In addition, the sediment buildup behind the large dams eventually lessens their usefulness, and also prevents these soils from replenishing downriver deltas which generally form the better agricultural areas. In contrast to large dams, smaller water diversion projects avoid these significant silting problems.[5]

In addition to sediment buildup, there were "cold war" problems associated with large dams. Atomic warfare (thermo-nuclear detonations) were capable of destroying large dams thereby flooding downriver populations. In addition, the 1950s and 1960s held the promise of nuclear power plants, and also the alternatives of natural gas and petroleum power plants were expanding. These alternatives to hydropower are more flexible in providing convenient locations close to densely populated areas. Construction costs are also much lower for these alternative systems.

[5] Small diversions include the use of pipes extended into the river to withdraw water for use. Silt can then be filtered from the water.

There were two other significant reasons for the decline in popularity of the large Western dams. Firstly, there was the declining number of appropriate sites for large dams. These sites are best found at the stable rock canyons present at some locations on the Colorado River—especially within the Grand Canyon. There have been perhaps a few other sites on the Columbia River, and/or the Snake River Canyons. The Hetch Hetchy Valley in Yosemite provided such a site as reviewed above. The second of these significant reasons was the political resistance of the Eastern US for further subsidies to Western projects. Many of the 1930s' projects were considered less than economically justified except for the need for employment projects during the Great Depression which were formed to counter the high unemployment levels. Between 1950 and 1976, however, the West received more than half of the American water project expenditures by the US Army Corps of Engineers and the Bureau of Reclamation—the two federal agencies responsible for these sorts of projects. The American South received 28 percent of these expenditures, and the Northeast received only 6 percent of these expenditures. The Northeastern and Midwestern sections of the US believed they were subsidizing the growth of Western cities. For example, California and Arizona received greater per-acre agricultural subsidies than the crop acreage in all other states. (See Pisani, 2002 reviews these subsidies.)

In defense of the Bureau's efforts, Floyd Dominy argued before Congressional committees that the Bureau's irrigation projects provided the larger portion of the US agricultural produce (lettuce, melons, carrots, and many other vegetables) during the winter months. But in addition, Dominy pointed out that the ten most visited Bureau reservoirs—especially Lake Mead—attracted more vacationers than the ten most visited National Parks. These agricultural and recreational areas also were the primary generators of many communities' economic health in the Western US. As a result, these communities grew at rates in excess of the rest of the US.

The Imperial Valley and the All-American Canal

Because of Southern California's "early and therefore priority establishing" use of the Colorado River's water, the *Bureau of Reclamation* has had an enormous impact on that region's development. The key to Southern Cal's diversion of the River's water is the Imperial Valley and its

All-American Canal. This Valley has been one of the US' more productive agricultural areas particularly during winter months when it produces a large portion of the Nation's vegetables. The Imperial Valley is bordered by the Colorado River on the east, and in part the lifeless Salton Sea to the west and north. The Salton Sea is a dead body of agricultural wastewater from the drainage of the Valley's irrigated fields. The Imperial Valley's southern border is Mexicali, Mexico. The Valley therefore extends 50 miles from the southern end of the Salton Sea to the US-Mexican Border. Yuma, Arizona, is on the east side of the Colorado River opposite the Valley. The Imperial Valley is hot desert with extreme temperatures during the summer. It is almost entirely below sea level. It was once part of the Gulf of Mexico, but the region was formed as the northern end of the Colorado River's delta where it flows into the Gulf of Mexico.

The Valley was uninhabited until 1901, when the Alamo Canal was created to divert the River's water through Mexico and into what was to be irrigated fields. A flood in 1905, however, destroyed the irrigation canals. In 1911, the Imperial Valley Irrigation District (IID) was created as a State of California entity. It captured and distributed the diversions from the Colorado River as it flowed through Mexico via the Alamo Canal and back into the southeastern tip of California.

In 1917, the Imperial Irrigation District (IID) requested that the *Reclamation Service*—later to become the *Bureau of Reclamation*—build a new diversion canal that would be entirely within the domain of the US. The *All-American Canal* was authorized by the *Boulder Canyon Project Act* of 1928, along with the Hoover Dam and the Imperial Dam. The latter is farther downstream on the Colorado River and forms a lake from which the *Canal* draws its water. The construction and operation of all these entities is (and was) under the Bureau's domain.

The Hoover Dam, Imperial Dam, and the All-American Canal were constructed during the 1930s. All are owned by the Bureau. The Canal began delivering water to the Imperial Valley in 1940. It also conveys water to the Yuma Project—an agricultural area in Yuma, Arizona—and to nine cities in Southern California. The canal is 82 miles in length, and is the largest aqueduct in the world with a capacity of 26,155 cubic feet per second. It also has eight hydroelectric plants with a combined generating capacity of 58 megawatts.

Today, eighty-three years after the Bureau first diverted irrigation and municipal waters to Southern California, the Imperial Valley is at the center of the struggle for revisioning the Colorado River. Water usage in the Western US tends to be distributed by "priority rights" of who first diverted or used the Colorado's waters with all contenders perceiving the struggle as a zero-sum game. This "Law of the River" system is archaic with struggles concerning bureaucratic power, legal adjudications, and ease of administration at its heart. The practical functionality of this Law, however, has been destroyed by our global warming and consequent drought since this archaic system can only lead to the zero-sum game of distributing the dust of the arid Southwest rather than any semblance of abundant water. As is explored below, the "Law of the River's" usage system must change or the Colorado River's System will become only a series of "dead pools" that ceases all usefulness for California, Arizona, Nevada, and Mexico. An extreme "tragedy-of-the-commons" scenario is emerging from the current heavily subsidized "the water is free" system with the heads of its participants stuck in the very dry desert sands of the Southwest. Much more about this tragic scenario is explored below.

Storing Water for What Purpose? The Colorado River Storage Project Again

The last significant project produced by the Bureau under Floyd Dominy's direction was the *Central Arizona Project* (CAP) in 1968. This project consists of a series of aqueducts that annually delivers 1.4 million acre-feet (maf) of Colorado River water to irrigate nearly one million acres of agricultural land in south-central Arizona (Maricopa, Penal, and Pima counties), and to also provide municipal water for various Arizona communities that include the larger southern Arizona cities of Phoenix, Tucson, and Yuma. To facilitate the aqueduct's flow, fourteen pumps lift the aqueduct's water 2,900 feet. These pumps are the largest user of electric power in Arizona. It is essential to our narrative to also understand the impact of the CAP on Southern California's considerable water diversions.

The CAP was authorized through $4 billion of federal expenditures in 1968—the most expensive project ever for the *Bureau of Reclamation*. Arizona's Senators Barry Goldwater and Carl Hayden, plus Congressional Representative Morris Udall, and his brother the US Secretary of the Interior Stewart Udall (the Udalls also being Arizona natives), posed a formidable team of federal leaders who established this project. The main stem of the CAP's system of aqueducts runs 336 miles through the Sonoran Desert—from Lake Havasu on the Colorado River to the CAP's terminus southwest of Tucson. (Fig. 4.1 maps this area.) Between

Lake Havasu and Phoenix, the aqueduct flows through three tunnels of concrete pipes of 22 feet in diameter.

At the time of the *Colorado River Compact's* formation, it was estimated that on average the River's flow was 17 million acre feet (maf) per year. The Upper Basin agreed to allow a flow of at least 7.5 maf to the Lower Basin. Each Basin was then to allocate to its respective states with the exception that at least 1.5 maf was to be allowed to flow into Mexico. (Fig. 4.1 indicates this flow to Mexico.) Subsequent tree-ring studies, however, concluded that the long-term average water flow of the Colorado is significantly less than 17 maf per year, which was originally estimated during the unusually high precipitation period of 1905–1922. The true long-term average is considerably less, by perhaps at least one-third. It is actually below 13.5 maf. The current "drought" has lasted 21 years (up to 2023), but the evidence indicates that this drought has flows below 7.5 maf per year. This is actually the approximate current normal average. This realization of lower water flows led to significant reductions in commitments to the Lower Basin states. The current commitments are now contingent upon the water level of Lake Mead. This agreement concerning adaptive allocations to droughts consists of:

- *Light Shortage*: When the surface elevation of Lake Mead is 1,075 – 1,050 ft. above sea level, the Lower Basin states are to receive only 7.2 maf per year, with 4.4 maf allocated to California, and 2.5 maf to Arizona, and 0.2 maf to Nevada.
- *Heavy Shortage*: When the surface elevation of Lake Mead is 1,050 – 1,025 ft. above sea level, the Lower Basin states are to receive only 7.0 maf per year, with 4.4 maf allocated to California, 2.4 maf to Arizona, and 0.2 maf to Nevada.
- *Extreme Shortage*: When the surface elevation of Lake Mead is below 1,025 ft. above sea level, the Lower Basin states are to receive less than 7.0 maf per year, with 4.0 maf allocated to California, 2.3 maf to Arizona, and 0.2 maf to Nevada.

The political struggles in reaching and adjusting these drought allocations have been significant, and largely involve conflicts between California and Arizona, which were the two faster growing states in the post-World War II period. The current drought is now classified as more severe than even the "extreme shortage" scenario. It is due to global warming, and threatens Lake Mead with being below 1,000 feet above sea level, a critical state for the interests of the Lower Basin.

The important background context of the Compact's negotiations includes the June, 1922, US Supreme Court decision (prior to the interstate agreement). That decision established "the rule of prior appropriation" as applying to Colorado River extractions regardless of state law. At that time, this "rule" meant that the fast-growing Southern California, with its early claims to Colorado River extractions, might dominate the future claims of the rest of the Basin. When this decision was rendered, however, interstate compacts had not been tested, but given the Court's precedent-setting decision, the states that were to form the *Colorado River Compact* were paranoid about federal government power and actions. They did not want "the law of prior appropriation" to apply, but rather the Compact's agreed upon allocations. In particular, the Hoover Dam is located close to California and was thought to likely facilitate greater diversions to Southern California unless the Compact could control allocations. In addition, the *Bureau of Reclamation* planned (and completed) the eighty-mile "All-American Canal" in Southeastern California, which originates at the Bureau's "Imperial Dam," just north of Yuma, and diverts water from the Colorado into California's Imperial Valley and then ultimately to Los Angeles. It would be the only significant water source for the Imperial Valley with its enormous agricultural irrigation demands.

For Arizona to fully utilize its allocation of Colorado River water, it needed a diversion project of its own, one similar to California's "Imperial Valley." Since Arizona was not utilizing its full allocation (and has never done so), the excess above its allocations flowed to California which extracted this excess under the "law of prior appropriation." As a result, Arizona developed the *Central Arizona Project* (CAP) for the expressed purpose of reclaiming its share from California. In 1963, ignoring the "Compact's" allocation system, in *Arizona vs. California* the US Supreme Court settled the issue using the "law of prior appropriation" with California being allocated 4.4 maf, Arizona 2.8 maf, and Nevada 0.3 maf. Federal funding, however, was appropriated for the CAP's construction in 1968. Construction began in 1973, and the project was substantially completed in the 1990s. But *Arizona vs. California* also empowered the Secretary of the Interior (Bruce Babbitt at the time) to act as the water master for the Lower Basin with the authority to apportion future surpluses and shortages among the states. Note that as of 2020, Arizona was still not fully utilizing its 2.8 maf. It has only used 2.5 maf, but California and Nevada have over extracted their allocations.

The multi-state Compact, therefore, has not resolved or managed the allocation problem, but instead has demonstrated a type of "tragedy-of-the-commons" where each state's attitude is, "We want our water so we will develop projects that will maximize our share, and even do this in the context of existing in a thirsty desert." This is essentially the point argued by Reisner, 1987.

The Upper Basin has not experienced the growth-stress of the Lower Basin. It allocates its portion of Colorado River water according to a percentage system agreed to in 1948: 51.75 percent of the forecasted flow allocated to Colorado, 23 percent to Utah, 14 percent to Wyoming, and 11.25 percent to New Mexico. The states of the Upper Basin anticipate (or forecast) the total supply-flow from the River and allocate according to these percentages. The Lower Basin, however, lies in a desert with insatiable thirst associated with great development potential if only the water is available. Its attitude has been "us versus them." It argues that its allocation is an absolute "priority right" not to be compromised due to the current drought emergency.

The Current Drought

The current severe drought (as of late 2023) is thought to possibly be the worst in the last 1,250 years. It is due to global warming, and therefore coupled with evaporation. This has resulted in struggles for withdrawal levels that the River cannot support. The level of Lake Mead is now down by more than two-thirds as compared to the early 1990s; the level of Lake Powell is down three-quarters. In addition, the US Geological Survey (USGS) estimates that by 2050, the Colorado's water flow will be one-quarter less than it currently is. The effects of global warming include lower levels of snow-packs in the Rocky Mountains, and greater evaporation. But warmer temperatures have stimulated plant growth in the Mountains, and this further absorbs snowmelt runoff that previously would flow to the Colorado River. (Demircan, 2020, reviews this problem.) The Colorado River is therefore severely over allocated, but still has almost insatiable demands from the Lower Basin States.

There have been 29 dams built along the Colorado and its tributaries, and numerous diversion projects that feed municipal and agricultural demands. The resulting sediment buildup behind these is substantial, and so is the associated increase in water temperatures and evaporation. The consequential environmental degradation follows. The Colorado River is not managed for environmental protection, but for development. It should be judged as an environmental tragedy, one resulting not from a lack of governmental supervision, but from near total purposeful governmental alteration from the River's natural state, and conversion of it into a resource that has attempted to serve only development interests.

In recent years, the growth in Arizona and Nevada reduced the amount of water available for California. The Upper Basin states, however, have still not used their full entitlement, which therefore allowed California to use more than its legal allocation according to the Compact. In August of 2000, all the "Compact" states, including California, agreed to a fifteen-year phase out of overuse of the Colorado, particularly the overuse by California. Within this "phase out period," California was to eventually reduce its usage to its allocation of 4.4 million acre feet (maf). This should allow the 1.5 maf flow to the Gulf of California's Colorado River Delta in Mexico, which should be capable of facilitating the environmental restoration of the delta. (Sudman, 2020, reviews this possibility.) The problem is that this water is the runoff from the irrigated fields in the Imperial Valley and Yuma. Consequently, it is heavily polluted and is capable of feeding noxious algae blooms. Environmentalists have argued that the US has an obligation to provide more water to the Mexican Delta to assist endangered species that reside in this wetland. US officials contend, however, that the "Endangered Species Act" ends at the Mexican border. (Sudman, 2020, reviews this legal debate.) Note that Southern California has yet to reduce its extractions. In the meantime, global warming means greater evaporation of Colorado River water.

A further updated agreement among the Lower Basin and Upper Basin States was reached on May 23, 2023. It involves the Lower-Basin states cutting additional consumption by 3 million acre-feet (maf) over the years 2023–2026 as compared to 2022, i.e. a 14 percent reduction in total. The time sequence of these cuts requires that at least 1.5 maf must be cut by the Lower-Basin States by 2024, and that California alone must cut 1.6 maf by 2026. These cuts, however, are to be compensated by approximately $1.2 billion in federal *Inflation Reduction Act* funding. But according to the Bureau, all these cuts are only half of the minimum amount required to prevent critical emergencies at the Colorado River's Dams, where the "critical level" is that necessary to continue hydropower generation. (Jones, 2023, reviews this problem.) To avoid being below this "critical level," Bureau Commissioner Camille Touton indicates an additional 2 maf must be cut over the 2023–26 period. Professor Jack

Schmidt of Utah State University estimates, however, that greater cuts should be required to begin to restore the lakes of the *Storage Project* back to the levels of the 1990s.

Seepage, Evaporation, Drought, and Subsidies

Climate change has had global effects on precipitation and mountain snowpacks. The world's great rivers generally depend on winter buildups of snowpacks followed by spring melts. In essence, these snowpacks provide water storage for spring and summer runoffs which provide agricultural irrigation for the growing seasons. The winter buildup of snowpacks have been deficient in recent years due to global warming. Also, the warmer temperatures and increased humidity have increased vegetation at higher altitudes, and this vegetation absorbs some of the runoff from snowmelts. With respect to the Rocky Mountains and the Colorado River, hydrologists estimate that global warning has caused at least a twenty-percent reduction in the River's average flow. (James and Hennessy-Fisk, June 26, 2023, reviews this research.) As a result, the levels of Lake Mead and Lake Powell are now at one-third and one-quarter of their peak levels respectively (as of July, 2023). Lake Mead, for example, is estimated to currently lose 6 feet per year due to evaporation. (James, February 9, 2023, reviews this evaporation problem.) The current levels of Lake Mead and Lake Powell are close to being below the hydropower intake levels which when breached, will create "dead pools" of these reservoirs, i.e. a halting of downriver flows that feed the CAP and the irrigated fields of Yuma, the Imperial Valley, plus the municipal water supplies to Arizona, Nevada, and California.

This emergency exacerbates conflicts among the users, especially concerning who has priority rights for applicable allocation cuts. Addressing an irrigation conference in 1893, Wesley Powell—the initial explorer of the Grand Canyon—presciently stated, "I tell you, gentlemen, you are piling up a heritage of conflict and litigation over water rights, for there is not sufficient water to supply these lands." (Powell, 1876). This "heritage of conflict" crisis is now greatly exacerbated by the consequences of global warming.

The *Colorado River Compact* of 1922 apportioned the river "in perpetuity," allocating 7.5 maf to the Upper Basin, and 7.5 maf to the Lower Basin. The *Compact* did not allocate "half to the Upper, and half to the Lower Basin;" it allocated absolute amounts as though these were known minimums of a probability distribution for the River's flow. Perhaps these absolute commitments resulted from cavalier ignorance, or cavalier lack of due diligence, or purposeful delay of the inevitable crisis, but one should recall that an unfulfillable contract that would be judged as such if due diligence were performed is akin to a "lying promise." Before the signing of the *Compact*, scientists warned there was insufficient flows to justify the 7.5 maf commitments. An immoral kinship between the Lower Basin and Upper Basin States was therefore established in 1922. The inevitable conflicts that followed up to today have their roots established a century ago. The abundance of water was never real; the reality was always of "scarce water in the desert."

The current struggle over where cuts in allocation are to be made is largely between Arizona with its CAP, and California with its Imperial Valley Irrigation District (IID). The IID claims priority due to its earlier usage of water diversions, and therefore it argues for smaller cuts to its allocation. It should be noted that 80% of the Colorado River's withdrawals are for irrigated agriculture, and the All-American Canal gets the larger share of this irrigation water. It should also be noted that the irrigation use in the IID is greater than the allocations to the municipalities of Los Angeles, Phoenix, and Las Vegas combined.

One key to solving the water crisis involves alteration of the old methods used for irrigation, i.e. a change from "flood irrigation" where water is released through canal gates and into the agricultural fields, and to more efficient methods that would avoid high evaporation. The farms are currently charged per acre of the plot irrigated, and not for the actual amounts of water applied onto the fields. This inefficient method exacerbates evaporation. Sprinkler systems, for example, are methods that significantly reduce evaporation, and they have begun to be applied. Drip irrigation would even be better. But the farmers in these areas are demanding federal subsidies in order to install the piping and sprinkler heads necessary. The farmers are also demanding subsidies to fix any leakages in the piping systems that would also save significant amounts of water. (James, January 27, 2023, reviews this leakage problem.) Currently the *Bureau of Reclamation* is offering $4 billion in drought mitigation funding to the IID to eliminate seepage to individual farmers. This would include replacing the lining of its earthen canals with concrete, as well as subsidizing the necessary changes to irrigation systems.

As inducements to promote these irrigation and other efficiencies, the Lower and Upper Basin States except for California want all the state allocations to include the evaporation and seepage losses as part of their allotments, thereby inducing areas such as the IID to install evaporation and seepage mitigations. Without these mitigations, California would continue to have the highest evaporation and seepage. California, however, is resisting the inclusion of evaporation and seepage as part of its allocation unless, of course, federal subsidies are provided to pay for their mitigations. Meanwhile, Lakes Mead and Powell continue to approach their "dead pool" levels that if reached, would totally cease all Colorado River flows below Lake Mead (James, February 9, 2023, reviews these conflicts over evaporation and seepage.)

The Coachella Valley as Desert Oasis

The Coachella Valley lies north of the Salton Sea. It includes the resort and luxury housing area of Palm Springs, and the municipalities of Indio, Coachella, La Quinta, and Cathedral City. The Little San Bernadino Mountains are on the Valley's east, and the Santa Rosa Mountains are on the west. The Coachella Valley receives a relatively large water allocation from the Colorado River in that although its population is much smaller than Las Vegas, it receives a larger allocation.

The Coachella Valley first received irrigation water in 1949 when the Coachella Canal—a northern branch of the All-American Canal—was completed. Currently, agriculture consumes 72, percent of the Coachella Canal's water to produce grape, dates, peppers, lemons, and vegetables. Since 2003, the Coachella Valley has purchased water from the *Imperial Valley Irrigation District* (IID).

There are 120 golf courses in the Coachella Valley which absorb 18, percent of the Valley's water use. Lagoons and surf parks have also been developed as recreational facilities. Both of these are water intensive establishments. Grass lawns around luxury homes and swimming pools are common in this area. Gated real estate developments constructed around golf courses are common between La Quinta (toward the Valley's southeastern edge) and Palm Springs (on its northwest edge). They were developed and promoted as desert mirages with palm trees, grass greenery, swimming pools, and artificial lakes and waterfalls, i.e. a playground oasis with abundant water. The water supply comes either from the Colorado River via the Coachella Canal, or from an aquifer that is replenished from the Canal's water. (James, January 29, 2023 reviews the sources of the Valley's water.) With climate change and global warming, the associated water crisis likely dooms this desert mistakenly viewed as oasis.

The Drought and Colorado's Water: Diversions at the Source

The Colorado River has its source on the west side of the Continental Divide of the Rocky Mountains, northwest of Denver. Our current global debacle of drought affects the cities of the State of Colorado. Storage projects for the City of Denver, which is located to the east of the Rockies in Central Colorado, and the much smaller city of Durango, which is located to the west of the Colorado Divide in southeastern part of the State, are currently being constructed. They might be the last of these sorts of projects within the highly stressed Colorado River System.

There are two new water storage projects now under construction that will serve Colorado's urban Front Range directly west of Denver: the Chimney Hollow Reservoir, and the Gross Reservoir expansion. Both of these projects are more than 20 years in their planning and construction. If and when completed, they will store an additional 167,000 maf. The municipal utility companies of Northern Water and Denver Water are responsible for their respective constructions. Northern water delivers municipal and agricultural water to approximately one million customers and 1.6 million acres of farmland. It brings 220,000-acre feet of water annually from the west slope of the Rockies to the Front Range just east of the mountains. About half of this water is for irrigated agriculture, and half for fast-growing municipal use. This considerable engineering feat involves two pumping stations, five hydropower plants, seven reservoirs, and twenty miles of tunnels to transport the water to the Chimney Hollow Reservoir and the Front Range west of Denver. In this sense, it is a small-scale version and component of the much greater Colorado River System. The Utility company Northern Water intends to initiate filling the Chimney Hollow Reservoir in 2025. It intends for the Reservoir to take three years to fill at a rate of 30,000 acre-feet per year. Given climate change and associated drought, however, this reservoir might never be filled.

Denver Water serves 1.5 million customers in the vicinity of Colorado's largest City. Denver Water's Gross Reservoir is less than 50 miles south of Northern Water's Chimney Hollow. Gross Hollow was designed and built in the 1950s with expansion in mind. Recent construction will increase the height of the dam so that its storage capacity will more than double. Water from the Colorado River Basin is diverted and tunneled under the Continental Divide to the Gross Hollow Reservoir to serve Denver's customers. Denver Water anticipates its customers will grow from 1.5 million to 2.5 million by 2040.

Denver Water's facilities are all along the Continental Divide of the Rocky Mountains above the City. The climate models are ambiguous as to the extent of the precipitation in this critical area. The scenarios of drought and severe storms appear to be equally probable. As with Northern Water, building new facilities covers the needs of population expansion if the reservoirs fill and the increased water is demanded. The risk is, therefore, that the climate change will result in drought and the increase in the area's population will be impossible to serve.

Save the Colorado is one environmental advocacy organization that opposes any new dam diversions of the Colorado River, and consequently *Save the Colorado* is suing both Denver Water and Northern Water due to their planned diversions at the headwaters of the River. In light of our shrinking Colorado River flow, environmentalists and hydrologists are critical of building more downstream capacity when upstream reservoirs are struggling to fill and ecosystems are severely stressed. The Director of *Save the Colorado* explains that, "These two projects will further drain and destroy the Colorado River at the exact moment the system is collapsing." (Emmer, 2022)

Durango is an ideal small town in Southwest Colorado on the Western Slope of the Rockies. It is close to the ancient dwellings of Mesa Verde, and also near to Moab and Southeastern Utah's National Parks such as Canyonlands and Arches. It is a boomtown for real estate development in that its location is in considerable demand. A small house in Durango now costs over $1 million. But like other small towns in Colorado, Durango has less than 10 days of water stored. It relies on its ability to draw from two nearby rivers—the Florida River and the Animas River—but climate change and drought now place these water sources in doubt. Precipitation in Durango's watersheds has declined by twenty percent since the late 1980s, and mountain runoff to the River has dropped by more than a third. Note that Southwestern Colorado has always been drier than the rest of the State, and Durango borders on being in the American Southwest Desert. To resolve its water problems, and to avoid a potential collapse in its hot real estate market, Durango now seeks a pipeline to nearby Lake Nighthorse. This lake was built by the *Bureau of Reclamation* for recreation and a water source for the nearby *Southern Ute Indian Reservation*. Now it will be used to save Durango.

The Drought and Glen Canyon Dam

The Glen Canyon Dam was completed in 1963, but Lake Powell did not reach its enormous reservoir capacity until June of 1980. In its storage capacity it is second only to Lake Mead. Once it was filled, Glen Canyon's pocket canyons of sandstone cliffs and spires, and its ancient and sacred remains of the Navajo were submerged. According to environmentalists, Glen Canyon became "America's lost National Park." In 1968, along with the photographer Eliot Porter, David Brower published a brilliant book of photographs *The Place No One Knew: Glen Canyon on the Colorado River*. It brought the vivid colors of the Canyon to the public. Lake Powell, however, became a popular tourist and recreational destination in the form of a 1.25 million-acre *Glen Canyon National Recreation Area*. The lake's marina had elaborate houseboats. It accommodated two million visiting water-sport enthusiasts per year. In 2011, the National Park Service reported that 2.3 million visitors spent $238 million in nearby communities, which supported 2,819 jobs.

But after the Glen Canyon Dam was completed, the *US Geological Survey* estimated that Hoover Dam and its Lake Mead had sufficient unused capacity to store all of the Lake Powell reservoir so that the Glen Canyon's storage capability was actually unnecessary. Lake Mead could have handled it all. Nevertheless, the dam already submerged the Canyon's fabled grottos, ancient cliff dwellings and scenic pockets. The environmental damage appeared to be complete, and given the enormous concrete structure in place, it did not appear to be fixable.

In 1975, author Edward Abbey published the popular novel *The Monkey Wrench Gang* which depicted the use of sabotage to oppose the environmental destruction of the Southwestern US. This novel concerned a small group of environmentally motivated saboteurs who had particular outrage for the Glen Canyon Dam. Based on this novel, the term "monkey-wrench" became popular among activists. In 1981, the radical but small group *Earth First* faked a crack in Glen Canyon Dam by unfurling a 300-foot banner of a black slash down the front surface of the Dam's structure that depicted a split in the concrete. The atmosphere of active resistance was growing stormy among the Southwest's environmentalists.

According to the *US Census*, the population of the Southwest States (Arizona, Nevada, New Mexico, and Utah) grew by more than one-third between 1990 and 2010. The Colorado River's flow, however, shrank over these decades, and continues to shrink up to the current year (late 2023). The American West has experienced an historic drought within our era of global warming (In August, 2024, the daily average global temperature continues to set record levels). This has caused the draining of ancient aquifers throughout the Southwest in order to service irrigated agriculture. Meanwhile, Lake Powell and Lake Mead have been shrinking for the last twenty-five years (up to the Winter of 2024).

Eric Balken of the *Glen Canyon Institute* argues that the Colorado River can no longer support two massive dams and storage reservoirs. (Langlois, 2017, reviews this storage problem.) Given the current critically low levels of Lake Powell and Lake Mead, Balken calls for a "Fill Lake Mead First" strategy which calls for the *Bureau of Reclamation* to drain Lake Powell in order to at least partly restore Lake Mead. This strategy has two water conservation aspects.

1. The total surface area of the restored Lake Mead would be less than the combined surface of the current combination of Lake Mead and Lake Powell. This reduction in surface area would reduce evaporation thereby conserving large water amounts.
2. The leakage of water through seepage into Glen Canyon's sandstone is substantial, but Lake Mead's surrounding canyon is hard rock with far less seepage. Eliminating Lake Powell's seepage by draining it into Lake Mead would substantially conserve water.

Lake Powell is already at critically low levels. Balken's strategy would keep the dam in place, but allow the flow to continue naturally through the Grand Canyon to Lake Mead. If needed in the future, the dam would still remain and could be reutilized.

The *Glen Canyon Institute's* 2013 study was conducted by hydrologist Tom Meyers. It argues that Lake Powell loses enough water through evaporation and seepage to fulfill all of Nevada's allocation from the Colorado River. It is worthy to note, however, that hydrologist Jack Schmidt of Utah State University questions the old data used in Myers' study in that given the current critical conditions, the seepage data Myers used may not reflect what might be currently saved.

All dams have limited lives, even the very large concrete structures such as the Hoover and Glen Canyon Dams. In popular lore, these structures are "to last like the pyramids," but seepage along the reservoir's walls, or in most cases the eventual large buildup of sediment behind the dam, defeats the dam's purpose. Sediment buildup occurs behind all dams, including hydropower dams, and will eventually create a marshy area behind the dam, and/or create substantial problems for electric generators. The dam then loses its purpose and should be decommissioned and deconstructed. The Colorado River is particularly prone to dam silting problems because it carries large amounts of soil. Note that in its natural state prior to damming, the Colorado never ran clear. It was always red, or brown, or another color, and was always opaque. But the flow coming from the Colorado's dams is now usually close to clear, which indicates that the dams are blocking the silt from flowing downstream. Note that the siltier water flows lower in the water column so dams allow the clearer flows to pass.

Over its 50 years, the Glen Canyon Dam has blocked millions of tons of sediment which now lies at the bottom of Lake Powell; 100,000 cubic yards of additional silty sediment accumulates every year. (Markosian, 2021, presents this estimate.) This sediment is contaminated by agricultural runoff and mining wastes. In a natural state, the silt creates oases in the coves and bends along the river. These naturally verdant areas of native vegetation have been disappearing from the Grand Canyon because the Glen Canyon Dam blocks the natural silt from being carried downriver. With Lake Powell being at such low current levels, however, these natural coves are again appearing and some native vegetation is again growing in these reemerging areas.

As of February of 2023, Lake Powell is only at 23 percent of its capacity. During a recent (February 7) virtual meeting of the *Bureau* with its various constituents, the options for restructuring the Glen Canyon Dam were reviewed and discussed. The immediate concern is that the reservoir will drop below the level necessary for power generation. (James, 2023, February 9, reviews this February meeting.) If that were to happen, then water would only flow through four smaller tubes which would reduce the dam's water releasing capacity to flow through the Grand Canyon down to Lake Mead. This reduction could further lead to Lake Mead becoming a "dead pool" with no flow past Hoover Dam.

Three proposals were discussed at the February 7 meeting:

1. New lower-level intakes could be drilled through the dam's concrete.
2. A new lower-level power plant could be constructed.

3. Shafts could be tunneled on the sides of the dam to establish new power plants.

Meeting participant Kyle Roerink of the *Great Basin Water Network* observed, "They're not sugarcoating that things have to change pretty quickly" (Ibid.). Note that any structural penetration involves risks of weakening the dam. Also, all three of the above options are expensive ($0.5 billion to $3 billion) and time consuming according to the Executive Director of the *Colorado River Energy Distributors' Association*, a meeting participant.

Nonprofit cooperatives purchase the electric power from the Bureau and then distributes the power to their respective localities. The increased costs due to the dam's modifications would be passed on to consumers. During 2022, Glen Canyon powered approximately 240,000 homes and businesses. But the *Bureau* suggests that the apparent inevitable loss of hydropower at Glen Canyon can be replaced by either solar or wind turbine alternatives.

This hydro crisis is caused by global warming which is driving the aridification of the Southwest. The scientific analysis indicates that a global average increase of 1.8 degrees Fahrenheit reduces the average flow of the Colorado River by approximately 9 percent. The hydropower future of Glen Canyon Dam is therefore now highly questionable. Kyle Roerink states, "I think we need an option that is just a bypass option without a power plant at the end of it." (James, 2023, February 9, includes this quote.) With respect to the reengineering options for the dam, Roerink says, "Are the taxpayers' dollars worth it? How long will it be until this just proves to be a futile exercise?" (Ibid)

The Emergence of the *Glen Canyon National Park*

John Weisheit, a cofounder of the environmental advocacy organization *Living Rivers*, poses a relevant question. "I think it is imperative that we know exactly what the life span of the dam is" (Ibid.). For example, the sediment buildup behind the dam has already reduced the dam's storage capacity by at least seven percent compared to its pervious maximum capacity, and one-third of its current low level. Also, the solutions to the intake problems will likely reduce the life of the dam's electric generators and concrete structure. The proposed expenditures for the various alternative options to salvage the dam range from $0.5 billion to $3 billion.

Congressional approval and environmental reviews are required for these, and all of this plus the various proposed constructions requires time. John Weisheit argues that rather than spending $3 billion on attempts to salvage Glen Canyon's "shrinking amount of hydropower capacity," investing these funds in solar cells and wind turbines would offer more power generation. Weisheit's solution is to deconstruct the dam because it is not the right technology given global warming and consequent drought.

One should note that wind and solar facilities are flexible in their location. Dams are not flexible in location, and therefore require longer transmission infrastructure which in turn dissipates the electric power, a sort of leakage of this power. Locating the renewable energy facilities closer to Phoenix, Tucson, and the other dense populations of south-central Arizona inherently conserves both energy and the expenditures on longer transmission infrastructure. Furthermore, consider the irony of favoring hydropower in a geography that severely lacks water resources, and not favoring the renewable sources that are clearly natural for a sun-drenched desert geography. The latter is the inevitable technology of choice. In this circumstance, expending significant public resources on saving the apparent outdated technology appears to be a foolish public choice. What could possibly favor such a selection? Only the forces of current vested interests would oppose the change to renewable energy.

The *Glen Canyon National Recreation Area* is now almost entirely gone given the receding Lake Powell. A relevant question is, "Can a new *Glen Canyon National Park* emerge where hikers can visit the oases and slot canyons of Glen Canyon?" Annette McGivney presents a narrative of her excursions into the emerging canyon starting in May of 2003. Her narrative is published in *The Guardian*. (McGivney, June 11, 2023). Her description of her hike through the now revealed canyon is riveting.

> This resurrection of a river that had been dammed to create the reservoir was a beautiful yet unsettling sight. The climate crisis was exposing flaws in a water system that – after years of denial – Western states have finally been forced to confront. Over the following decade, I returned to Lake Powell many times to hike the slot canyons and tributaries emerging as drought shrunk the lake, watching a world long thought buried coming back to life.

McGivney's account describes hiking in the scoured-out narrows along "Twilight Canyon" which was once 110 feet below Lake Powell's surface. She also describes paddleboarding through the "Cathedral in the Desert," a side canyon that has emerged from the receding waters. Her description fits Ed Abbey's characterization of Glen Canyon as being "a portion of Earth's original paradise."

McGivney also described her other ventures into the emerging parklands.

> The first years of drought left a dystopian scene of impassible mud flats, abandoned houseboat parts and thorny native plants. But over time, flash floods swept that away, allowing streams to trickle over scoured slickrock, springs to seep from fern-decked canyon walls and waterfalls to tumble over ledges. As I hiked through more than a dozen of the same tributaries over a period of years, I saw native willow and cottonwoods saplings take root and all manner of wildlife species return, from dragonflies to beavers to bobcats. Fifty feet above my head, the white stripe of the reservoir's high-water mark lined canyon walls, a bizarre reminder of where boats once floated. (Ibid.)

More than 100,000 acres of Glen Canyon have now emerged from Lake Powell as dry land. This is an area larger than *Arches National Park*, where equally magnificent vistas, arches, waterfalls, oases, and monumental spires are commonly found.

Biologist Seth Arens' presentation to the *Utah Native Plant Society* states,

> In locations that have been exposed since 2000 or 2011, substantial restoration of riparian ecosystems has already occurred. Hanging gardens fed by springs seeping from canyon walls are growing in isolated locations and thriving riparian ecosystems with extensive stands of willow and cottonwood exist. (Ibid.)

In addition to the reemerging canyons and the like, 40 miles of "Cataract Canyon," a legendary whitewater section of the Colorado River with rapids similar to those in the Grand Canyon, have already emerged. Eric Balken of the *Glen Canyon Institute* sums up the Glen Canyon narrative.

> So much of the Colorado River story is about depletion and decline, but in Glen Canyon we have this opportunity for rebirth and getting something back. (Ibid.)

The debate about what to generally do with the Colorado River, and specifically about Glen Canyon, is becoming increasingly feeble. Global warming is entirely determining the debate's direction with options rapidly narrowing. As of this writing (July of 2023), record temperatures (over 107 degrees Fahrenheit) make visiting any of the Grand Canyon areas during mid-summer uncomfortable at best, and dangerous at worst. It is highly likely that the hydropower presence along the river will be much reduced within the next few years. The Lake Powell water recreation area will likely disappear as we now know it. A National Park area similar to Arches, or Monument Valley, or Canyon Lands is emerging for significant numbers of visitors during the months of moderate temperatures. Glen Canyon Dam is highly likely to be either entirely deconstructed, or so modified as to no longer resemble its water-park past in function. The natural environment, along with advances in technology, will likely determine the entire design of the Colorado River's future features. Deep in the reemerged canyons of Abbey's "original paradise, springs are flowering, wildlife is returning, and cottonwood saplings are reaching toward the sun" (Ibid.). The mistake of Glen Canyon Dam is being corrected by nature and human intelligence.

The Original Sin and Its Resolution

We can now perceive the extent of the consequences of the "great lying promise" posed by the 1922 *Compact*. The Lower-Basin States do not have reasonable prospects for reaching generally accepted agreements for distributing what will likely be far less than 7.5 maf per year for much of the distant future. Arbitrary, unfair, environmentally destructive, and economically inefficient solutions—such as the "law of prior appropriation"—will likely be foisted on the Southwest. These solutions might be legally convenient for the courts, but they are seriously problematic economically and culturally, and for any other likely consideration except legal convenience. The overexpansion of agriculture and population in the Lower Basin persist in posing insatiable demands to "feed the desert." Note that water shortages will likely inhibit these expansions, but the attitude has apparently been, "Since people are moving there, we must provide them with water." This apparent attitude has the direction of causation backwards, i.e. provide the water first and people will come to get it. The Upper Basin States now have their own water distribution problems caused by climate change, and resolution of these problems will likely exacerbate the problems of the Lower Basin. The 1922 *Compact* cannot resolve these severe problems. It must be discarded as a remnant of some irrational paranoid uninformed distantly-past conference. That conference was unconnected to the realities of our current severe drought and the probable future that climatologists and hydrologists predict will most likely persist for a lengthy period.

The problem of Glen Canyon is not one of an environmental treasure to be viewed in isolation. Its destruction occurred due to the sin of the 1922 *Compact*. Its resurrection can only occur in a less sinful culture of political rationality, or else perhaps nature will move in its own way to provide the solution independently of our decisions.

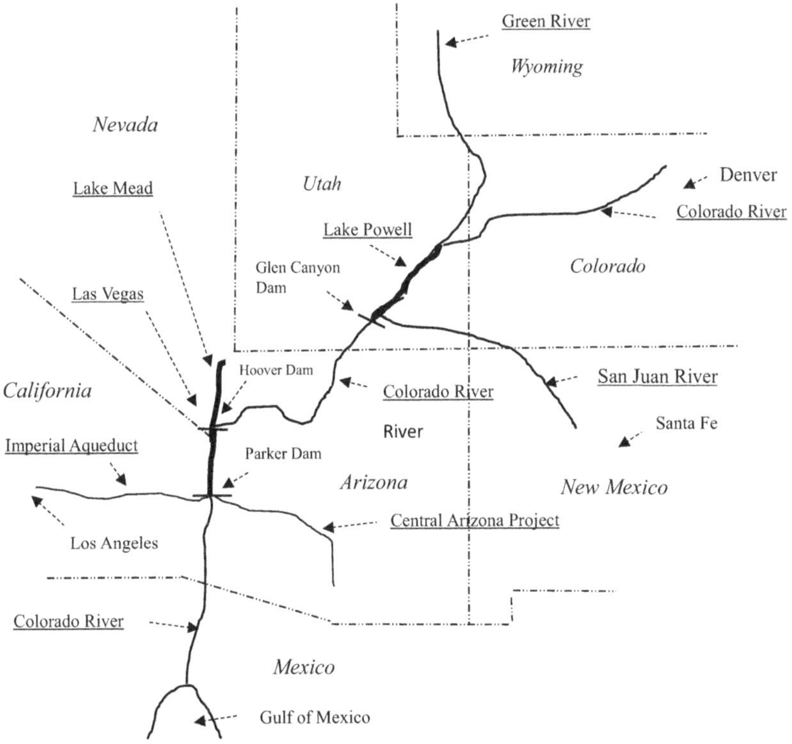

Fig. 4.1 Colorado River Basin

References

Abbey, E. (1975). *The Monkey Wrench Gang*. Lippincott.

Adler, R. W. (2007). *Restoring the Colorado River Ecosystems: A Troubled Sense of Immensity*. Island Press.

Brower, D., & Porter, E. (1968). *The Place No One Knew: Glen Canyon on the Colorado River*. The Sierra Club/Ballantine Books.

Demircan, H. (2020). Colorado River: Lifeline of the Southwest. *CASSE*, October 1, 2020. www.steadystate.org/colorado-river-lifeline-of-the-southwest-effects-of-economic-growth-and-climate-change/

Emmer, O. (2022, August 17). Two New Colorado River Reservoirs are Rising on the Front Range, are They the Last of their Kind?" The Water Desk, by *Fresh Water News*.

James, I. (2023, February 18). As the Colorado River Shrinks, Federal Officials Consider Overhauling Glen Canyon Dam. *Los Angeles Times*. www.latimes.com/environment/story/2023-02-18/federal-officials-consider-overhauling-glen-canyon-dam

James, I. (2023, January 9). Why Desert Golf Courses and Artificial Lakes Remain Untouched by the Colorado RiverCcrisis. *Los Angeles Times*, Los Angeles, CA.

James, I. (2023, February 9). As Colorado River Shrinks, Water Evaporation Becomes Critical to California's Future Supplies. *Los Angeles Times*, Los Angeles, CA.

James, I. (2023, January 27). In California's Imperial Valley, Farmers Brace for a Future with Less. *Los Angeles Times*, Los Angeles, CA.

James, I., & Molly, H.-F. (2023, January 26). The Colorado River Is Overused and Shrinking. Inside the Crisis Transforming the Southwest. *Los Angeles Times*, Los Angeles, CA.

James, I., & Sean, G. (2023, February 7). Why California Is So Far Apart From Other States in Colorado River Water Cuts Plan. *Los Angeles Times*, Los Angeles, CA.

James, I. (2023, January 26). A Rriver Guide's View of Lake Powell's Decline and the Depths of the Colorado River Crisis. *Los Angeles Times*, Los Angeles, CA.

Jones, B. (2023, May 23). Why the New Colorado River Agreement Is A Big Deal. *Vox*. www.vox.com/2023/5/23/23734404/colorado-river-cuts-lake-mead-deal

Langlois, K. (2017, September 4). Down with the Glen Canyon Dam? *High Country News*. www.hcn.org/49.15/rivers-down-with-the-glen-canyon-dam

Lin, R. G., & James, I. (2023, February 5). Colorado River Crisis Is So Bad, Lakes Mead and Powell are Unlikely to Refill in Our Lifetimes. *Los Angeles Times*, Los Angeles, CA.

Markosian, R. (2021). Water in the West: Should We Demolish Glen Canyon Dam?" *Utah Stories*, February 17. https://utahstories.com/2021/02/water-in-the-west-should-we-demolish-glen-canyon-dam/

McGivney, A. (2023, June 11). A Portion of Paradise: How the Drought Is Bringing a Lost US Canyon Back to Life. *The Guardian*. www.theguardian.com/us-news/2023/jun/11/lake-powell-glen-canyon-drought-reservoir

Meyer, J. M. (1997). Gifford Pinchot, John Muir, and the Boundaries of Politics in American Thought. *Polity 30*(2), Winter, 267–284.

Palumbi, A. (2023). *At the Base of the Giant's Throat: The Past and Future of America's Great Dams*. University of Nebraska Press.

Pisani, D. J. (2002). A Tale of Two Commissioners: Frederick H. Newell and Floyd Dominy. Presented at the *History of the Bureau of Reclamation: A Symposium*, Las Vegas, NV.

Powell, J. W. (1876). *Report on the Lands of the Arid Region of the United States*. at https://pubs.usgs.gov/publication/70039240.

Reisner, M. (1987). *Cadillac Desert: The American West and Its Disappearing Water*. Penguin Books.

Smith, H. (2023, February 6). Why Southern California Water Restrictions Remain Despite So Much Rain. *Los Angeles Times*, Los Angeles, CA.

Smioth, J. (2023, July 10). New Study Shows Durango's Water Supplies Declining Dramatically as Climate Change, Drought Hit Home. The Water Desk, by *Fresh Water News*.

Sudman, R. (2020). Colorado River Compromise. *Water Education Foundation*. www.watereducation.org/western-water-except/colorado-river-compromise/

Tallmadge, J. (1997). *Meeting the Tree of Life, A Teacher's Path*. University of Utah's Press.

CHAPTER 5

After the Gold Rush: Klamath River Justice

Introduction: Correcting an Accident of People and Geography?[1]

The story of California is one of water, how to get it from where it is abundant (the far northern coastal region) to the State's arid south. The State's population density is in the south; its rich agricultural soils extend into the south. In the first three decades of the twentieth century, Californians thought they only needed to defy nature and replumb their rivers, particularly if they could use federal funding to do so. The problem was thought to be as simple as rearranging the drainage in an urban neighborhood, but just on a much larger scale. All that would be needed was to put in more drainage tiles and send the water in the proper direction! But the problem of replumbing such a large and complex state is not simple. For one thing, the ecologies of northern and southern California radically differ, and replumbing on the scale attempted substantially destroys the ecologies of both. For another thing, the statewide replumbing substantially rearranges the populations of the state. It particularly supports the population centers in the southern half of the Central Valley, centers

[1] This is a modification of a quote from California's thirty-second Governor, Pat Brown who served from 1959 to 1967. His quote was aimed at explaining the Central Valley Project's and the California State Water Project's extensive replumbing of Northern California's rivers to irrigate the southern sections of the Central Valley.

R. M. Robinson, *Restoring America's Rivers*,
https://doi.org/10.1007/978-3-031-81758-8_5

formed around big agriculture. For a third thing, the replumbing substantially opened the political influence door to big agriculture, and the resulting political struggles impacted the water allocations in ways that were economically very inefficient. Big agriculture must ultimately suffer in an arid climate where the water pumped in is from an unreliable source due to frequent long droughts. For a fourth issue, replumbing badly hurt some of the in-place populations of the north which gave up their water for the temporary benefit of those in the south. These four points are explored in this Chapter.

Nevertheless, such a large engineering task required federal funding, and therefore federal bureaucracy. It required a bureaucratic entity that could provide the massive funding necessary to redesign this mostly arid State. The US Bureau of Reclamation became that federal entity, and the Depression of the 1930s gave it the opportunity. The particular story told here concerns how this federal institution piped the waters of the Klamath and Trinity Rivers into the Sacramento River and did so for the purpose of feeding the municipal and irrigation demands of the Central Valley.

California's Central Valley is an elongated flat 450-mile long agriculturally rich land mass that extends from just north of the town of Redding down to just south of the town of Bakersfield. (Fig. 5.1 provides a map of the Central Valley.) It covers 18,000 square miles (eleven percent of the entire State) between the Sierra Mountains on its east side and the Coastal Range on its west side. It varies from approximately 40 to 60 miles in width along its length. It has almost seven million acres of irrigated agriculture (sixty-one percent of the Central Valley) that grows almost half of the fruits, vegetables, and nuts produced in the US measured in dollar value. The Central Valley is irrigated since it has little rainfall. There was once a considerable amount of aquifer water available, but this has now been mostly pumped dry.

In the 1930s, the federal government via the Bureau of Reclamation, along with the State of California, began an extensive replumbing of the State's rivers by taking water from the rain forest of Northern California (in particular the Shasta, Klamath, and Trinity Rivers) to irrigate the central and southern portions of the Central Valley. The Bureau accomplished the task by channeling water from the Shasta River and the Trinity River into the Sacramento River which flows southerly from the north into the central portion of the State's Central Valley. From there, the water is pumped over low mountains and into the southern sections of the Central Valley. The environmental effects of these water diversions are substantial, particularly on the once prolific salmon and steelhead runs of Northern California. The consequential impacts on Indigenous Tribes have also been substantial. The history of these diversions and their effects are examined in this chapter including the justice issues associated with dam removals and their impacts on the Indigenous Tribes of Northern California and Southern Oregon.

The Central Valley

California's Central Valley consists of four sections: (1) the Sacramento Valley in the north, (2) the Delta and Eastside Streams in the center just to the east of San Francisco Bay, (3) the San Joaquin Basin which is a large area south of the Delta, and (4) the Tulare Basin in the far south. (Figure 5.1 indicates these sections.) The Sacramento River flows 400 miles to reach the San Joaquin River delta, and then the San Francisco Bay. It drains 26,500 square miles (sixteen percent of the land area of this very large State) of mountainous area and fertile agricultural floodplain bounded by the Sierra Mountains on the east and the Coastal Ranges on the west which forms the Sacramento Valley. Since the 1950s, this River and its tributaries have been dammed for hydroelectric power, and for providing municipal water to half the State's population, and also to the irrigated agriculture of the Sacramento Valley and the rest of the Central Valley.

The Sacramento River originates at Shasta Lake immediately north of Redding. (Figure 5.2 provides a map of this River.) The Upper Shasta River, the McCloud River, and the Pitt River all flow into Shasta Lake. The Upper Shasta River originates by draining Mount Shasta which is seventy miles directly north of the small City of Redding. Just north of Redding, the Shasta Dam impounds the River to form Shasta Lake. This dam provides a primary water storage and hydropower generating facility of California's Central Valley Project (CVP). It also provides flood control, hydropower, and irrigation to the Upper Sacramento Valley. After flowing five miles to the south, the River encounters Keswick Dam where it receives an average of 1.2 million acre feet annually from the nearby Trinity River. (This diversion is reviewed in the section below.) After the River flows through Redding, it carves the Iron Canyon at Red Bluff where a pumping station diverts water for nearby irrigation. Beyond Red Bluff, the River reaches the low floodplain of the Valley. Between Red Bluff and the City of Sacramento (California's State Capital) the

River receives Butte Creek at Calusa, then the Feather River and the American River just north of Sacramento. Thirty miles farther south, the Sacramento River enters the San Joaquin Delta.

In 1917, the federal government funded the "Sacramento River Flood Control Project." Its main feature was the establishment of a series of weirs and bypasses that divert floodwaters when they reach a cut-off level. The purpose was to send flood water into natural floodplains and then allow this water to naturally recede. These were established at Calusa and surrounding towns. The irrigation, hydropower, and municipal water supplies are therefore protected, and the historical floodplain preserved for dissipating emergency flood conditions.

The San Joaquin River originates in the south-central Sierra Nevada Mountains, south of the San Joaquin Delta. It flows 355 miles, first westerly. It then turns northerly a few miles north of the City of Fresno as depicted by Fig. 5.1. On this route to the San Joaquin Delta, it is joined by the Merced, Tuolumne, Stanislaus, and Calaveras Tributaries, all of which drain the Sierra Mountains and join the San Joaquin on its east side as depicted by Fig. 5.3. A few miles north of Stockton, the San Joaquin reaches its Delta, a myriad of canals and streams that forms a braid of variegated islands before the freshwater flows into the estuary of Suisun Bay, and then the San Francisco Bay. The San Joaquin flows into the Delta from the south where the Friant Dam diverts irrigation water from the River into the California Aqueduct. This Aqueduct then transports water south into the Central Valley to be distributed through a system of federal and state canals.

The San Joaquin Basin includes the Valley north of Fresno and up to the Delta. This area is hot and dry during the summers, and moderate during the winters but with periods of heavy ground fog. It produces approximately thirteen percent of California's agricultural produce by dollar value. For example, the area around Stockton produces most of the asparagus consumed in the US, and the area around Fresno produces most of the raisins consumed in the US. But this Basin suffered greatly from the recent megadrought. Its groundwater aquifers are now at significantly lower levels. It also has poverty rates among the highest in the US.

The southern portion of the San Joaquin Valley is the Tulare Basin. This Basin is named after the once freshwater Tulare Lake, which since the late 1800s has usually been dry except during periods of unusually high precipitation, such as occurred in 2023. During high precipitation periods, it becomes marshy wetland. It is, however, usually dry since the rivers that feed it are extensively drained for irrigation water.

Despite the difficulties with drought, plus the draining of aquifers and other difficulties with water sources, the irrigated agriculture of the Central Valley produced eight percent of the US agricultural output in 2013 ($43.5 billion). Almost all of the non-tropical crops have been produced in the Central Valley, including tomatoes, table grapes, cotton, apricots, asparagus, and almonds. About seventy percent of the world's supply of almonds (600 million pounds) are produced in this Valley.

It is the extensiveness of this system of agriculture carved from the central valley's semi-desert—a system with particularly high thirst for irrigation water—that generated the demands for damming Northern California's rivers.

The Central Valley Project and the Northern Rivers

In 1933, the "Central Valley Project" (CVP) was federally created and funded. It was begun under President Franklin Roosevelt's Administration during the worst days of the "Great Depression" as a public works project aimed at providing irrigation and municipal water to the Central Valley. This Project was originally intended to be a California State effort, but the State's revenue decreased during the Depression. This forced California to ask the federal government to assume the Project in both funding and management. The Bureau of Reclamation took over the CVP in 1933. The Project's purpose was to send water south from the northern California rain forest to the water starved San Joaquin Valley, and to do so by constructing a series of canals, aqueducts, and pumps. Some of this series has been shared by the "California State Water Project" (SWP) which was a later state-level effort (reviewed below). The CVP was designed, however, to be more than only a redistribution of water. Using its extensive system of more than twenty dams and reservoirs, it also generates a capacity of over 2,000 megawatts of hydroelectric power, and provides flood control for California's heavy November and December rains, plus California's heavy spring snow melts in the Sierras. This federal effort transformed the semi-arid San Joaquin Valley into productive farmland. As a consequence of the CVP's dams, the anadromous fish of the Sacramento River Basin, the Klamath River Basin, and the Trinity River Basin were all decimated. The culture and economy of California's Northern Indigenous Tribes were also consequently decimated. These consequences are reviewed below.

The CVP can store thirteen million acre-feet of water in twenty-one reservoirs in the Sierra Nevada and Klamath Mountains. It can pass 7.4 million acre-feet downstream annually through its canals. Five million acre-feet of water can irrigate three million acres of the Central Valley, but six-hundred-thousand acre-feet of this is used as municipal water in the Valley. The material reviewed below indicates the extensive degree of what should only be called the "over engineering" of the rivers that flow from the Sierras and surrounding mountains.

Mount Shasta is a 14,162-foot peak in north-central California, fifty miles north of Redding. The Sacramento River drains Mount Shasta's southern slopes, and flows into Shasta Lake north of Redding as indicated by Fig. 5.2. The McCloud River drains the eastern slope of Mount Shasta, and flows into the northern tip of Shasta Lake. The Pit River generally drains the mountainous area to the northeast of Redding, and joins Shasta Lake on its eastern edge. This reservoir is the largest in California, and has a surface area of 30,000 acres. It is impounded by Shasta Dam, a gigantic structure of 602 feet in height and approximately two-thirds of a mile in length. It can hold 4.5 million acre feet of water on the northern edge of the Central Valley which ultimately provides flood control, hydropower, and irrigation for the Sacramento River Valley and the entire CVP. It was begun in 1938, and completed in1945.

The Keswick Dam is immediately downriver from the Shasta Dam. There the Sacramento receives water diverted from the Trinity River as explained below. The Shasta Dam and the Keswick dam together regulate the flow of the Sacramento River so that downstream diversion dams can capture this flow and send it into canals where pumps then distribute the water into agricultural areas.

In 1942, the Bureau of Reclamation began to investigate the "Trinity River Project" of damming the Trinity River for the purpose of diverting water into the Sacramento River. But in 1945, the Shasta dam was completed and this sufficiently served the water storage and flood control purpose of that time. Undeterred and not wanting to leave a river undammed, the Bureau proposed the project again in 1951. It was authorized in 1953 as the "Trinity Diversion of the Central Valley Project." While the original "Congressionally authorized annual diversion" into the Sacramento River was 704,000 acre-feet, the Bureau increased that diversion amount to 865,000 acre-feet.

The Trinity diversion was completed in 1963. The Upper Trinity River has two dams, the Trinity Dam that creates the very large Clair Engle Lake, and the slightly downstream Lewiston Dam which diverts water through a 10.7-mile Clear Creek Tunnel under the Trinity Mountains to the Whiskeytown Lake, a reservoir a few miles east on Clear Creek. From there it flows a bit farther east through another 2.4-mile tunnel to join the Sacramento River at the Keswick Dam near Redding. The dams along the Upper Trinity River block salmon runs from their 109 miles of upriver spawning habitat. This destroyed the fishing economy of the Trinity River Valley.

After the Trinity River Dams were completed and its diversions began, the Bureau of Reclamation did not adhere to the limits set by Congress. It regularly diverted seventy-two percent of total Trinity River flow, and sometimes as high as ninety percent, which decreased the salmon runs by eighty percent as compared to the 1950s. This abuse began to be corrected in 2005. (See more details below.)

The Feather River is the longest tributary of the Sacramento River as depicted by Fig. 5.3. Its North Fork, Middle Fork, South Fork, and West Branch all drain the Sierra Nevada and flow into Lake Oroville, which is created by the Nation's tallest dam—Oroville Dam. The main stem of the Feather River flows seventy-three miles from the Oroville Dam to the confluence with the Sacramento River, twenty miles north of the City of Sacramento. The Feather River's branches originate on the east side of the Sierra Mountains, and flow through low passes to reach the west side.[2]

At 770 feet high, the Oroville Dam is the tallest in the US. Lake Oroville is the second-largest reservoir in California with a storage capacity of 3.5 million acre feet. It was constructed by the California Department of Water Resources, and is the main storage and hydroelectric generating facility of the California State Water Project. The Oroville Dam is earthen with a concrete spillway, but it had considerable structural problems that needed emergency repairs in 2016–2017 when Northern California experienced its wettest winter in more than a century. (Chapter 11 reviews this emergency).

The Oroville Dam was constructed by the California Department of Water Resources and completed in 1968. (Figure 5.2 indicates this dam.)

[2] The total drainage area of the Feather River is 6,200 square miles with 3,604 square miles being upriver of Lake Oroville. Most of this drainage area is in public lands, mostly in the Plumas National Forest.

It is a primary facility of the CVP for its water storage and hydropower generation. Downstream of the Oroville Dam, the Feather River is heavily diverted into irrigation canals, so much so that the River's flow is significantly reduced. Its major tributary, the Yuba River, has two dams (Bullard's Bar and Englebright), and the Feather River's other major tributary, the Bear River, is impounded by Rollins Dam. Together, these dams reduce the flow of the Feather River by eighty percent.[3]

Of the Sacramento River's significant tributaries, the American River is the most southern, and the most downstream. The Sierra Madre around Lake Tahoe drains into Folsom Lake, from where the American River's main stem flows thirty miles south-westerly to join the Sacramento River in the City of Sacramento. Folsom Lake is the reservoir formed by Folsom Dam which was constructed in 1955 by the Corps of Engineers as a principal component of the CVP.

Folsom Dam provides flood control and hydropower generation for the Sacramento area.[4] This dam is operated by the Bureau of Reclamation, although it was constructed by the Corps. Seven miles downstream from the Folsom dam is the Nimbus Dam that diverts irrigation water into Folsom South Canal which is part of the CVP. This canal irrigates 7,000 acres of farmlands.[5]

The South Fork of the upstream American River has eight hydroelectric dams operated by the Sacramento Municipal Utility District (SMUD). The environmental advocacy organization American River Conservancy is attempting to restore the South Fork's stream beds and riparian habitat so that it will support anadromous fish.

The Sacramento River flows into the San Joaquin Delta, which has an extensive and complex system of aqueducts and pumping stations that sends water to the southern half of the California's Central Valley.

[3] Downriver from Oroville Dam, between 1902 and 1967 (before the Oroville Dam was constructed), the average annual stream gage flow was 5,834 cubic feet per second. From 1969 to 2009, the same stream gage's average annual flow was only 1,086 cubic feet per second, a reduction of more than eighty percent.

[4] Folsom Lake has a surface area of 11,930 acres, and a capacity to store almost a million acre feet. The Dam's hydropower capacity is close to 200 megawatts.

[5] The Nimbus Dam is small by CVP standards. It stands only 87 feet high, but is 1,093 feet long. It is part of the CVP and completed in 1955 as a supplement to the Folsom Dam and Project. Nimbus Dam creates Lake Natoma, a 8,700-acre feet of storage reservoir. Its hydropower capacity is 13.5 megawatts. The Nimbus Dam diverts irrigation and municipal water 27 miles south through the Folsom South Canal.

As indicated in the section above, the San Joaquin River originates in the Sierra Mountains south of the San Joaquin Delta. It flows 355 miles (mostly north) to reach the Delta as depicted by Fig. 5.3. Along its course, it is joined by its tributaries: the Merced River, the Tuolumne River, the Stanislaus River, and the Calaveras River. Since they all drain the western slopes of the Sierras, they all flow into the San Joaquin from the east. The CVP has several dams on the San Joaquin that divert water to the Central Valley Aqueducts. For example, the Friant Dam—completed in 1942—sends water south through Tulare Lake and to Bakersfield which is at the southern tip of the Central Valley. Water is diverted for irrigation along the way.

The Merced River flows 145 miles from above Yosemite Valley (Yosemite National Park) where it is a fast-flowing mountainous stream that becomes a meandering slow moving river when it reaches the San Joaquin Valley. It has been dammed in several locations so that at times, irrigation diversions reduce it to a dry bed. It drains 1,726 square miles in the central Sierras. The upper section is classified as a National Wild and Scenic River. The New Exchequer Dam is the largest of the Merced River dams. It forms Lake McClure which holds more than a million acre-feet for flood control, irrigation, and hydropower generation. Merced Falls Dam is downstream of the New Exchequer and forms Lake McSwain for irrigation diversion. These diversions remove half the natural flow of the Merced, and even more of a portion during drought. In 2022 (a drought year), all of the flow of the Merced was diverted for irrigation, and for several months the lower Merced River was dry.

The Tuolumne River originates as a fast-flowing stream in Yosemite National Park. After the San Francisco earthquake and devastating fire in 1906, the City decided that the municipal water supply was inadequate for firefighting. In great controversy, the City proposed damming the Hetch Hetchy Valley in Yosemite (the O'Shaughnessy Dam). The leader of the newly formed Sierra Club—John Muir—led the opposition through legal challenges, but this opposition effort failed. This failure led to the environmental movement becoming better organized to oppose similar desecrations. Along the Tuolumne, the "New Don Pedro Dam" was constructed a mile and a half downstream from the "Old Don Pedro

Dam." This earthen dam was completed in 1971, and it replaced the Old Don Pedro which was completed in 1924.[6]

The Stanislaus River is the farthest downstream tributary of the San Joaquin River. It is south of Stockton and the San Joaquin Delta. It originates where its North, Middle, and South Forks of the High Sierra Mountains join at the New Melones Reservoir. It flows 96 miles to reach the San Joaquin Valley to which it furnishes agricultural irrigation.

There are twenty-eight dams on the Stanislaus River and its tributaries. Together they have a combined storage capacity of 2.8 million acre feet. They irrigate 213,000 acres of farmland.

Fourteen of these Stanislaus dams are hydroelectric. Over an upriver course of sixty miles, it drops 10,000 feet, which yields an unusually steep drop of 167 feet per mile, perfect for locating hydroelectric generating facilities.[7] The average annual flow of the Stanislaus is 1.1 million acre feet. All of these specifics appear to indicate an ideal asset for meeting the demands of a major metropolitan area like the San Francisco Bay area. But this River flows into California's Central Valley with its rapacious agricultural demands, and as reviewed below, new environmental demands were also being placed on the Stanislaus in the 1970s.

The water authorities that control the Stanislaus Watershed included local irrigation districts along with the US Bureau of Reclamation. They have committed 19.7 million acre feet to various irrigation needs, which is far beyond the River's average annual flow of 11.1 million acres. This extensive overallocation led to sections of riverbeds being dry for months at a time especially during Central California's frequent drought periods.

The Bureau of Reclamation has long recognized the insatiable demands for irrigation and municipal water use of the Stanislaus Valley. That is why the Bureau first proposed the "Tri-Dam Project" in 1948. This led to three dams being completed in 1957 and 1958 (the Dannells Dam, the Beardsley Dam and the Tullock Dam). At the time, these three tripled the storage capacity of the Stanislaus system, but that was far from adequate. Finally, the "New Melones Dam" replaced an old structure on

[6] The New Don Pedro is 585 feet high, the tenth highest in the US. Lake Don Pedro has a surface area of 12,960 acres. It has a storage capacity of slightly more than two million acre-feet. Its hydro capacity is 203 megawatts.

[7] The dams of the Stanislaus have a combined capacity of 780 megawatts.

the main stem in 1972.[8] The "New" dam was constructed just upstream of the Tullock Reservoir. But this large construction and reservoir initiated reactions from environmentalists particularly since the New Melones Reservoir flooded some prime whitewater sections that flowed through steep and picturesque canyons. This occurred at the time that whitewater rafting and kayaking became popular activities during the 1970s in the San Francisco Bay area. As a result, the Bureau of Reclamation began to leave commitments to irrigation districts unfilled in order to meet environmental demands for water flow. The public consciousness concerning the environmental effects of big dams began to have political consequences.

In 1973, the California "State Water Resources Control Board" (SWRCB) ordered the Stanislaus irrigation districts to provide adequate water flow to protect the environmental habitat of the River and its tributaries. In a 1978 decision by the US Supreme Court held that California has the constitutional right to make these demands on the federal Bureau of Reclamation. Furthermore, the 1982 Reclamation Reform Act set higher standards for the cost–benefit analysis for water projects such as dam constructions. Environmental consideration now had to be considered. Also that year, and three years after the New Melones dam was completed, a planned hydropower dam on the Tuolumne River was canceled. The new environmental consciousness was having an effect.

The Mokelumne River is a ninety-five-mile long body flowing from the central Sierra Nevada Mountains (157 miles if the length of its North Fork in included) as depicted by Fig. 5.4. It flows into the San Joaquin River very close to where they form the Delta area with the confluence of the Sacramento River. Its watershed is immediately to the east of San Francisco Bay. This River is formed by the confluence of its North and South Forks in the high Sierras. The 62-mile long North Fork is

[8] The New Melones Dam was built in 1979 as a 625-foot high earthen embankment structure, the second-largest earthen dam in the US after the Oroville Dam. It is the sixth-largest US dam. It impounds 2.4-million acre feet of water with a reservoir surface of 12,500 acres. Its hydropower capacity is 300 megawatts. The fourteen hydropower dams in the Stanislaus Watershed produce a combined 780 megawatts. It was constructed by the Corps of Engineers for flood control protection for the downstream towns of Oakdale, Ripon, and Riverbank.

the longer of the two forks, and it contains Bear Lake and Salt Spring Reservoir, both of which are created by their respective dams.[9]

In 1999, PG&E's agreement with FERC requires that three upstream small tributary dams be removed. In addition, 37 miles of the Mokelumne above the Pardee Reservoir is being considered for "Wild and Scenic" designation as of 2018. If approved, this will prevent any further development along this section of the River. As is required by the National Park Service for this designation, the classification is supported by the State of California, local governments, and local environmental advocacy organizations (Foothills Conservancy and Friends of the River).[10]

The number of dams, reservoirs, and water diversions constructed by the CVP since the 1930s indicate what should be called "an over engineered Central Valley," but the State's efforts make the already complex system even further complex. This is briefly reviewed in the next section.

The California State Water Project

The California State Water Project (SWP), which is frequently called the "Water Management Project," must be understood as an extension of the federally led and financed "Central Water Project" (CVP). The SWP "adds onto projects" begun by the CVP. The origins of the SWP, however, stem from the Bureau of Reclamation's 1951 "United Western Investigation Report," which assessed the potential for Western "inter-basin water transfers." It particularly assessed the potential for damming the waters of Northern California's rivers and piping these waters into the Sacramento River for diversion southward into the Central Valley. Coincidentally, California's State Engineer first proposed what became the "Feather River Project." (See the review of this project presented above.) Following this, several propositions and evaluations led to calls for a comprehensive statewide water management plan, and in that direction

[9] In 1931, Pacific Gas and Electric (PG&E) completed the hydroelectric Salt Springs Dam, Plus the Bear River Dam, and the Tiger Creek Dam. They produce 206 megawatts combined. Below the Pardee Dam, and 34 miles above the Mokelumne's confluence with the San Joaquin River, PG&E constructed the Comanche Dam and Reservoir in 1963 (See Fig. 5.4). The combination of the Pardee Dam and Comanche Dam provide the flood control and water storage for the East Bay. The Comanche Reservoir has a capacity to store 431,000-acre feet.

[10] In addition, the expansion of the Pardee Reservoir has been successfully prevented by local environmental organizations.

the California Department of Water Resources was formed in 1956. This Department formed the State's "Water Management Plan." This included the damming and diverting of water from all the major rivers of the northern section of the State. California's significantly large agribusiness lobbied for the "Plan" to be enacted, which included:

- the Oroville Dam of the Feather River,
- the California Aqueduct that takes water from the Sacramento and San Joaquin Rivers and diverts it into the southern section of the Central Valley,
- the Delta pumping stations that accomplish the above diversions,
- the Damming of the Klamath River,
- the damming and diversion of the Trinity River,
- and many more projects.

The more than 400-mile California Aqueduct is the principal feature of the SWP. It begins at the southwest corner of the Sacramento—San Joaquin River Delta where it flows south splitting into three branches of canals, tunnels, and pipelines. It is one of the largest aqueduct systems in the world. Ultimately, it conveys 650 million gallons of water up to 705 miles to serve 27 million people and 750,000 acres of farmland. The Aqueduct's water is pumped 1,926 feet over the Tehachapi Mountains after which it splits into the three separate, mostly concrete segments. The Central Branch of the Aqueduct ends at Lake Cachuma which is ten miles north of Santa Barbara. The West Branch conveys water to Castaic Lake in Los Angeles County, providing water to the northern section of this County. The East Branch ends at Silverwood Lake in San Bernadino County. The California Aqueduct provides municipal water to the Santa Barbara, northern Los Angeles, and San Bernadino areas. But along its northern route, it also provides agricultural irrigation.

Ultimately, the SWP constructed 21 dams designed to divert 2.4 million acre-feet of water annually to the south. Note that the "Plan" was to divert 4.23 million acre-feet, but slightly more than half of this planned diversion was accomplished before environmental considerations interfered. Also, for this Sacramento River water to reach the San Joaquin Valley, it must be pumped 2,882 feet over the Tehachapi Mountains, and these pumps were installed and funded by the SWP. The State estimates that over $400 billion in agricultural and municipal water revenue has been created by the SWP for California.

When one considers the extraordinary extensiveness of the system of dams and diversions, one must come to the recognition that this is a massive altering of nature that serves some socio-economic interests at the considerable expense of others. Was the agricultural infrastructure and production of the southern Central valley worth the public's expense? Is there any probability that this "replumbing" is environmentally sustainable, or even economically sustainable? These questions must be addressed, and are at least approached by the next section.

The Contentious Klamath

The sections of this chapter that concern the CVP and SWP are included to explain the extraordinary demand for agricultural irrigation and municipal use extracted from every river and stream source available. This is the critical fact that confronts the indigenous tribes of Northern California and Southern Oregon. The waters from the Klamath, Shasta, and Trinity Rivers meet some of these insatiable demands created by America's Southwest urban and agricultural empire of California.

For twenty-five years the Klamath River has been a central focal point for America's dam removals. In the late Spring and Summer of 2024, the four upriver dams located close to the Oregon-California border were finally being deconstructed. But this culmination occurred only after a series of nearly completed, but nevertheless frustrated agreements failed because of fact-resisting interfering politicians. What finally brought these dams down? The answer is that the many decades of persistent work by the five native tribes of the area, plus the hard work of other environmentalists, and the hideous growth of poisonous algae in the shallow reservoirs behind the four dams, were all contributing factors that led to accomplishing the task of removal. The Chinook and Coho salmon of the Klamath will likely be saved because of these forces.

The Klamath River flows 257 miles through the Cascade Mountain areas of Southern Oregon and Northern California. The Klamath Basin was a marshy area twenty-five miles directly south of Crater Lake, and twelve miles north of the California border. The Williamson River and the Sprague River feed this marshy shallow lake the outer borders of which have been largely filled to create farming acreage. (Figure 5.5 indicates locations.) This Basin is on the east side of the Cascade Mountains, out of the temperate rain forest of the west side. This is dry sage brush country with low density pine forest and junipers that grow in the sandy soils to the east of the spine of the Cascades. This is the Klamath Basin from which the Klamath River flows south through the Link River Dam and then the nearby Keno Dam, both of which are hydroelectric facilities. The former dam was constructed in 1921, and the latter was constructed in 1966, both by the California Oregon Power Company (COPCO). The Klamath then flows southerly through the J.C. Boyle Dam, the COPCO 1 and 2 Dams, and then the Iron Gate Dam, all four of which are hydroelectric structures, and all four will be deconstructed prior to the end of 2024. As of July, 2024, all four deconstructions are ahead of schedule.

After the Iron Gate Dam, the Klamath flows through the Klamath Mountains in Northern California. Here the flow leaves the arid high desert around the Upper Klamath Lake and enters the temperate rain forest on the west side of the Cascade Range. The Klamath then flows west along the south side of the Siskiyou Mountains until it turns sharply toward the south at Happy Camp. It then flows through a whitewater rapids section of the Klamath National Forest and Six Rivers National Forest where it meets the confluence with California's Salmon River. At the town of Weitchpec, the Klamath turns sharply to the northwest to flow into its sixteen-mile tidal estuary. The mouth of the Klamath is at Requa on Northern California's Pacific Coast (south of Crescent City and in the middle of the Redwood National Park).

The Klamath River's Watershed is 15,689 square miles, almost nine-tenths of the size of California's Central Valley. It is mostly rugged mountains, forests, and canyons. Its main tributaries include the Shasta River, the Scott River, the Salmon River, and the Trinity River. The Shasta River drains the area around Mount Shasta in north-central California. It flows directly north to its confluence with the Klamath close to the town of Yreka approximately ten miles south of the Oregon border, and twenty miles downriver from where the Klamath enters California. The Scott River flows parallel to the Shasta River thirty miles to the west. Farther downstream, the Salmon River's confluence is thirty miles north of Weitchpec (the point where the Klamath turns sharply north and the Hupa Valley Tribe's Reservation is located). The Trinity River is the Klamath's largest tributary. Flowing from the south, it has its confluence with the Klamath at this Weitchpec turning point.

The Trinity River is (or was prior to its flow diversion into Central California) the largest Klamath River tributary. It flows 165 miles through the Klamath Mountains and the Pacific Coastal Range. Its headwaters are at the confluence of High Camp Creek and Chilcoot Creek as depicted by Fig. 5.5. It is classified by the US National Park Service as a "Wild and Scenic River" given that most of its course flows through steep picturesque canyons and high mountain meadows. It once sustained prolific runs of Chinook, Coho, and steelhead which sustained the indigenous tribes (particularly the Hupa Tribe) of the Klamath Mountains prior to the "gold rush" of the mid-1800s. (The Native Tribes of this area are reviewed below.) This mid-century event brought the gold seekers and disease to the Hupa, and the following half-century brought logging, ranching, and runoffs from the gold mines, all of which degraded the River's ecology and its area's indigenous population.

The story of the indigenous tribes of Northern California, however, must be intertwined with the story of the State's massive plumbing of its natural water flows, i.e. "The Central Valley Project." The term "massive plumbing" certainly applies to the case of the Trinity River. In 1963, ninety percent of the Trinity's upriver flow was replumbed by the Lewiston Dam (reviewed below) into the Sacramento Valley. After this engineering achievement, only 150 cubic feet per second were allowed to flow downriver. In 1991, however, a minimum flow of 470 cubic feet per second was established, but this is too small of a flow to sustain the salmon runs of what was previously the third-largest producer of salmon on the west coast.[11] Because of the destruction caused by these dams, Coho salmon are now listed as endangered.

Dams and Philosophy

How is it possible that our society dams the Klamath and Trinity Rivers, and so diverts the flow of the latter that it largely destroys the riparian culture of the indigenous tribes that have occupied their banks for hundreds of years? The rationale for these forced river transformations stem from the methodology of "cost–benefit analysis" which prior to 1965 had little standardization as a government decision process. As such, it invited bureaucratic abuse. But this problem has substantially improved in the decades since. This type of project analysis is lightly reviewed

[11] See Eric Stene (1994), "Central Valley Project: Trinity River Division," US Bureau of Reclamation History Program at www.usbr.gov/dataweb/html/trinity.html.

here along with its substantial defects for the river development issues in question.

It is only natural that prior to pursuing any public project, the costs and benefits should be estimated. Once they are measured and reported, however, the automatic decision to pursue all those projects with benefits greater than costs is rooted in the philosophy of *utilitarianism*, a decision methodology that has some fundamental defects. This section reviews these "defects" and our general methods for public project evaluation.

Utilitarianism as the Great Mistake, or "What Do We Do About Klamath Culture?"

The "Klamath River Basin's Indigenous Tribes" include the Klamath, Yurok, Karuk, Hupa, and Shasta peoples. Their cultures are similar, riparian, and based on their tribal locations along the Klamath and Trinity Rivers of Northern California and Southern Oregon. The anadromous fish of these rivers are central to this Klamath Basin culture. The destruction of these rivers therefore directly attacks this culture. We should therefore ask, "What socio-economic philosophy would justify the destructive actions of damming and diverting the Klamath and Trinity Rivers?" To answer this, we need to examine whether any acceptable philosophy would justify these actions. To properly pursue this examination, we need to first examine the socio-economic philosophy of *utilitarianism*. Was some version of this philosophy used to justify the clear destruction of this culture?

This utilitarian notion began with the early nineteenth-century philosophy of Jeremy Bentham (1748–1832), an English philosopher from University College, London. He originated "modern utilitarianism," a theory that moral principles should be those that most benefit overall society. This is a hedonistic-consequentialist theory where a particular notion of human good is to be served, and this human good consists of the sum of individuals' happiness.[12] Our Western society does not accept this theory in that it expresses the belief that there are fundamental individual rights, i.e. *inalienable rights,* which must be respected despite the

[12] A "consequentialist philosophy" is one that judges the rightness of an action on the consequences it produces. By "hedonistic" we mean material well-being. Therefore, a "hedonistic-consequentialist" philosophy is one that produces the best materialistic outcome. The measure of "the best" is, however, vague and debatable.

overall total sum of the happiness of society's members. At least theoretically, our society believes that no one is to be directly sacrificed in order to serve the enhancement of society's overall happiness.[13] The weakness of the utilitarian theory is examined extensively below where it is argued that this philosophy is not especially useful as a practical method for organizing our policies or actions.

John Stuart Mill (1806–1873) was also a British philosopher, perhaps the greatest of the nineteenth century. He argued that common sense morality, which we learn as children, represents a lengthy historical accumulation of wisdom about desirable and undesirable consequences of actions. He also argued in *Utilitarianism* (1863) that basing moral principles on a desire to maximize societal satisfaction would not lead to common sense solutions to ethical problems. As an example, the wealthy miser might still derive great pleasure from an additional gift of a small amount of cash, while the poor person might not. Does it fit common sense to have the poor redistribute to the wealthy?

Utilitarianism is a consequentialist branch of ethical philosophy that claims that our moral rules should be so designed as to maximize the total sum of society's happiness. By this we mean the total sum of the happiness of all individuals in that society. This philosophy, therefore, does justify the sacrifice of the well-being of some provided this sufficiently benefits others. This definition of utilitarianism is actually rather limited in that in its extreme form it could justify the gross violation of the rights of some in order to benefit the whole of society. As such, we would never accept this extreme form of consequentialist theory as the basis for our morality.

For an example of our rejection of the utilitarian philosophy, consider the 1961 movie *Judgment at Nuremberg* which concerns a semi-fictional post-World War II trial of German jurists. These judges and prosecutors had warped the legal structure of 1930s Nazi Germany so as to allow the wrongful persecution of various political and ethnic minorities, a persecution that evolved into what we ultimately describe as the *holocaust*.[14] (The trials of similar German jurists actually did take place in post-war Germany. They ended in 1949.) The two main characters in this cinema-play are Judge Ernst Janning, played by actor Burt Lancaster, who is

[13] Society does accept the *a prior* risk of possible individual sacrifice as in military actions.

[14] This movie was based on the 1957 Abby Mann television play of the same name.

being tried for crimes against humanity, and presiding Judge Haywood, played by Spencer Tracy. One of the critical acts that Judge Janning is accused of concerns a particular trial that occurred early in the Nazi era (in 1935). Janning is accused of knowingly allowing an innocent man to be convicted, and presiding in such a way as to enable that conviction. Judge Janning is found guilty, and when Judge Haywood reads the sentence, he states, "We declare now that we stand for truth, justice and the value of a single human being!" At the end of the movie, the two judges meet in Judge Janning's jail cell. Janning pleads he had not known that his undermining of the judicial system would lead to the horrors of the *holocaust*. Judge Haywood responds, "Judge Janning, it became that when you knowingly first convicted an innocent man!".

This *Judgment at Nuremberg* story actually illustrates both the fault with the extreme consequentialist form of utilitarianism, and its two variations indicated immediately below, each of which is more practical but not necessarily applicable and useful. In Western society, theoretically the rights of the individual are held as a sacred ideal. This philosophy of individual rights is the foundation of Western democratic constitutional arrangements, i.e. there are limits to the democratic rights of the majority.

The three branches of utilitarianism referred to above are:

- Hedonistic utilitarianism: People seek pleasure and avoid pain so as to maximize the difference between these sensations. Whatever actions maximize the sum of these differences across society evolve into our moral rules. This might explain *ex post* how our rules have evolved and therefore can be used to justify their existence.
- Rules-based utilitarianism: We a priori adopt whatever rules we believe will maximizes our expected overall societal happiness.
- Social-welfare based utilitarianism: We should adopt actions and rules we believe will maximize our expectation of a societal welfare function, which considers not only societal happiness, but also by some specified extent, the rights of individuals. In this case, the specified rights of individuals can be considered as constraints on our attempts to maximize societal happiness.[15]

[15] The first two utilitarian theories originated with Jeremy Bentham (1748–1832), and were extended by John Stuart Mill (1806–1873) and Henry Sidgwick (1838–1900), all three of whom were English philosophers.

At first glance, utilitarianism can appear to be based upon common sense. It fits common slogans such as "the greatest good for the greatest number," which some argue should be our guide for action and rule-making. The problem is that in application, this sort of slogan is problematic in measurement of this "good." Are we to try to apply utilitarianism as a norm in establishing rules and actions, or are we to use the theory as an explanation of how our rules evolved, i.e. did they evolve so as to maximize some conception of societal welfare? If it is the former, then we must answer "Who is the *we* who will be deciding the rules and actions?" Will this be by democratic vote? Will we sacrifice the interests of minorities to benefit the majority? Will we sacrifice the welfare of future generations to benefit the welfare of the current generation? These questions are particularly important for addressing environmental concerns.

We can illustrate the problems with utilitarianism by considering the conundrum of income redistribution. To what extent is it in the interests of social welfare to redistribute income from the wealthy to the poor? There is no doubt that if we tax the income of the wealthy, they will be less happy as a class although perhaps some individual wealthy people will not be less happy. If we redistribute this income to the poor, there is little doubt that as a class, the poor will be happier, but how much happier as compared to the loss of the happiness of the wealthy? How much if any redistribution should we undertake? Our society decided that some redistribution is in order since we have progressive income taxes, but should we not redistribute income so that it is equal? If the relation between income and happiness were identical for all individuals, then this sort of income equality might maximize current overall societal happiness. Is this what we want?

Of course, the relation between income and happiness varies greatly across individuals. Some prefer more leisure and less time spent on earning income. Others, for example, pursue what we classify as more cultural pursuits rather than income; or they prefer to be engaged in social service rather than pursuing higher income.

One principle is for sure: our society strongly believes in inalienable rights, and this principle applies to individuals, and to groups. Note our support for minority rights as an example. A positive claim that our current rules and actions evolved so as to maximize anticipated social welfare might be accurate, but it is hardly testable because we have no clear notion of "social welfare." We therefore have no method of judging the extent that we follow this philosophy. As such, this social philosophy claim is of little interest. Note that our society is generally interested in exploring new rule modifications since this is the democratic process, but any new ethical maxims should be consistent with our principle of inalienable rights, principles we know we hold sacred according to long-standing law and belief. Note as an example, however, that our society does claim the right to eminent domain for constructing various public projects, but always with what we deem as adequate (fair) compensation. Nonetheless, contention over that compensation is an issue worthy of investigation.

Our Water Project Decision Institutions[16]

The "Flood Control Act" of 1936 gave the Army Corps of Engineers the responsibility for preventing flood damage along our navigable rivers. But the Act specified that flood control projects should be undertaken "if the benefits to whomever they may accrue are in excess of the estimated costs, and if the lives and social security of people are otherwise adversely affected." This established the criteria that economic benefits must exceed economic costs, but that other social costs, benefits, and impacts should be considered. This criteria of economic benefits exceeding costs was extended to other federal agencies that potentially undertook water projects: the Bureau of Reclamation, the National Resources and Conservation Service of the Department of Agriculture, and the Tennessee Valley Authority (TVA). In 1950, a Federal Inter-Agency River Basin Committee was formed to produce a homogeneous set of standards for the above listed federal agencies. Their report was known as the "Green Book," which was revised in 1958.[17] These reports set the standards "... to provide a guide for the effective use of the required economic resources ..." It states that the objective for undertaken a federal water project should be "to maximize net economic returns and human satisfaction from the economic resources used for the project." One should, however,

[16] Part of this section is based on Susan Durden and Jim Fredericks (2009, June), "Economics primer," *US Army Corps of Engineers, Institute for Water Resources,* Report 09-R-3 at https://planning.erde.dren.mil/toolbox/library/IWRServer/iwrreport-09-R-3.pdf.

[17] The Green Book's title was "Proposed Practices for Economic Analysis of River Basin Projects."

note the potential conflicts between these objectives. "Human satisfaction" from economic resources begs the question as to the very nature of this "satisfaction" and its linkage with "maximizing net economic returns." As we will explore, this conception of "net economic returns" may be measured from the standpoint of citizens acting only as individual consumers and not as a collective citizenry. But what vision might we collectively have for a river or watershed? The answer might be different from the consensus of individuals each acting as lone isolated consumers. The collective vision might require preservation or restoration, but this might differ from the old-style cost–benefit type rules.

This "maximize net economic returns" decision rule, according to the Corps' interpretation, established the principles of the subsequent National Economic Development Analysis. Because of the ambiguous conflicts in the objectives listed in the Green Book, in 1962, Congress passed "Senate Document 97" which specifies that "Preservation (proper stewardship of the nation's bounty) and the Well Being of the People" should also be objectives. This sharpened the potential conflicts in pursuit of maximizing National Economic Development (NED) net benefits, the interests of preservation, and also the meaning and intention of the "well-being" objectives.[18]

These conflicting objectives led to the "Water Resources Planning Act of 1965," which created the *Water Resources Council* (WRC) with the purpose of resolving these conflicts by clarifying and standardizing the project evaluation process. The WRC was a committee of representatives of federal agencies involved in water projects. They conducted an open process that solicited input from all interested parties, and in September of 1972, the WRC issued its *Principles and Standards for Planning Water and Related Land Resources* (P&S). The P&S specified how the "National Economic Development Benefits" would be calculated, and the results of which would be kept in one of two separated accounts with respect to any project examined, i.e. the NED Accounts. The first account was to measure the gross value of increments to the national value of goods

[18] See Delaney Edge (2021, December 14), "The Role of Cost–Benefit Analysis in Public Policy Decision-Making," *Berkeley Public Policy Journal*, Spring/Summer 2024, at www.bppj.studentorg.berkley.edu/2021/12/14/the-role-of-cost-benefit-analysis-in-public-decision-making. For the US Army Corps of Engineers cost–benefit procedures see www.asafm.army.mil/Portals/72/Documents/Office/CE/US%20Cost%20Benefit%20Analysis.pdf.

and services produced, i.e. the NED benefits. The second account would measure the impact the water project would have on the regional values of goods and services since these might be impactfed while the national level might not be impacted (there might just be a shift in productive resources from one region to another with a zero net national increase in production). Nevertheless, these regional effects could be important. Those in poverty in one region might benefit while there is no national increase. But the 1972 P&S also required that a third account be recorded, i.e. the impact the project would have on environmental quality. The conception was that America's water assets had been substantially degraded, and now needed preservation or restoration rather than further degradation. A project with positive net benefits might be environmentally degrading and therefore not be pursued.

As indicated above, evaluation of "National Economic Development Benefits" is an attempt to measure dollar value increments to our national income resulting from our government projects generated by federal actions such as development of facilities along rivers. Measuring the expected benefits to be derived from these projects pose a variety of problems reviewed here. These are the problems that must be examined in detail in order to explain California's water development, the degradation of its rivers, and their possible restorations. These problems concern conceptions such as "market valuations" and "the vision problem" as explained below.

With respect to our river-related projects (or those projects that otherwise have environmental impacts), our governmental agency analyses beg three important questions:

i. Is it fair to expend federal resources on projects that benefit only a local few?
ii. Might the Nation prefer placing greater weight on projects that positively impact poor under-developed or economically stricken areas even though it does not produce significant *National Economic Development Benefits* (NED benefits)?
iii. Would the Nation wish to negatively affect already marginalized groups in order to generate significant new NED benefits?

These three questions essentially concern resource transfers or compensations to disadvantaged groups via environmentally impacting projects.

The old cost–benefit analysis answers these three questions by separating the costs of resolving problems to the disadvantaged from the national benefits associated with pursuing public projects. If the NED benefits are strong even though the disadvantaged have been negatively affected, then appropriate compensation might be arranged. Politically, however, these are not easily separable solutions and compensation is often not arrangeable. But the answers to the first two questions above are generally "Yes!" The *US Army Corps of Engineers* identifies disparate impacts on the disadvantaged in analyses of all of its projects including those environmentally related. With particular relation to question (iii), the efficient answer is to generally undertake all projects with positive NED benefits, but to compensate those disadvantaged by those projects. The explicit problems of compensation are not examined here but Robinson, 2021, Chapter 6, reviews this and other broad questions concerning this cost–benefit analysis.

The public discourse and examination required by the issues reviewed above include:

i. the clear identification of those groups impacted,
ii. as clear as possible identification of the disparate costs and benefits impacting these groups, and
iii. clear identification of the methods and amounts that could be used to compensate those negatively affected.

These three points describe the information that should be provided to the decision-makers in the political process. The clarity pursued in the above can potentially contribute to a "reasoned public discourse."

Cost–Benefit Analysis and Public Discourse

American rivers exhibit the three characteristics of public goods[19]:

i. they are capable of being used by all in their locale,
ii. they do not have well defined property rights, and
iii. they can only be controlled by a collective (meaning "public") authority.

[19] There are rivers in Britain that are actually privately owned and controlled.

The problems posed by these characteristics concern how, or even if, our "collective authority" will be exercised. For example, will we pursue the restoration of some river if few can access the river's locale, or if the property on both sides of the river are owned and controlled by a private entity? But usually these three characteristics are strongly present so that preservations, restorations, or development can be pursued.

Following the Congressional passage of the *Water Resources Development Act* of 2007, and the Executive Branch's *Principles, Requirements, and Guidelines* (PR&G) of 2015, all federal agencies revised their "Agency Specific Procedures" (ASP) for evaluation of water projects. In April of 2024, the Army Corps of Engineers became the most recent agency to complete their ASP revision. Evolving from the old cost–benefit analysis (CBA), current federal water project evaluations must now consider (i) the project's effects on environmental preservations and restorations, (ii) the project's distribution of benefits, costs, and other effects across populations, (iii) the possibility of nonstructural problem solutions (no significant construction but perhaps significant deconstruction), and (iv) the possibility of broadly integrated approaches to resolution of water problems.

With respect to the first above consideration under the 2015 PR&G (the environmental impact account), note the difficulty in making comparisons of the NED Benefits with the environmental impacts. The latter is particularly difficult to measure, it usually could only be indirectly quantified through index-type measurements such as its potential impact on fish and wildlife. The decision-maker is thereby left to make their own subjective judgment with respect to a comparison of values. But the measurements of the market values of goods and services that the NED Benefits Account generated by the project appears to be an objective quantitative measure.

In pursuit of this evaluation, four accounts are therefore recorded:

- The National Economic Development Account (NED) that measures the expected change in the National value of goods and services if the project is undertaken.
- The Regional Economic Development Account (RED) that records the localized economic impact on goods and services.
- The Environmental Quality Account (EQ) that records the expected effects on natural and cultural resources.

- Other Social Effects Account (OSE) that records the expected effects not reflected in the above listed categories.

We can point out two particular problems with this approach: (1) public projects must evaluate benefits that are both marketable and non-marketable, and the latter can be difficult to conceptualize and measure while the former usually has identifiable market prices where the latter does not, and (2) there are often substantial opportunity costs associated with public water project development. These alternative "opportunities" should be explored and analyzed. Since usually these alternatives tend to be mutually exclusive only one opportunity can be selected, and that might be to do nothing.

To explore these sources of difficulties in valuation, consider the potential damming of a river for deriving both hydropower and agricultural irrigation. Allow us to suppose that the costs of constructing the dam, the installations of hydroelectric generators and the necessary transmission lines are all readily obtained. The market value of the electric hydropower should also be readily forecasted. But what about the opportunity value of leaving the river in its natural state? What about the recreation values, and the cultural values of the river left alone? How much would people be willing to pay to leave the river undammed? Note the important essential question. We need not ask anyone to actually pay to have the river left in its natural state, we are only asking, "What is the value of this resource to society if left in its natural state?" Could this value exceed the value of the dammed river? To put this question somewhat differently, would society prefer that the river be left "as is?".

In the past, the very question we pose above was generally left unexamined. Consider the makeup of the institutional bureaucracies (the Corps of Engineers, the Bureau of Reclamation, EPA, State Water Resources Department, or many other agencies) that typically evaluated the development of Northern California's rivers (the Sacramento, Shasta, Klamath, and the Trinity Rivers and their tributaries). The direct costs of constructing the infrastructure were relatively easy to measure by the Corps, or other technical experts. The agricultural revenues to be generated, the electric power revenues to be generated, and flood control benefits were also forecasted by economic analysis. The market value of the projected increase in agricultural production, the market value of the generated power, and the expected value of the property left undamaged can and were all estimated. But there were no serious examinations of the alternative of leaving these rivers "as is" although there have been recent evaluations of restoration of these rivers.

There have been, therefore, two fundamental interrelated methodological defects associated with the old cost–benefit analyses: (1) valuations were oriented toward the market values of individual consumers, and not toward those same consumers acting as a collective to produce an alternative, and (2) as such, it did not consider any posed visions of alternatives these same collective consumers might have preferred. For example, consider a public forest area where the decision is to either harvest the trees or leave it intact as a park so that it can be used for recreation and/or watershed protection. The net-market value of the timber can be estimated (gross revenues minus the costs of harvesting). That value considers what the individual consumers of timber place in total on the harvest option. But how much would the collective of these individuals be willing to pay for the forest area left intact? Forests can protect watersheds and wildlife that can be explored through trails and the like. There is no easily measurable market value to place on this "leave it alone" option, but we know there is a likely value to the community that would use it for nature-based recreation or watershed protection. In addition, if the forest is left intact, then we do not preclude other future options that might emerge for its use.

Can we collectively examine alternative futures for our environmental assets including leaving the asset "as is" for future available options. For another example, the Northern California rivers could have been considered as components of a large area where alternative agricultural, flood control, and power generation possibilities might be arranged. In not posing the scenario that future alternatives might be arranged differently, the arrogance of the present destroys those possibilities. There is value in preserving options even though we know there are times when we are ready for hard decisions to be made. But not allowing time for different visions to be formed before "not easily alterable long-term decisions" are made was a serious political defect in the older project analysis approach. Now, in our newer cost–benefit approach, we are careful to allow sufficient time for these new collective alternatives to be formed and properly posed before we commit. Consider the case of the quickly decided Glen Canyon Dam trade for preserving Echo Park Dam as reviewed in Chapter 4. Why did the decision involving these dams have to made so quickly?

Another fundamental issue in our project analysis is whether to monetize the costs and benefits for our proposed water projects. When public funding for a project (such as a hydropower dam's construction) is required, it is only natural to measure costs in dollar metrics because the public's tax revenues used to finance the project are, of course, measured in dollars. It is therefore natural to measure the costs of providing the projects' benefits in dollars along with the costs of the various options involved such as the possible sizes of the dams, the possible sizes of the hydropower generators, the costs of providing new fish passage technology. It is therefore also only natural to attempt to monetize the estimates of various benefits to be derived, e.g. the dollar estimates of revenues from power sales, the estimates of dollar values of the likely flood damage to be avoided, and also of the recreational benefits if any, etc. But we know there are also costs and benefits that are difficult to quantify, such as any negative sight-impacts the project might cause, the reservation demands for preserved natural areas, the option value of natural preservation rather than development. These effects can only be estimated with great difficulty and uncertainty, especially with respect to monetized benefits. Nevertheless, the various expected costs and benefits must be indicated and if possible, quantified by some measures and/or indices. They provide important information for decision-makers who perhaps must subjectively evaluate the importance of these metrics. All of this should be part of the process of discourse and consideration. But it is also natural for bureaucracies to want to rely on simple single benefit/cost ratios that appear to be an objective decision-oriented metric. But we know this "objectivity" is ultimately illusory because any complete monetization of costs and benefits will offer imperfect measures. They therefore must be subjectively evaluated as such by the decision-makers. The question then remains, how do we design processes that can facilitate our analyses while properly incorporating and considering all the relevant expertise and information available and relevant?

Our decision method for consideration of water projects should consist of a *highly structured stakeholder collaborative process* whether it is examining a new construction, the removal of an old structure, or the preservation of an existing natural water body.[20] The public must perceive this process as being *fair, reasoned, and fully informed*. These criteria

[20] See Gowan, Stephenson, and Shabman (2006) concerning the FERC process for nonfederal hydropower dams.

compose the subject of Robinson (2021), which explored in depth the reasoning process for our environmental decisions. That volume used our Western philosophical foundations to argue that our public process for environmental decisions should ultimately be judged according to the criteria of being (i) fully informed, (ii) inclusive of appropriate input from those affected, (iii) free of corruption and conflicts of interest, and (iv) deemed as logical under objective review by competent judges.[21] These should be the key elements for our process of project review and decision. These elements are reviewed here.

As indicated above, there were two fundamental interrelated methodological defects associated with the old cost–benefit analyses: (1) it was oriented toward the market values of individual consumers, and not toward those same consumers acting as a collective, and (2) as such, it did not consider any posed visions of alternatives these same consumers might have preferred and established as a collective. We consider the second "methodological defect" here.

Not allowing time for different visions to be formed before "not easily alterable long-term decisions" are made was a serious political defect in the older cost–benefit approach. Now, in our newer cost–benefit project-analysis approach, we are careful to theoretically allow sufficient time for these new collective alternatives to be formed and properly posed before we commit.

We should realize that our previous cost–benefit analysis (pre-1972 reforms) were strictly crude expressions of utilitarianism with all its philosophical defects. For example, at the time of the damming of the Trinity River and the diversion of its water to the Sacramento River for further diversion south (authorized in 1953 and completed by the Bureau of Reclamation in 1963), the cultural value of the River to the Northern Indigenous Tribes was not even considered. If it were, the issue of "how much would they be willing to pay" would not have been appropriate given the relative poverty of the Northern Tribes. The tribes might have been willing to give a considerable portion of their wealth, but given their relative poverty, this might not have been much. Furthermore, issues of alternative flood control, alternative irrigation systems, and alternative power generation and/or alternative conservation, were not examined.

[21] See Robinson (2021, Chapter 5).

The factors that were examined under the old cost–benefit analyses included only the market values of the increment in agricultural and energy production because these were easily measured. The values of alternatives require envisioning "what could be established by the collective," and what non-marketable environmental characteristics people are willing to pay for either in the form of the preserved or restored environment.

Consider the problem of valuing the preserved or restored river. How do we place a value on what the Northern Tribes would be willing to pay (not what they would be forced to pay) to restore the Trinity River to what it was before damming and water diversions? How much would the rest of our society be willing to pay for the restored Trinity River? Also, what is the value of keeping the Trinity dammed and diverted? Perhaps the more relevant question is, "Did the Northern Tribes have an inalienable right to the preserved Trinity River?" If so, perhaps it should have been preserved for this reason, or it should now be restored.

Consider, for example, the potential restoration of the Klamath River in Northern California and Southern Oregon. Restoration of this River requires the removal of its dams and restoration of its ecology. This requires the elimination of a small amount of hydropower, and the bearing of the costs of dam removals. Since this involves the restoration of the very culture of the Klamath Basin Indigenous Tribes, the benefits provided to future generations—who will eventually form a collective entity—could be considerable. But there is also an essential associated issue that we consider below. "Is it morally fair for us to make decisions concerning environmental benefits for future generations?" The answer is that there is no ethical justification for such a procedure, but we are stuck with these decisions nonetheless. Given that future generations cannot participate in our reasoned social discourse concerning river restoration, should we also marginalize their interests by prioritizing our own? Our current society's method for cost–benefit analysis does not recognize this moral obligation. The benefits measured only include those of current individual consumers who purchase hydropower and possibly the flood control. They do not consider the collective cultural benefits to future tribal members, benefits that stem from the riparian culture as reviewed below.

After the Gold Rush: Klamath River Justice

After the "gold rush" of 1849, California became a State in 1850. It signed treaties with the Karuk that established aboriginal territories, but these treaties were not ratified by the State Senate. So the Karuk were without a reservation. But in 1864, the Native Tribes of the Klamath Basin ceded twenty million acres to the US which forced the Karuk, Yurok, and Hupa onto the small "Klamath Indian Reservation." But this reservation was modified in 1887 by the Dawes Act which allotted individual claims to tribal members. Many of these claims were subsequently sold to timber companies. In 1954, however, Congress removed federal recognition of these tribes, but their recognition was restored in the 1970s when the Klamath, Yurok, Karuk, Hupa, and Shasta tribes were recognized as the "Klamath River Basin Tribes."

The Yurok tribe has almost 5,000 members. Although the Yurok reservation was first created in 1855, through the late 1880s the Tribe was moved through several reservations that were eventually closed. Its current reservation is 63,000 acres on the Lower Klamath. The Yurok's tribal culture and ceremonies emphasized their connection to the Klamath River and its salmon.[22]

The Karuk was federally recognized in 1979. Its members have been working to regain its Klamath Forest Reserve lands since 1905. Its strong culture centers on the concept of an annual "World Renewal." This annual ceremony (every August) is viewed as essential for maintaining the Karuk's dedication to environmental stewardship and its relations with nature.

The Hupa Valley Tribe is federally recognized with 3,000 members. Its reservation includes 80,000 acres and is the largest in California. It is located on the Lower Klamath at Weitchpec. The Tribal ceremonies are similar to the Yurok and Karuk people, and also rely on salmon as the essential ceremonial element.

The Shasta Tribe is not federally recognized given the California Senate's rejection of granting its reservation in 1851. The Shasta's culture celebrates the first salmon run of the year with ceremonies similar to the other Klamath Basin Tribes.

[22] See Dr. Kathleen Sloan (2011, February), "Yurok and the Klamath River," at https://media.fisheries.noaa.gov/dam-migration/yurok_klamath_doi-2011.pdf.

The Klamath also include the Modoc and Yahooskin-Paiute. They were federally recognized in 1986 with 4,800 members. They are dispersed through Southern Oregon and Northern California, but they hold sacred ceremonies along the Sprague River in Oregon.

The Contentious Klamath Watershed again

In 1905, the US Forest Service (part of the US Department of Agriculture) began to manage traditional Klamath Tribal lands. The Forest Service encouraged extensive logging, mining and dam construction. Six dams on the Upper Klamath River and two on the Trinity River were constructed.[23] This decimated salmon and steelhead runs and destroyed much of the California's Redwood Forests.[24]

With the "Wilderness Act of 1964," the US established its wilderness protections. In 1964, the first federally designated "Wilderness Areas" within the Klamath Basin were created: Marble Mountain Wilderness in California, and the Mountain Wilderness in Oregon. The Russian Wilderness, the Siskiyou Wilderness, and Sky Lakes Wilderness were added in 1984. These "wilderness designations" banned timber harvests, road or other constructions, and also motorized vehicles, but it also prevented tribal fishing and tribal management on what the indigenous tribes perceived as their traditional lands. In the late 1990s and early 2000s, the Forest Service and Bureau of Land Management, however, began collaborating with the Native Tribes to use their knowledge of land management such as their brush burning practices. Nevertheless, native tribal autonomy over traditional land management was perceived as further eroded by this wilderness creation.

Between the 1918 and 1966, four hydroelectric dams were constructed by the California Oregon Power Company (COPCO) on the Upper Klamath below Klamath Lake as depicted on Fig. 5.5. These constructions blocked salmon and steelhead from large stretches of their spawning beds. Their possible removal posed the key contentious issue in recent decades. Despite intensive lobbying by the Klamath Basin Tribes along

[23] See James Agee (2007), *Stewards Fork: A Sustainable Future for the Klamath Mountains,*" University of California Press, p. 121.

[24] See "Oregon Clearcuts Endanger Salmon Even More Than You Think," at *Oregonwild.org,* at https://oregonwild.org/about/blog/oregon-clearcuts-endanger-salmon-even-more-than-you-think.

with fishermen and environmental groups, the 2004 renewal application for a fifty-year license for the dams was granted by FERC even though there were no provisions for salmon and steelhead to return past the dams and reach the 300 miles of their upriver spawning habitat. (Chapter 1 reviews the FERC licensing procedures.) But in the 2007 FERC agreement, the PacifiCorp (the company that followed COPCO as the dams' owners) was required to equip all four dams with effective fish passages at a cost of $300 million.

In addition to these issues involving the dams, in the early 2000s there was a very significant controversy involving the allocation of irrigation water on the Klamath River. A severe drought in 2001 endangered the survival of the anadromous fish in the Klamath Basin. But in that year, the Klamath Basin farmers sued to obtain more irrigation water which would occur at the expense of the tribal right and demand for protection for salmon runs in the River. The Klamath Tribes won this suit, and their river allocation was upheld again in 2013. But in 2001, Vice President Dick Cheney personally intervened to ensure full water deliveries to irrigators rather than to maintain minimum adequate stream flows for migrating salmon and steelhead.[25] This diversion to irrigators clearly violated the treaty rights of the Klamath Tribes, rights that predated and hence had priority over those of any irrigators. "First users" still have priority rights under the California law. The Klamath River consequently suffered the largest fish kill known up to that time. Vice President Cheney was investigated by Congress for releasing extra water to ranches for political gain.[26] In addition, "gill-rot disease" attacked the salmon below the confluence with the Trinity River due to the low water-level flows.

Another significant problem is that the Upper Klamath has persistently suffered from high phosphorous, nitrogen, overly warm water temperatures, poisonous algae blooms, and low oxygen. These problems are associated with irrigated agriculture and the warming water in the reservoirs behind the dams. The EPA, Native Tribes, and California State

[25] "The Truth About Klamath Project Irrigation Water Deliveries in 2001," *Klamath Basin Coalition*, May 15, 2003, at www.klamathbasin.info/2001waterfacts.pdf.

[26] See Kovner, Guy (2007, July 9), "Cheney's role in fish kill probed 2002 Klamath River disaster," *The Press Democrat*. Santa Rosa, California, at www.pressdemocrat.com/csp/mediapool/sites/PressDemocrat/News/story.csp?cid=2165338&sid=555&fid=181.

Government agencies attempted to address these problems.[27] The only solution is dam removals since the algae problem in the reservoirs had become extremely bad, and the water needed aeration to provide oxygen and cooling.

When the Klamath River's flow is warm and low, the juvenile salmon cannot migrate downstream to reach the Pacific, and the adult salmon cannot migrate upriver to spawning grounds. In February of 2010, the poor condition of the River led to the "Klamath Basin Restoration Agreement" between PacifiCorp, the federal government, and the States of Oregon and California. But this "agreement" only specified that the federal government was to assess the costs and benefits of removing the four upriver dams (Iron Gate, COPCO 1 and 2, and John C. Boyle). A final determination concerning removal was to be made by March of 2012.

A local political group—the Klamath County Tea Party Patriots—formed in opposition to the agreement. It succeeded in unseating some local elected officials who supported the removals. But in April of 2013, the US Department of the Interior presented a final analysis report that recommended removal of all four dams, plus $1 billion for river restoration funding. Following the recommendation, the "Klamath Basin Water Recovery and Economic Restoration Act" passed in 2014. The four dams would be completely removed by September of 2024.

Table 5.1 presents various specifics concerning the four dams deconstructed along the Upper Klamath River.

There are still two dams upstream of the four dams currently being deconstructed. These two dams (the Keno Dam and the Link River Dam) were constructed with fish passages that are now deemed inadequate. The Link Dam creates Klamath Lake, and was constructed in 1921. In 2022, NOAA, the environmental advocacy group Trout Unlimited, and the Pacific States Marine Fisheries Commission formed a "Reservoir Reach Restoration Plan" that details 200 projects to improve salmon habitat and passage. Note that the "Reserve Reach" refers to the stretch of the Klamath River that includes the four deconstructed dams, plus the Keno

[27] See "Action plan for the Klamath River total maximum daily loads and Lost River implementation plan," *California Environmental Protection Agency, State Water Resources Control Board,* March 2010, pp. 1–2. at https://www.waterboards.ca.gpv/northcoast/water_issues/programs/tmdls/klamath_river/1009270/03_BasinPlanLanguage_Klamath_Lost.pdf.

Table 5.1 The Four Deconstructed Dams of the Upper Klamath River

Dams	*Year constructed*	*Hydropower Capacity in Megawatts (MW)*	*Reservoir Size (acre feet)*	*Construction*	*Size*
JC Boyle	1958	98 MW	3,495 a.f	Earthen	68 by 693 feet
COPCO 1	1918	20 MW	46,900 a.f	Concrete	126 by 415 feet
COPCO 2	1928	27 MW	73 a.f	Concrete	33 by 278 feet
Iron Gate	1962	18 MW	944 a.f	Earthen	173 by 740 feet

and Link River Dams. The NOAA "Plan" recommends $20 million for the effort. The Yurok, other tribes, and NOAA will identify how and where the restorations will take place and the necessary improvements in fish passage facilities. The fish passage options range from a more salmon-friendly ladder to complete dam removal.

Reconnecting the Klamath to its upper reaches will help salmon and steelhead to survive the warming climate and disasters such as drought and forest fires. Migrating adults will reach spawning grounds that have cold-water springs. This will increase the survival rate of juveniles. NOAA scientists expect small numbers of salmon to begin upstream passage as soon as this autumn. Salmon and steelhead have 3- to 5-year life cycles so out-migrating juveniles won't return for several years. It will therefore take more than a decade for large populations to return. As indicated by NOAA Fisheries Biologist Tommy Williams, repopulation may occur more quickly than 12 to 25-years. "We're 10 years past the removal of major dams on the Elwha River in Washington, which was blocked for more than 100 years. It's now seeing large numbers of Fall Chinook. That's two and a half generations. This gives us confidence in what these fish can do."[28]

NOAA will fund extensive monitoring to understand how fish repopulate the newly opened habitat. Similar to the procedures used for monitoring the restoration after the Elwha Dam removal, the monitoring

[28] See NOAA Fisheries, July 23, 2024.

project will use a sonar camera and in-river sampling to measure the fish passage. Radio tags will be used to evaluate the effectiveness of the restoration effort.

Is Mother Nature Still on the Run?

In the autumn of 1971, I first probed the Red Wood Forest of Northern California. I picked up two young men in their early twenties hitch-hiking along the coastal Route 101. The practice of hitch-hiking was still popular then. They were locals. In conversation, they volunteered their preference for cutting the redwoods for timber. The local industry was timber cutting, and that industry was in retreat in California so sawmill jobs were declining. They also volunteered a rather strange judgment that the only forest worth preserving was the Black Forest of Germany. An uncle of one of the two had explained this assured opinion to both, and they accepted the judgment as valid. I calmly indicated that I was in awe of these giant redwoods. Then I turned the radio on to silence further discussion. Neil Young's "After the Gold Rush" played. It had recently been released. This was certainly ironic given Young's rather sad lyrics of "flying mother nature's silver seed to a new home in the sun." In that California locale, the timber was now largely depleted, and people were moving on to the next "gold rush."

Later that day, I took a small hike through the redwood forest. The trail was beside a perfect rock filled but fast-moving stream, a gift of nature. Two other young men—teenagers of 16 or 17, younger than the two hitch-hikers—were approaching from the opposite direction on a fairly quick pace. One handed a small flat-bottle of what appeared to be whiskey to the other who finished the bottle and flipped it into the stream. I could hear the bottle shatter, an extraordinary sacrilege if I ever witnessed one. I was surprised, but pleased I was there to gather the shattered glass pieces and take them back to the trash barrel at the parking lot. It is good, a gift perhaps, to have these fortunate opportunities; to be at the right place at the right time to help Mother Nature.

There are engineers who are trained in dam building. If you give them the resources and the political clout of "Big Agriculture," they will build dams, a lot of dams. California's "Central Valley Project" is clearly the prime example of over engineering a large important section of the world. This State has ample precipitation in the rain forests of the north, but the south is largely arid desert, plus California had the determination to replumb extensively. The Indigenous Tribes of Northern California were sacrificed. The fish passages of their rivers were drained dry to provide a temporary small quenching of the great thirst of the desert of Southern California. No river in this State was left unplumbed.

In this Fall of 2024, the Northern California Rivers (Klamath, Shasta, and Trinity) are being restored through dam deconstructions. Restorations are being monitored. It is likely that more dams will "come down" in the years to come. There is no doubt that the Klamath Basin Tribes will likely rebound to once again practice their sacred ceremonies of celebrating salmon returns.

We ask, "Is Mother Nature still on the run?" The answer is "No! She is on the ground trying to resuscitate. She might still survive." In the coming months, and next few years, we will see.

How do we explain California's "Central Valley project?" It is the result of the political exercise of California's basic industry, "Big Agriculture." The US needed big public works projects in the 1930s, and federal funding for dam constructions were available in the Post-World War II era. So California was reengineered, and the northern rivers were casualties along with the culture of the Indigenous Tribes. But all of this was facilitated by an old-style cost–benefit methodology with its market paradigm bias. This old-style methodology has now changed, and this is why we see some potential recovery. Now we recognize that the natural river and fish runs have value. This is a massive social change in attitude, and it has taken place in the last half-century. Mother Nature should be honored!

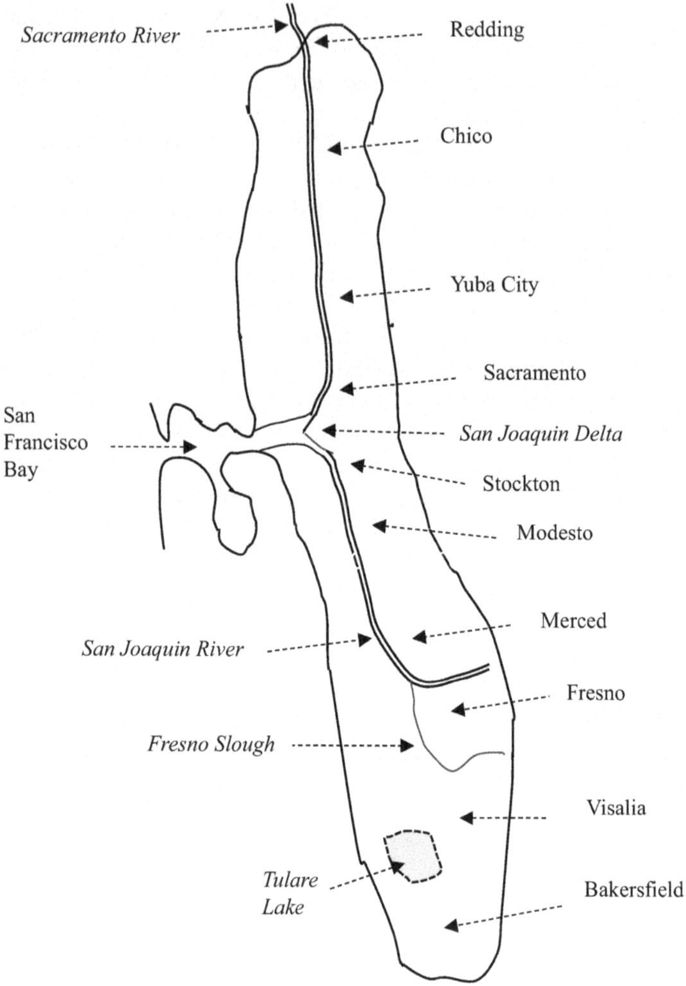

Fig. 5.1 California's Central Valley

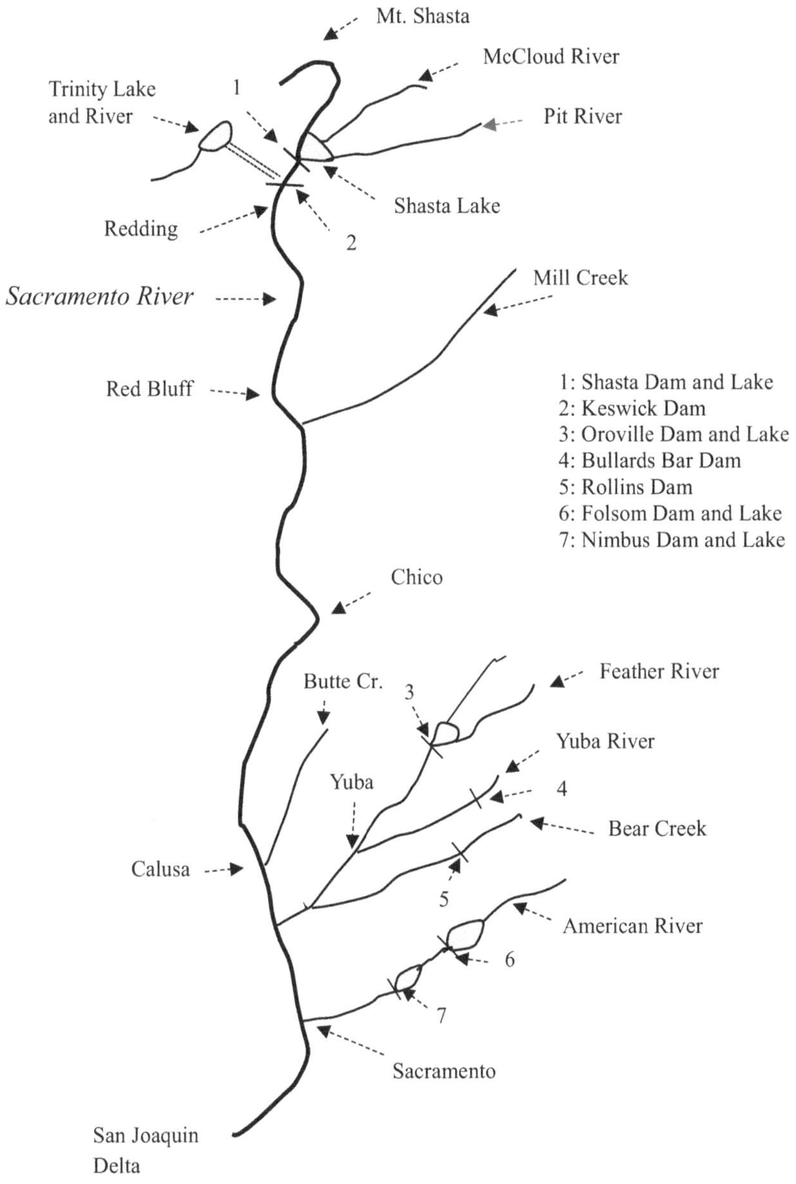

Fig. 5.2 Sacramento River and Significant Dams

Some San Joaquin facts:

- The Stanislaus River and its tributaries have 28 dams with 2.8 million acre-feet of storage. The River's average annual flow is 1.1 million acre-feet, but its irrigation districts have commitments of almost 19 times this amount.
- The Merced River has 1 million acre-feet of storage behind its dams, but its flow ran dry in 2022.
- At the Friant Dam, the Friant-Kern canal diverts water south, and the Madera Canal diverts water north, both for agricultural irrigation.
- Below the Friant Dam, the San Joaquin River runs dry during the Summers except during high flow years.

Fig. 5.3 San Joaquin River, Tributaries, and Significant Dams

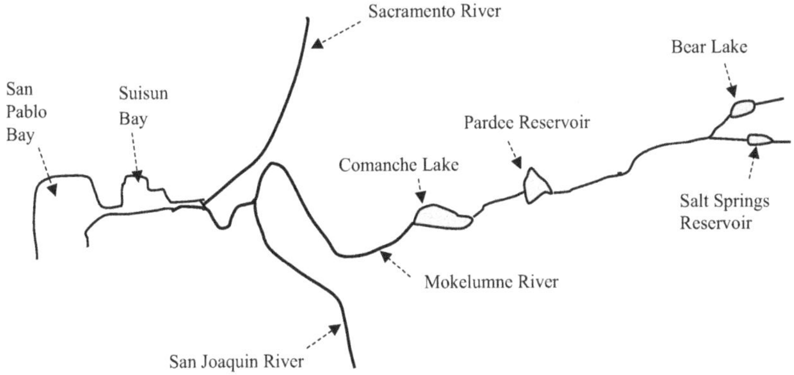

Fig. 5.4 Sacramento—San Joaquin Delta

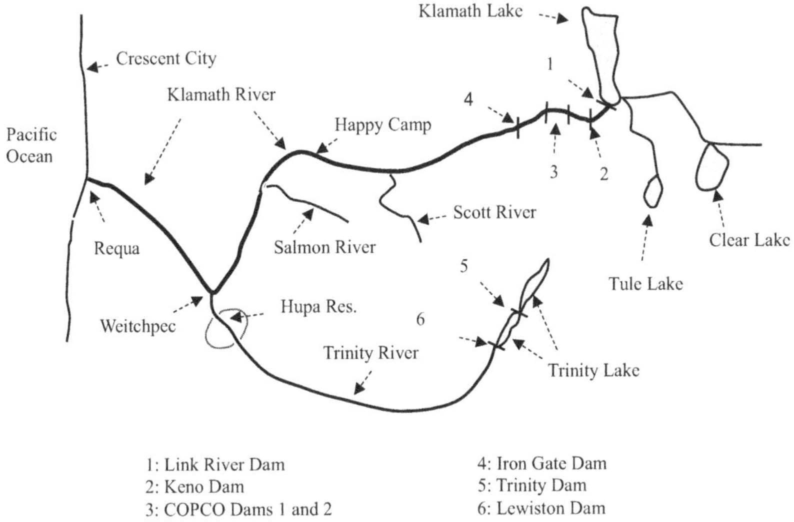

Fig. 5.5 Klamath River and Trinity River

References

Gowan, C., Stephenson, K., & Shabman, L. (2006, April 1). The Role of Ecosystem Valuation in Environmental Decision Making: Hydropower

Licensing and Dam Removal on the Elwha River. *Ecological Economics, 56*(4), 508–523.

Greeley-Polhemus Group (1991). National Economic Development Procedures Manual: Overview Manual for Conducting National Economic Development Analysis, prepared for the US Army Corps of Engineers, Water Resources Support Center Institute for Water Resources, Fort Belvoir, VA.

NOAA Fisheries (July 23, 2024). World's Biggest Dam Removal Project to open 420 Miles of Salmon Habitat this Fall. www.fisheries.noaa.gov/feature-story/worlds-biggest-dam-removal-project-open-420-miles-salmon-habitat-fall

Robinson, R. (2021). *Environmental Organizations and Reasoned Discourse.* Palgrave Macmillan.

CHAPTER 6

Columbia River and Salmon Passage

Introduction: "Speak with One Voice"

In November of 2023, sixteen Native American tribes held a "summit meeting" at the Tulalip Resort (a Tribal Owned Resort on the Puget Sound north of Seattle). The "summit" included the Nez Perce, the Yakima, the Confederated Tribes of the Umatilla, the Shoshone-Bannock Tribe, and other Northwest Tribes from Idaho, Oregon, and Washington. The gathering was called "Rise Up in Unity." The purpose of the meeting was to "speak with one voice" concerning the formulation of plans to breach the "four dams" on the Lower Snake River—the Columbia River's most significant tributary. It had become apparent that federal litigation would now inevitably lead to either a complete deconstruction of the four dams or a substantial breaching of them. The purpose of the deconstruction was to facilitate the survival of Pacific salmon—a very much endangered species listed as such under the Endangered Species Act (ESA)—and to consequently restore this essential and fundamental element of the culture of the Northwest Tribes.

R. M. Robinson, *Restoring America's Rivers*,
https://doi.org/10.1007/978-3-031-81758-8_6

Big Water from the Canadian Rockies

What precipitated this November, 2023, gathering of native tribe representatives? How did the severe degradation of the great northwest river come about? These questions concern the fate of the Pacific salmon of the Columbia River.

The Columbia River is the major artery of the Pacific Northwest. It forms in the Canadian Rocky Mountains, and then flows 1,243 miles on its course through British Columbia (BC) and the states of Washington and Oregon. Its watershed is 258,000 square miles, and also includes Idaho, Western Montana, and a small northeast section of Nevada. The Columbia River's headwaters are in the Rocky Mountain Trench of BC. The "Trench" is a deep valley between the Canadian Rockies and the Columbia Mountains of British Columbia (BC). The River first flows 200 miles from its headwaters in a northwesterly direction before turning sharply south. The Pond Oreille River joins the Columbia just two miles north of the Washington/British Columbia border. Figure 6.1 maps this area. On its downriver course, the river again turns at the Spokane River confluence, but this time westerly. But farther downstream it again turns south after its confluence with the Okanagan River. Close to its confluence with the Snake and Yakima Rivers (and close to the cities of Richland, Pasco, and Kennewick), the river turns west to establish the border between Oregon and Washington. Figure 6.1 also shows the significant tributaries of the Columbia. This major River finally reaches the Pacific Ocean near Astoria in Oregon where it forms the Columbia Bar, a large and moving sandbar that creates hazardous-to-navigate waves that are typically 30 feet high.

The Columbia River has both heavy flow and significant elevation drop of 2.16 feet per mile on average. For comparison, the Mississippi River drops less than 0.65 feet per mile on average. This creates significant hydroelectric potential. The Columbia has one-third of the hydropower potential of the US, and it delivers 44 percent of the actual US hydroelectric generation.[1] The main stem of the Columbia has fourteen dams—three in Canada and eleven in the US. There are, however, more than 400 dams used for hydropower and irrigation within the Columbia Basin. Four of the dams on the Columbia's significant tributaries have been removed.

[1] This is a 2012 estimate.

Most of the Columbia Basin's watershed of 258,000 square miles is between the Rocky Mountains on the Basin's east and the Cascade Mountains on the west. Since the Rockies form North America's "Continental Divide," its western slopes drain into the Columbia Basin. This includes the western slopes of Grand Teton National Park and Yellowstone National Park which are both in Wyoming, and also includes Glacier National Park in Montana.

As of 2016, over eight-million people live in the Columbia Basin. This includes 3.5 million in Oregon, 2.1 million in Washington, 1.7 million in Idaho, and 0.5 million in British Columbia. Note that the Puget Sound region of Northwest Washington is not in the Columbia Basin, and this includes the high population regions of Seattle, Bellevue, Tacoma, and Olympia. The northwest Pacific Coastal Region of Washington and the southern Pacific Coastal Region of Oregon are also not in the Basin.

Two of the more distinct geological features along the Columbia are the "Grand Coulee" and the "Gorge." During the Missoula Floods era of 8,000 to 13,000 BC, the flood water took a more direct southerly path than we currently observe in Central Washington between Spokane and Kennewick. There the floods formed an ancient river bed called Grand Coulee. When the flood waters retreated, the Columbia took the "C-shaped" curve of this "Big Bend" section. Here, the Grand Coulee Dam was constructed that created Lake Roosevelt. This large dam (fully reviewed below) powers pumps that send water into the ancient dry-lake bed now called Banks Lake.[2]

The Columbia River Gorge was created by the River cutting through the Cascade Mountains just north of the 11,243-foot Mt. Hood and along the border between Oregon and Washington. This extremely scenic area is also a major transportation route for both river and railroad traffic, the latter being on the Oregon side of the River. The Gorge is just east of Portland and the Willamette River's confluence with the Columbia.

The Columbia River carries large amounts of sediment that is now largely blocked by dams. This sediment created a delta of sandbars and wetlands that extends from Portland, past Longview in Washington, and then to the large Columbia River Bar at the River's mouth. At this mouth, the average discharge is approximately 265,000 cubic feet per second,

[2] See www.usbr.gov/projects/Project.jsp?_Name-Columbia%Basin%20Project.

which makes the River the largest by discharge into the Pacific Ocean from the Americas.

The Columbia River is situated as a major transportation artery for agricultural goods from the Willamette Valley (grains, fruits, and wines), the Snake River area (potatoes, fruit and grains), the Palouse area of Idaho (wheat), plus the Central Washington and Mt. Hood areas (fruit). Coal and other mining extractions from Montana are also transported along the Columbia and its nearby railroad. Lumber products from the Pacific Northwest rainforests—west of the Cascade and North Cascade Mountains—have always been a mainstay of shipments along the Columbia. All of this enables Pacific Northwest goods to reach the Pacific Ocean and thereby Western US and Asia.

Beginning in the 1880s, shipping channels for barge traffic were opened by dredging the Columbia, and also by the construction of canals and locks. In 1891, a 17-foot channel was first opened between the Pacific Ocean and the Ports of Portland and Longview. This channel was deepened to 40 feet in 1976. The Cascade Locks and Canal were constructed in 1896 to bypass the Cascade Rapids near the Columbia River Gorge.

In 1999, Congress authorized deepening the shipping channel to 43 feet so that large container ships could reach the Ports of Portland and Longview. The authorization of this project required that the US Army Corps of Engineers (ACE) restore large areas of wetlands in the areas of the dredging. This project was completed in 2010.

Damming the Columbia River

As reviewed in Chapter 2, the decade of the 1920s initiated mass electrification in the US, especially in urban areas. The political forces, however, strongly favored private ownership of electric utilities, and up to the early years of the Great Depression, the ownership of these utilities became heavily concentrated in holding companies. But the politics changed with the inauguration of President Franklin Roosevelt in March of 1933. In that year, Congress passed and Roosevelt signed the authorization for creating the Bonneville Dam and Grand Coulee Dam as public works projects. This legislation was led through Congress by Senator Charles McNary, a proponent of publicly owned electric utilities. The large McNary Dam between Bonneville and Grand Coulee Dams is named after the Senator.

The Grand Coulee Dam is the center piece of the Columbia Basin Project in Central Washington. Its primary purpose was agricultural irrigation, but it has also evolved into a significant hydroelectric generating dam. It was constructed by the federal *Bureau of Reclamation* as the largest water reclamation project in the US.[3] It supplies irrigation to more than 670,000 acres, and supposedly has potential expansion to an additional 340,000 acres. This Columbia Basin Project consists of the irrigation network created by the Grand Coulee Dam.

The story of the Grand Coulee begins in 1933 when Congress approved a proposal from the *Bureau of Reclamation* for constructing a relatively low dam of 290 feet on the Columbia River. But upon visiting the Grand Coulee construction site in August of 1934, President Roosevelt endorsed building a dam of almost twice the height (550 feet). At that height the dam would provide sufficient hydroelectric power to pump water from Lake Roosevelt (the new reservoir behind the dam) into Banks Lake which was formed on the ancient Grand Coulee dry bed. Congress approved the higher dam in 1935, and construction was completed in 1942.

The Grand Coulee Dam's reservoir is called Roosevelt Lake. Its creation forced the relocation of 3,000 people. The Grand Coulee Dam does not have a fish ladder and therefore it prevents salmon from reaching the upper Columbia including Canada. This was the first dam with total fish passage obstruction, but following its construction, the downstream Chief Joseph Dam and Wells Dam also totally prevented fish passage. These dams denied the Spokane and Colville Indian Nations their historical primary nutrition, and interrupted their cultural traditions.

Although the Grand Coulee dam was originally intended to serve the agricultural irrigation demands of Central Washington, this purpose was diverted by the needs of World War II. In the war years, the Dam's hydroelectric generation powered the aluminum smelters in Portland and Longview. It also powered the airplane construction industry in Seattle, and the shipbuilding industries in Portland and Seattle. The Grand Coulee Dam's power, however, was in greater demand at Hanford for its top-secret plutonium production (reviewed below).

President Roosevelt began financing the Grand Coulee construction with National Industrial Recover Act funds in 1933. Its later funding was

[3] Chapters 2 and 4 review this federal agency.

authorized in 1935 and 1939, and then by the Columbia Basin Project Act of 1943. The Dam, however, was actually completed in 1942 in time to provide hydroelectric power to the Hanford Nuclear Laboratory for creating and processing plutonium for the nuclear fission bombs of World War II. The Hanford site was selected because it was along the Columbia River in Central Washington, and because of the site's combination of remote location, plus large cooling water flows for nuclear reactor piles, and abundant electrical power for both the nuclear facility and the nearby homes of workers. The *Hanford Nuclear Reservation* was a 400,000-acre militarily guarded area of high security.

Hanford's Irrigation

In the late 1970s, the author of this book was invited onto the Hanford Nuclear Reservation—one of our National Laboratories similar to Los Alamos in New Mexico. I was invited to present a possibly relevant paper to a group of economists and engineers. I was working at the time for the Army Core of Engineers in Portland, Oregon, where I was investigating some technical issues related to cost–benefit analysis for agricultural irrigation in the Columbia River Basin. Without my knowing it, these technical issues were considered possibly relevant for Hanford's nuclear waste disposal. This was a revelation I was soon to discover. I flew from Portland to Richland in Central Washington. Richland was close to Hanford.

On the way to the highly guarded grounds of Hanford, I was accompanied by other economic analysts working there as we drove through the security gates of the Reservation. Once past the gates, I saw America's first nuclear reactor building several miles north across the flat tumbleweed desert. I also saw the largest center-pivot irrigation system I had ever seen, and I had seen many very large systems. I naively asked what was being irrigated since I could see no crops or pasture there. The area being irrigated appeared to be only tumbleweed desert. The response to my question was total silence, a very suspicious response. After driving a few miles through the desert, we came to a modern reinforced-concrete building that stood on concrete pillars above a deep reflecting pool of water. The design of this modern building was attractive, but given its isolated location in flat tumbleweed desert, I again was suspicious. In a small auditorium filled with economists and engineers, I gave my presentation. Afterward, the questions and discussions were such that before the end of the day I understood that the purpose of Hanford's irrigation was to hold down the otherwise wind-blown dust that contained the radioactive waste poured onto the desert or buried just below the surface. The winds across the Central Washington desert were strong, and the existence of this wind-blown radioactive dust was supposed to remain some sort of classified secret. The reflecting pool of water was to protect those working in this modern building from nuclear radiation emerging from the ground under the building. The task of the economists and engineers was to analyze how to manage Hanford's nuclear waste disposal.

After the day's presentations and discussions, I was asked if I might be interested in applying for a position at Hanford. If so, I would need a security clearance from the US authorities. Proper application and background investigation would be necessary. But I was already spooked by the thought of plutonium waste in the air, and while trying to be polite, I begged off saying I would give the possibility some thought. This was Hanford Nuclear where for decades plutonium was produced for atomic bombs. I suspected then that the problem of nuclear waste might be large, in fact so large that the expansive tumbleweed desert I viewed was severely contaminated. The extent of the problem was not publicly known for another few years, but it turned out to include leakage of large amounts of nuclear waste into the Columbia River. The problem became public with a *60 Minutes* nationally broadcasted story. This dangerous and extensive man-made disaster is further reviewed below.

When I flew back to Portland, I was seated next to a kindly older gentleman wearing a well-fitted business suit. He began to ask questions as to why I was in Richland, and who I might know there, and the like. I was already spooked by even being at Hanford Nuclear for a day. He was friendly, but persistent. I avoided conversation and withdrew into silence. The problems of Hanford were to be suppressed, not a subject for my discussion or even reflection, at least not at that time. The very thought of Hanford and working there was dangerous to my sense of a peaceful self. At that time, I did not want to discuss Hanford with anyone.

The serious work of cleaning the Hanford site began in the 1980s. When I visited the site, the analysis of how the cleanup might proceed was just starting. The actual cleanup process is ongoing even up to today. The site is often referred to as "America's most polluted."

The process of making the chemical element plutonium from the element uranium involved a nuclear reactor cooled by pumping a large amount of cold Columbia River water. The old reactor building was consequently located close to the river. Aluminum cannisters of uranium—or "rods"—were inserted into the reactor "pile" and some of this uranium was transformed into fissionable plutonium which then was used as the explosive ingredient in the atomic bombs. The plutonium, however, needed to be extracted from the cannisters. This process produced massive amounts of both radioactive liquid and solid wastes, and then only a very small amount of plutonium was made. This waste disposal became a monstrous problem. For forty years it was disposed of on-site in the desert. By the mid-1980s, it had become painfully apparent to the National authorities that the Hanford Nuclear Reservation must be deactivated, decommissioned, decontaminated, and demolished.[4]

Today, there are approximately 10,000 workers involved in the cleanup activities. At "the site," as it is referred to, there are leaking underground storage tanks, large amounts of severely contaminated soils and suspended radioactive liquids, old pipelines and other infrastructure, old reactor buildings, discarded worker clothing, and other contaminated debris all in need of careful disposal. To this end, engineers are constructing large on-site landfills called "Environmental Restoration Disposal Facilities" (ERDF) that are partly servicing the very complex problems of disposal.

During the production of this plutonium, hundreds of billions of gallons of radioactive liquid wastes were produced and disposed of by pouring it onto the desert ground, or into trenches, or into holding ponds. The extremely radioactive nuclear waste generated by plutonium extraction from reactor fuel-rods were placed in underground storage tanks. Record keeping was poor concerning how much was stored and where it was stored. Large amounts of solid wastes were placed in pits, or steel drums, or wooden boxes before being buried, but some of this solid waste was just buried without containers. Record keeping was also poor concerning these disposals.

Most of the recovered solid wastes, contaminated soils, and building debris are now being placed in Hanford's on-site ERDF, which is supervised by the US EPA. This "Facility" has eight cells of 500 feet by 1,000 feet by 70 feet in depth. There are also two cells each of which is twice the capacity of the other eight. Within these cells, 2.8 million tons of

[4] See "The Hanford Site," US Department of Energy at www.hanford.gov.

solid waste is being buried and covered with appropriate clay-type soil to remain "safely and permanently."[5]

In addition to all of this radioactive material, solid transuranic elements mixed with plutonium are also a significant byproduct of the plutonium production process. These mixtures remain radioactive for hundreds of thousands of years. These particularly poisonous wastes are being transported to the "Waste Isolation Pilot Plant" in New Mexico for burial.

The liquid wastes from the plutonium extraction rods that were stored in underground tanks are now being transformed by mixing them with glass-forming materials, and then heating these mixtures to 2,100 degrees Fahrenheit. The resulting cooled glass cylinders are either stored on site or shipped to other storage facilities. The liquid wastes that were poured over the desert, however, have contaminated the groundwater of shallow aquifers. Barriers, chemical processes, and even implanted microorganisms are being injected in attempts to prevent migrations of these contaminants to the Columbia River.

Hanford was the creation of the World War II emergency of having to beat the Axis Powers in developing the atomic bomb. The hydroelectric power of Grand Coulee Dam was not created to power the Hanford Nuclear Reservation, nor was it created to power the aluminum aviation industry needed for World War II. But these industries were located where there was abundant hydropower potential. Grand Coulee Dam was further developed to meet these strategic industrial emergency needs, and Hanford continued to produce plutonium after World War II because of the perceived emergency of the Cold War. Today it is difficult to envision or understand the then cavalier attitudes of the scientists and engineers responsible for creating the most severely polluted site in the US. But this should not surprise us. As reviewed previously, the alterations (damming) of the Tennessee River, the Colorado River, and the Columbia River are all products of what was then viewed as the employment crisis of the Great Depression. The alteration of the Grand Coulee from primarily an irrigation dam to a hydropower dam was perceived as necessary for the Manhattan Project's need for plutonium, and also World War II's production of military aviation aluminum. National emergencies appear to have generated the cavalier attitudes necessary to produce environmental disasters.

[5] Ibid.

Some Environmental Consequences of the Grand Coulee Dam

Grand Coulee Dam is extraordinarily high at 550 feet. This height facilitated its significant hydropower generation in time for World War II armaments production. But the required speed of this dam's construction discouraged any accompanying fish ladders or other passage mechanisms. Since its construction, the Columbia River's salmon—the mainstay of the upriver native American tribes—have not been able to migrate into the Upper Columbia River (including Canada) to spawn. The Spokane and Colville Native American Tribes consequently have not held their sacred salmon ceremonies since 1940.

The town of Inchelium in Washington was submerged by Roosevelt Lake (the Grand Coulee's Reservoir), and two-hundred and fifty Colville Tribe Native Americans had to be relocated from their ancestral homeland. Also, Kettle Falls—once the primary fishing location for the native tribes—was submerged. Their average catch of 600,000 fish per year was eliminated by Roosevelt Lake. The Army Corps of Engineers estimated that annual catches of more than one million salmon have been eliminated by the Grand Coulee Dam.[6] In June of 1940, the *Confederated Tribes of the Colville Reservation* held a three-day event they called the "Ceremony of Tears" which memorialized the loss of their Kettle Falls prime fishing site.

Constructing the Grand Coulee Dam eliminated the traditional life of the Colville and Spokane tribes. In the 1990s, the US compensated the Colville Tribes with a single $53 million payment plus $15 million per year thereafter. In 2019, the US began compensating the Spokane Tribe with $6 to $8 million annual payments. In addition, three modern fish hatcheries have been created on the Columbia River above the Grand Coulee in an attempt to partly restore the salmon in the upper sections of the river. Note that these modern hatcheries will not produce the large salmon that previously reached the upper Columbia (up to eighty pounds), but will produce smaller fish (only two to four pounds).

[6] See www.dams.org/kbase/studies/us_exec.htm.

Other Dams Along the Columbia

There are fourteen hydropower dams along the main body of the Columbia River, three of which are in British Columbia: the Mica Dam, the Revelstoke Dam, and the Keenleyside Dam. The Rock Island Dam and Hydro Project was built between 1929 and 1933. It was the first dam to span the Columbia River, and it is 12 miles downstream from the town of Wenatchee in Washington. It is not a "big concrete dam" since it was built from the nearby basalt rock remnants of an ancient lava flow. It was built for and by private utility interests, and it is now owned and operated by the Chelan County Public Utility District. It is not a particularly high construction (135 feet) and it has a capacity of 660 megawatts. This hydropower was expanded over three construction periods that ended in 1979.

The NcNary Dam was constructed between 1947 and 1954. (Figure 6.1 indicates its locations.) It spans the Columbia between Umatilla County in Oregon and Benton County in Washington. It is operated by the Army Corps of Engineers, and is located close to the town of Umatilla in Oregon. It has two fish ladders for salmon and steelhead passage. Its reservoir, Lake Wallula, extends 64 miles up the Columbia and borders on the Hanford Nuclear Reservation.

The Chief Joseph Dam was also constructed by the Army Corps of Engineers between 1949 and 1955. It has a large capacity of hydroelectric power of 2,620 megawatts, and is the second-largest producer of hydropower in the US. It does not have fish ladders or other fish passage capability.

Bonneville Locks and Dams is actually a series of dams and locks that span the Columbia River near the Gorge. These structures serve both navigation through the locks, and the hydroelectric facilities that generate 1,190 megawatts of capacity. It does have a large fish hatchery next to the dam. The Bonneville structures were completed in 1938. It is operated by the Bonneville Power Administration that distributes power to both public utility districts and private interests.

The Dalles Lock and Dam was constructed by the Army Corps of Engineers between 1952 and 1957. It is two miles east of the city of the Dalles. The dam spans the Columbia River between Washington and Oregon. On March 10, 1957, hundreds of indigenous tribal members watched as the rising water behind the dam inundated the fishing platforms of Celilo Falls, and also the Village of Celilo Falls. Its ancient petroglyphs were relocated to a state park. The construction of the Dalles Dam, even with its fish ladders, dramatically reduced salmon passage. This dam's hydropower capacity is 2,150 megawatts.

The John Day Dam is the newest on the main stem of the Columbia. It is located close to the confluence of the John Day River with the Columbia, 28 miles east of the Dalles. It was constructed by the Army Corps of Engineers and completed in 1971. Its hydropower capacity is 2.48 megawatts. Its lock has the highest lift at 110 feet. Its reservoir—Lake Umatilla—extends 76.4 miles up to the McNary Dam. It has fish ladders to enable passage.

As reviewed below, all these main stem dams served navigation, irrigation, and electric power interests of the region. But they also largely disrupted the ecology of salmon and other anadromous fish, and thereby the culture and sustenance of indigenous people. The current dam removal movement is a natural reaction to the dam construction movement of the 1930s through 1960s.

Indigenous Americans and the Columbia

Native Americans and First Nation people have lived along the Columbia for 3,500 years. These include the Colville, Spokane, Coeur d'Alene, Yakima, Nez Perce, Cayuse, Plus, Umatilla, Cowlitz, Kootenay, and the Warm Springs Confederated Tribes. After the Spanish early explorations that left horses in the North America, the indigenous people domesticated horses and consequently traded over long distances with the indigenous of the mid-western plains. Along the Columbia, native tribes had a strong fishing culture. Fishing along the River generated a central conflict between the native tribes and commercial fisheries. Native fishing rights as established by various treaties—particularly the *Treaty of 1855*—were upheld in 1905, 1918, and 1974. The latter is commonly called the *Boldt Decision*.[7]

Prior to the commercial fishing on the Columbia (the 1880s and after), salmon runs are estimated to have been 10 to 16 million fish per year. Since 1938, the largest salmon run has been 3.2 million fish in 1986.[8]

[7] See Chapter 3.

[8] See "Columbia River Fish Runs and Fisheries," by the Washington Department of Fish and Wildlife and the Oregon Department of Fish and Wildlife, August, 2002, at http://wdfa.wa.gov/fish/columbia/2000_status_report_text.pdf.

The most productive fishing site was at Celilo Falls which in 1957 was submerged by the Dalles Dam. The affected tribes were compensated by $4 million and used this to establish the *Kah-Nee-Ta Warm Springs Resort*, a commercial enterprise a short distance southeast of Mount Hood.

Celilo Falls and the traditional rights of native tribes to fish there and at other sites were supposedly protected by the 1855 Treaty. But in the early 1950s, the Army Corps of Engineers wanted to develop another hydro power facility on the Columbia. The Dalles Dam was authorized under the Flood Control Act of 1950, and the Federal Government compensated the local tribes for destroying their village and sacred fishing grounds, and also promised to improve the tribal members' living conditions with a new village. The plan was for this new village to be finished by 2020. But as of April 2023, it has not been completed, and it appears that it might not be completed.

The Salmon Fishery of the Columbia River Basin[9]

The salmon of the Pacific Northwest are anadromous (spawn in freshwater, migrate to ocean water where they mature, and then return to spawn in the same stream where they were spawned). Salmon provided native peoples of the Northwest with their main source of protein. This species includes the Chinook, silver (or pink), coho and sockeye salmon. Steelhead trout are also anadromous and include rainbow trout that migrate to the ocean, and also cutthroat trout that migrate to the ocean's bays. These anadromous fish were large (Chinook would routinely grow to over 40 lbs) and were readily caught by spears, baskets, and nets at central locations where the rivers and tributaries formed falls.[10] As the primary food source of indigenous tribes, salmon also posed spiritual aspects in that salmon were envisioned as one of nature's essential

[9] This section is based upon Robinson (2021, Chapter 11) for which the author retains copyright.

[10] See "Salmon for All: History," at www.salmonforall.org/hstory/ and Joseph E, Taylor, "Salmon," at www.oregonencyclopedia.org/articles/#.YBCU0-SmUK.

gifts to be preserved and not abused by humanity.[11] Native tribes envision themselves as an integrated component of nature in a tribal religious sense. The salmon were also considered an integrated component, i.e. a not-to-be-abused gift from nature to the indigenous people.

Up to the mid-nineteenth century, salmon and steelhead trout were extremely abundant in the Pacific Northwest, but post 1880 the Columbia River salmon canneries were constructed. There were 38 canneries constructed on the lower Columbia with 22 in Astoria, Oregon—close to the mouth of the Columbia. In 1884, these canners formed the "Oregon and Washington Fish Propagation Company" to market the product. The fishermen organized the "Columbia River Fishermen's Protective Union" in 1884, an organization that still exists (affiliated with the American Federation of Labor). Prior to the turn of the century, the Union, together with local native tribes, and the cannery industry called for the formation of fish commissions and state actions to regulate the diminishing harvests.

Between 1900 and 1930, more than 100 dams were constructed in the Columbia River Basin, the most significant at Willamette Falls (1888), Bull Run (1895), Swan Falls (1902), and Spokane Falls (1920). These were constructed to serve flood control and agricultural irrigation interests. The larger hydroelectric dams at Rock Island (1931), Bonneville (1937), and Grand Coulee (1941) followed. (Figure 6.1 provides a partial depiction of significant dam locations.) In addition to power generation, these hydroelectric dams were also designed for flood control, and to serve agricultural irrigation interests.

As early as the 1870s, the states of Oregon and Washington attempted to regulate the commercial salmon industry, and in 1915 these two states formed the cross-border *Columbia River Compact* which Congress then recognized as the regulative agency for all the River's fisheries.

After the dam constructions of the 1920s, 30 s, and 40 s, the salmon and steelhead stocks greatly diminished because the fish could not return to their spawning grounds. Native tribes, recreational fishing groups, and the commercial salmon industry called for fish ladders that would allow salmon to bypass the dams in attempting to reach their spawning

[11] See the various quotes from Okanagan tribal members in "Historic Agreement Reached Between Columbia River Basin Indigenous Nations, Canada and British Columbia to Collaborate on Salmon Re-Introduction," Sylix Okanagan Nation Alliance, July 29, 2019. Also see Michaelsen (1995).

grounds. These ladders were constructed, but only had partial success in restoring the fish runs. Why? The overall management of this salmon and steelhead fishery did not utilize proper scientific-engineering information in constructing dams along the Columbia. The dams were overly large in height, designed to generate hydro-power, and not designed for fish passage. For example, long fish-diversion canals of gradual slope could have been constructed, but these would have added construction expense and weakened the flow through the hydro-power generators.[12]

As the management institution for the River's fisheries, the *Columbia River Compact* now includes the *Joint Management Staff* of the Oregon and Washington Departments of Fish & Wildlife. It consults with the *National Oceanographic and Atmospheric Administration* (NOAA), the US Fish & Wildlife Service, the Idaho Fish & Game Agency, and the *Columbia River Intertribal Fish Commission.* As a result, this salmon fishery is now highly regulated particularly due to attempts to comply with the standards of the *Endangered Species Act* (ESA), and also comply with the *US vs. Oregon* (1968) court decision that adjudicates the treaty fishing rights of the four tribes included in the *Treaty of 1855*. But as explored below, these attempts at compliance were only slightly successful in meeting the demands of the *Intertribal Fish Commission*, or of the *Endangered Species Act* (ESA). Why? The industrialization of the Pacific Northwest has continuously threatened degradation of the Columbia Basin. This twentieth-century development jeopardized the salmon fishery through dams, mining, logging, agricultural irrigation diversions, pulp-mill pollution, urban and suburban pollution, nuclear waste pollution, and the overall degradation and loss of salmon spawning streams.

Salmon and trout spawn only in clean sandy-gravel beds, and then only in clean cold water. Silting and pollution from mining, logging, and urban-suburbanization destroys these spawning beds. In addition, between 1900 and 1930, more than 100 dams were erected in the Columbia River Basin, mostly for flood control and irrigation. But the 1930s saw the New Deal and the establishment of the "Columbia River Federal Power System" which, for example, facilitated construction of the Bonneville Dam begun in 1933. Originally, this hydroelectric dam was to

[12] See John Waldman, (April 4, 2013), "Blocked Migration: Fish Ladders on US Dams are Not Effective," *YaleEnvironment360*. Also see Rebecca Kessler (May 6, 2014), "Mimicking Nature, New Designs Ease Fish Passage Around Dams," *YaleEnvironment360*.

be constructed without fish ladders. The political influence of the "Fishermen's Protective Union" and the cannery industry forced establishment of fish ladders in 1938. Construction of the 550-foot high Grand Coulee Dam on the upper Columbia, however, was too high for fish ladders to effectively function. This cut off the upper third of the Columbia from the salmon (and steelhead trout) reaching their spawning grounds.

This prevention of the anadromous fish from returning to their spawning grounds started the era of fish hatcheries for developing the young fish. These young fish were then transported back to the lower Columbia for facilitating their travel to the Pacific Ocean. Upon return for spawning, they (or their fertilized eggs) were netted and returned to the spawning tributaries. This mechanized transport, however, provided only slight success. Whereas prior to industrialization, it is estimated that 15 million fish were spawned and returned per year along the Columbia, the dams eliminated half of the Columbia Basin's spawning grounds.[13] Since 1938, fish runs on the ladders of the Bonneville dams have varied from 3.2 million to 0.9 million. As a result, more than 75 percent of these anadromous fish now must be hatchery bred.[14] Since 2000, all Columbia salmon have been listed as endangered species and federal and state efforts at reestablishment have been initiated. Success has been limited because the hatchery program has had difficulties as explained below.

In response to the decline in the environmental health of the Columbia Basin, Congress passed the *Northwest Power Act* in 1980 which specified a goal "to treat fish and wildlife as co-equal partners with other uses in the management and operation of the hydro projects of this region." This legislation also created the *Northwest Power Conservation Council* to conduct studies and advise agencies about how to meet the goals of the Northwest Power Act. As indicated below, attempts to find and initiate solutions for saving the anadromous fish have only resulted in very limited success given the significant number of dams throughout the Basin.

Since the 1970s, however, suits in federal courts have upheld the native tribes' rights (represented by the *Columbia River Inter-Tribal Fish Commission* formed by these tribes) to half of the allowed harvest under an 1855 treaty. These legal disputes continue to the current period where

[13] Ibid.

[14] See Ibid.

the *Power Conservation Council* continues to attempt to reach consensus plans that will satisfy its competing constituents.[15]

As indicated above, the establishment of fish hatcheries was the response to the numerous dams' effects on the anadromous fish populations. It should be emphasized that the hatchery movement has not been entirely successful. Breeding hatchery salmon has been criticized for narrowing the gene pool and thereby producing inferior fish that do not survive in the wild. Starting in the 1980s, the response has been to move to smaller "hatch-boxes" planted along tributary streams, and to establishing "acclimation ponds" that rear juvenile fish under more natural conditions, and also that pay particular attention to genetic variants for breeding in hatcheries. Despite these efforts, in the 1990s, the EPA listed most of the salmon species as threatened under the 1973 Endangered Species Act. (The supermarket purchased salmon we consume now are mostly Atlantic salmon that are mostly farm-raised in large sea-pens.)

The National Marine Fisheries Service—a division of the US Department of Commerce—indicated in its 2015 "Biological Opinion" that considerable distress to salmon and steelhead results from the increasingly warmer waters of the Columbia Basin. This is a highly significant threat. Following the recommendations of the "Fisheries Service," the EPA issued its "Columbia River Cold Water Refuges Plan," which calls for restoration of the 12 primary tributaries of the Columbia that normally provide colder water to the River.[16] Warm waters kill the younger salmon and steelhead, but global warming has changed the timing and extents of snowmelt in the Northwest's mountains. This affects the amount of water flow, and the temperature of that flow, especially during the critical Spring and Summer periods when fish migrate. Therefore, the anadromous fish are not migrating to sea or returning to spawn. For protecting native tribe's fish-harvesting rights, in 2017 a US District Court ruled that more water must flow over the hydro dams within the upper Basin in order to allow juvenile salmon and steelhead to migrate between April and June. But the snowpack has not been as great as in previous decades, and the Spring melt has not been as large. Court decisions and management intentions have failed to counteract the effects of global warming.

[15] See William L Lang, "Columbia River," *Oregon Encyclopedia*, at www.oregonencyclopedia.org/articles/columbia_river/#YBCNq0-SmUK.

[16] See "News Release from Region 10," at https://www.epa.gov/newsrelease/newly-released-epa-plan-will-aid-salmon-survival-columbia-river.

In October, 2020, the Governors of Washington, Oregon, Idaho, and Montana committed to a "unified public process" in order to find solutions to key issues for recovering the imperiled fish. This *Columbia Basin Partnership Task Force*, facilitated by NOAA in 2017, has set goals for two dozen different stocks of salmon and steelhead, each stock being from a different tributary. The stated goal is to "identify the actions and investments needed to recover harvestable salmon and steelhead populations, conserve fish and wildlife, honor and protect tribal needs and way of life, and strengthen the electricity and agricultural service that communities rely on."[17] The potential conflicts in these stated goals should be apparent: power generation and agricultural irrigation have been in conflict with the health and survival of the anadromous fish since the first dams were built in the Columbia Basin over a century ago. Up to this point, these conflicts have always been settled on the side of industrial development and agricultural irrigation. The "good will" of the "Task Force" might not be sufficient to change this.

What lessons can be concluded from the history of this important resource? The Pacific Northwest salmon fishery will certainly not be restored to its pre-industrial condition. For analyzing the problem, the relevant socio-economic information has been gathered through the "Inter-tribal Council," and various recreational and commercial fishing interests. The necessary scientific information has been provided through the US Fish and Wildlife Service and through NOAA. But the hydro dams and agricultural irrigation interests and other demands remain strong, and are not likely to easily allow the elimination of the numerous large dams in the Columbia Basin. In addition, despite the scientifically proper efforts to restore the fishery through modern hatchery efforts, global warming threatens the destruction of these anadromous fish. Perhaps this fishery represents an environmental resource that now has its future seriously degraded despite efforts of *environmental advocacy organizations* and tribal interests. The industrialization of the Basin occurred early in the twentieth century, and this has permanently altered the Northwest's environment. But other means (enhanced hatchery and transport programs for these fish) to facilitate the survival of this fish are being pursued.

[17] See John Harrison (2020, October 19), "Partnership Identifies Goals for Columbia River Basin Salmon Recovery," Northwest Power and Conservation Council, www.nwcouncil.org/news/partnership-identifies-goals-columbia-river-basin-salmon-recovery/

As an example of what could be the future of the Basin's salmon ecosystem, consider that in August, 2019, in a tribal cultural event, sixty hatchery-bred chinook salmon were released above the Grand Coulee and Chief Joseph dams. In September, 2020, Colville tribal biologists counted 36 spawning nests along an eight-mile stretch of the Sanpoil River, an upper tributary of the Columbia.[18] This is the first spawning of Chinook in the upper Columbia in 80 years. It generated significant hope for an ecological and perhaps spiritual regeneration. The problem remains, however, as to whether the juvenile Chinook will travel entirely on their own past the dams to reach the Pacific, or whether some degree of mechanical transportation will be necessary, or whether the Chinook will remain land locked. Whatever the result, if these regenerations can be successful, the native tribes' reintroduction of salmon to the upper Basin is underway, and this includes attempts in the Upper Snake River—the Columbia's most significant tributary. Hopeful regeneration efforts are therefore underway as led by the indigenous tribal groups.

Lessons from the White Salmon

The White Salmon River flows forty-five miles from the glaciers of Mount Adams in south-central Washington to the Columbia River. Mount Adams is a 12,307-foot peak in the North Cascade Mountains. The White Salmon's confluence with the Columbia is within the Columbia River Gorge National Scenic Area, about half-way between the Dalles Dam and the Bonneville Dam. (Figure 6.1 depicts the dams' locations.) More than twenty-five miles of the White Salmon River is federally classified as "Wild and Scenic." Between 1913 and 2011, the Condit Hydroelectric Dam blocked the River's flow at 3.3 miles above its Columbia River's confluence. This dam was removed on October 26, 2011.

After Chinook salmon migrate back to the river where they were spawned, they deposit their eggs in that river's sandy-gravel bed, and then they expire. After their death, their carcasses turn white. The Fall Chinook runs leave large numbers of white carcasses along the river's bed and banks, hence the descriptive name of the river. But after the Dam was erected in 1913, Chinook were unable to reach the river above the dam.

[18] See Eli Francovich, "Salmon Spawning in Upper Columbia River for the First Time in 80 Years," *The Spokesman Review*, December 17, 2020, at www.wenatcheeworld.com/news/.

The Condit Hydroelectric Project was named after its Chief Engineer, B.C. Condit. It generated electric power supplied primarily to Crown Willamette Paper Company in nearby Camas, Washington. This dam was "big concrete," 90 feet thick at the base and 125 feet high. Originally, the dam had fish ladders, but these were twice destroyed by floods. After the second destruction the fish passages were not rebuilt. Consequently, the Washington State Fisheries Department required the dam to have fish hatcheries installed, but survival rates for young salmon remained very poor. In 1996, as a condition for relicensing, the Federal Energy Regulatory Commission (FERC) required that fish ladders be reinstalled. This requirement was based upon the National Marine Fisheries Service analysis concerning the dramatic reduction in Columbia River salmon. The expense of compliance, together with the low level of the hydropower generated, finally persuaded the owner (PacificCorp) to decommission the dam and thereafter deconstruct it.

The Condit Dam's reservoir—Northwestern Lake—covered 92 acres of surface and extended two miles up the White Salmon River. The runoff from Mount Adams and its canyons produced large amounts of sediment that settled in the Lake—actually 2.4 million cubic yards of built-up silty sediment. Deconstruction could have involved either first dredging the accumulated 2.4 million cubic yards before breaching the dam, or allowing the rapid flow through the breach to dissipate the sediment downstream. PacificCorp chose the latter solution.

It was expected that it would take six hours for the flow and sediment to pass through the dam's breach, but it only took one-half hour—an extremely fast dissipation.[19] Eyewitnesses commented on the putrid smell of the black silty water that rapidly flowed through the breach. The considerable amount of the resulting downstream sediment, along with the remaining sediment in the bed above the breached dam, both required appropriate plantings of native vegetation. Because the dissipated sediment had been under water for 99 years, it became course, acidic, and high in iron and sulfur. As the lead manager for the Yakama Nation Fisheries, Jeanette Burkhardt commented that, "There was no playbook on how to restore former reservoir sites. We were learning as

[19] See McClusky (2021). Also see the video "The White Salmon River after Condit Dam." Oregon Field Guide, *Oregon Public Broadcasting* on *YouTube*. Also see the video "Spectacular Time Lapse Dam Removal," *National Geographic*, on *YouTube*.

we were going."[20] The dissipated sediment was essentially sterile without the natural buildup of organic material normally found along riverbanks. As a result, Burkhardt relied on spreading mulch from local tree bark plus other organic material prior to planting native Ponderosa Pine, oaks, various shrubs, and grasses. At the upper end of what was once the reservoir, but was now a muddy remnant, 500 volunteers planted 2.8 acres with 7,000 plants and shrubs. The restoration of the riverbanks appears to be successful.

The sediment also accumulated to form a sandbar at the mouth of the river. This blocked the boat ramp at the White Salmon's confluence with the Columbia. Approximately 15,000 cubic yards were dredged and relocated above the river's mouth. Native vegetation was successfully reestablished to create a four-foot high bank to protect the river's mouth.

Five years after the River's restoration, Ian Jezorek and Jill Hardiman of the Western Fisheries Research Center in Seattle investigated the extent of the return of salmon and steelhead to the White Salmon River between 2016 and 2021.[21] They found juvenile salmon and steelhead above the former dam site. Tagged fish had returned to the river. Since it takes four to six years for juveniles to migrate down the Columbia to the Pacific Ocean, and then to reach maturity and migrate back upriver and spawn, it was apparent that a second or third generation of fish is now present in the upper river. Also, a team from the Washington Department of Fish and Wildlife made repeated field trips to the White Salmon River above the former dam site during the Fall Chinook run. The team observed extensive spawning of fish. The dam removal has been so far successful in reestablishing Chinook to the upper White Salmon.[22] The White Salmon River now appears to provide a successful example of river restoration through dam removal and appropriate remediation activities.

Four Dams Along the Lower Snake River

The events described above brought us to the November, 2023, meeting of the sixteen Native American Tribes referred to in the first paragraph of this chapter.

[20] Ibid.

[21] This research was sponsored by the Yakama Nation Fisheries, the Mid-Columbia Fisheries Enhancement Group, the US Geological Survey, and the US Department of the Interior.

[22] See McClusky (2021, October 16.).

The Snake River is the Columbia's largest and most important tributary. It flows 1,078 miles through Idaho, Washington, and Oregon. Its headwaters are in the Grand Tetons of Wyoming from where it flows westerly through the Snake River Plain—a significant agricultural area of Southern Idaho. It then flows northerly through the rugged wilderness of Hells Canyon on the Oregon-Idaho border, and then through the rich wheatfields of the rolling hills of the Palouse area around Moscow in west-central Idaho. The Snake empties into the Columbia River near Pasco Washington at Lake Wallula, the reservoir created by the McNary Dam on the Columbia. (Figure 6.2 provides a mapping of the Lower Snake.)

The Snake River Basin drains parts of five Western States: Idaho, Washington, Oregon, Utah, and Wyoming. Native American Tribes have lived along the Snake for 11,000 years. Pacific Salmon once annually spawned by the millions on the Snake and its tributaries. The sustenance and culture of the Nez Perce and Shoshone (the dominant tribes of the area when Lewis and Clark made their early eighteenth-century expedition to the Northwest) depended on the salmon, particularly the large Chinook.

There are four dams of particular concern on the lower portion of the Snake River in Southeastern Washington State: (1) the Ice Harbor Dam, (2) the Lower Monumental Dam, (3) the Little Goose Dam, and (4) the Lower Granite Dam. (Figure 6.2 provides a mapping of these dams.) The Ice Harbor Dam and Locks is approximately only 10 miles upriver from the Snake's confluence with the Columbia, and 9 miles upriver from Pasco Washington. This dam forms Lake Sacajawea which in turn extends 32 miles upriver to the Lower Monumental Dam and Locks. This Dam then forms Lake Herbert G. West which extends 28 miles upriver to the third structure in the sequence, the Little Goose Dam and Locks. In turn, the Little Goose Dam creates Lake Bryan which extends through the 37 miles to the base of the fourth structure in the sequence, the Lower Granite Dam and Locks, the most upriver structure of this navigational system. This latter dam creates Lower Granite Lake which extends 39 miles to the upriver shipping port of Lewiston in Idaho. All four of these dams have navigational locks that enable the barges to travel through to the Lower Columbia River. These river barges can now reach the Ice Harbor Dam, and then navigate through Lake Sacajawea, Lake Herbert G West, Lake Bryan, and then Lower Granite Lake to finally reach the shipping port of Lewiston in Idaho. There the wheat from North America's richest wheat producing area (the Palouse area) is loaded and shipped down the Columbia River for domestic use and export.

More than fifty-two million tons are shipped by barge from the Lewiston port per year. This represents $21 billion in foreign shipments—in particular, ten percent of all the US wheat harvest was exported through the Lower Snake River Locks in 2017. This represented forty percent of all the US wheat exports. According to the Whitman County Washington Commissioner's Office, these barge shipments would otherwise require 135,000 semi-trailer truck trips, or 35,000 rail car loads traveling downriver through the "Gorge" to the Port of Portland, all of which would generate carbon emissions. But so do the tugboats that power the barge shipments.

In addition, 18,000 cruise-ship passengers travel through this navigational system in 2017.[23] This generated $15 million in economic benefits to the region. In addition, the Lower Snake Dams produce 1,700 to 2,000 megawatts of peak hydroelectric capacity, enough to power 1.3 million homes. Also, these dams and reservoirs facilitate the irrigation of 5.1 million acres of fruit, potatoes, vegetables, mint, and feed crops. Note that the wheat grown in the Palouse region is a highly valuable hard-red winter wheat that is not irrigated.

These four dams were all constructed between 1955 and 1975. They collectively devastated the salmon and steelhead spawning runs on the Snake River. Despite each dam having "fish ladders," few fish survive the trip over these four high structures. In addition, as exacerbated by global warming, the four navigational lakes of this system are now stagnant pools of warm water that breed toxic algae blooms. Salmon and steelhead demand faster flowing, cool and clean water. These anadromous fish only spawn in the sandy gravel of flowing streams, and do not successfully spawn in the silty warm lake beds of the four reservoirs on the Lower Snake. In addition, the electric power generators of these four hydropower structures kill the young "smolts" on their attempted downriver migrations to the Pacific Ocean where they can mature before their upriver journey to spawn. But in the recent hot summer of 2023, toxic algae blooms have been a particular enemy of migrating salmon and steelhead.

In the first week of October, 2023, satellite images showed that for 30 miles behind the Lower Granite Dam, a neon-green sheen spread throughout the Snake River which indicated a toxic algae bloom.

[23] Cruise trips up the Columbia River, through the "Gorge" area, and into Idaho are a popular recreational industry.

Whitman County Health Department indicated it had not previously seen a harmful bloom of this size on the Snake. Blue-green algae blooms in high concentrations that are toxic and will cause liver damage in people and pets. It also accumulates in salmon and other fish which consequently must not be consumed.

"It's pretty gross," said Maddy Lucas concerning the October, 2023, bloom. She gathers water samples for the Health Department. "In the areas where the scum is thick, it's this neon-green milkshake consistency across the river, and it smells pretty awful."[24] Lucas indicated that the smell occurs because the matter is decomposing and the bloom is nearing its end. This past Summer and early Fall saw more frequent harmful algal blooms than usual according to the Health Department's testing of water samples. Algae blooms particularly occur in stagnant warm water with high nutrient levels as found behind the Snake River Dams where heavy nutrient loads are present because the upriver Snake flows through the extensive agricultural area of Southern Idaho.

In February of 2021, the Congressman from Central-Eastern Idaho argued for a $33 billion federal fund to resolve the river and dam problems of the Columbia Basin, including those of the Lower Snake. The Congressman's plan includes total or partial breaching of the "four dams." In this sense, it appears that the era of dam removal may have now fully reached the Upper Columbia Basin. Alternative energy sources (wind, the new small nuclear, and solar) would first have to be developed to replace the hydropower lost by dam deconstruction. Notice that an estimated $17 billion has already been expended on attempts to save the salmon since the species was first listed as "endangered" in 1991. (See below for an explanation of the $17 billion expenditure.) But so far, this effort is generally recognized as a failure.[25] Washington Governor Jay Inslee and Senator Patty Murray are in opposition to merely removing the dams without addressing the wheat transportation issue together with the overall problems of restoring the salmon and steelhead runs. The Biden Administration, however, planned to expend approximately $845 million through the Bonneville Power Administration, the Department of the

[24] See Flatt (2023, October 11).

[25] See Geranios (2022).

Interior, NOAA, EPA, the Department of Transportation, and the Army Corps of Engineers, all on remediation of the salmon problem.[26]

Recent studies support the dam removals on the Lower Snake. A 2017 investigation found that removal would result in a doubling or tripling of salmon in the Basin.[27] A study by ECONorthwest "strongly justified" removing the Lower Snake Dams. To enjoy a "free-flowing" Snake River, Pacific Northwest residents were found to be willing to pay $40 per household for the dam removals, which was sufficient to justify their removal. The study's report maintains that the dams have reached the ends of their useful lives. They will require significant reinvestment for the dams to be operated another 30 years, and given their effects on fish, deconstruction is justified.

To some extent, the dams' viability depends upon Bonneville Power Administration's (BPA) decisions. BPA markets one-fourth of the Pacific Northwest's hydroelectricity, and owns three-fourths of its transmission lines. Beginning in the 1930s, Bonneville's distribution of low-cost power from Grand Coulee and other main stem Columbia River dams gave the Northwest area a competitive advantage for its economy. Now, however, BPA faces severe financial difficulties due to hydropower issues. The cost of renewable and natural gas generated power has dropped below BPA's price, particularly what it charges its 134 Northwest public utility cooperative customers. The price BPA charged these customers was previously low because it could previously charge higher prices to its California customers. But California is switching to less expensive alternative renewable energy sources. As a result, BPA now must charge increased rates from its Northwest public utility customers. BPA's public utility contracts largely expire in 2028, and many customers will likely opt for other lower cost sources. In addition, most of BPA's dams are at least fifty years old, and will require billions of dollars in maintenance expenditures to keep them commissioned.

Because of the ecological damage caused by its dams, BPA spent $727 million (one-quarter of its annual budget) between 2008 and 2017 on fish-damage mitigation. Much of this was spent on salmon hatcheries although the evidence now indicates that hatchery salmon negatively affect wild salmon survival. The salmon recovery effort has cost BPA $16

[26] See "Presidential Memorandum" (2023, September 27).

[27] Published by the Fish Passage Center, a Columbia Basin study group.

billion between 1980 and 2019.[28] But the Lower Snake Dams' removal would harm farmers who rely on the barge traffic. Approximately 1,000 farmers rely on this barge transportation rather than the more expensive rail or truck transports.[29] The barge system, however, requires that salmon from the Snake River encounter eight dams before reaching the Pacific Ocean. They must swim through 325 miles of inhospitable warm slack-water which leaves the fish smaller and less prepared to survive the ocean's challenges. Consequently, in 1993 environmental groups sued BPA and other federal agencies for violation of the Endangered Species Act because of the damage the dams caused to salmon and steelhead. In five separate court decisions, federal judges ordered agencies to evaluate the removal of the "four dams" on the Lower Snake River. Each time, the agencies responded with delays and diversions, once arguing that the dams were "immutable parts of the landscape;" as though it were an ocean that is not subject to habitat restoration or preservation for the purpose of saving the endangered salmon. Instead of dam deconstruction, the agencies have tried various mitigating schemes, none of which significantly helped the salmon recover.

On September 30, 2022, the National Oceanic and Atmospheric Administration (NOAA) issued its final evaluation report concerning methods for saving salmon in the Columbia Basin—"The Rebuilding of Interior Columbia Basin Salmon and Steelhead." This report included significant input from the US Fish and Wildlife Service. The report called for breaching the "four dams" of the Lower Snake plus other habitat restoration efforts that were designed to collectively not only preserve the salmon as a species, but to restore the species to abundance so as to satisfy the indigenous tribes' *Treaty of 1855*. The program report called for (1) facilitating the young fish to return healthy to the Pacific Ocean, and then return to spawn, and (2) expanding the upriver spawning grounds for which other restoration efforts would also be made.

Adjudication concerning the Northwest Tribes' *Treaty of 1855* rights began in 1993. The tribes won five lawsuits that sought to restore their rights to "forever fish for the salmon and steelhead" of the Columbia Basin. But this litigation has not resulted in saving the salmon from

[28] See Leslie (2019, October 10).

[29] This is according to the Washington Association of Wheat Growers who represent 1,800 wheat farmers.

being close to extinction. In 2021, a coalition of conservation groups represented by the nonprofit environmental law organization *Earthjustice* with support from the State of Oregon and the Nez Perce, agreed to pause their two-year-old lawsuit against five federal agencies—the US Army Corps of Engineers, the Bureau of Reclamation, Bonneville Power Administration, the US Fish and Wildlife Service, and the National Marine Fisheries Service. These agencies either operate or contribute to the operation of the "four dams" of the Lower Snake River. It appears that following the previous five litigations, the federal agencies are unable to defend the continuation of the "four dams." The pause, therefore, allows *Earthjustice*, its coalition, and the federal agencies to negotiate how and when to breach the dams, and how to finance the deconstructions. The pause is to enable this negotiation. All parties are to report an agreement to the court, or the negotiation will continue, or the court will assert a solution. The Biden Administration was amenable to requesting the Congress to approve the finance necessary for the deconstructions.

Parties to the negotiations have not been allowed by the court to publicly discuss proposed plans until the final agreement is reached. But lawyers for Earthjustice, and the Chairman of the Nez Perce have publicly indicated that the "four dams" should be removed. The continued salmon losses are particularly exacerbated by a lack of river water due to its use for irrigating the crops of an expanding farmland in the Basin. This irrigation will be largely eliminated once the reservoirs are gone. Other direct causes include overfishing in the ocean, the increase in water temperatures behind the dams, and the adverse impact of hatcheries and hatchery fish. As reviewed above, NOAA also indicated in its 2022 report that dam removals were the best way to rebuild salmon populations in the Basin. It appears that the day for dam deconstructions in the Columbia and Snake River Basins has arrived. One of the area's commentators, Randy Stapilus, opines that the Snake River Basin Adjudication, which actually began in 1987, "has now been completed—at least in general terms—and it only took a remarkably brief 26 years."[30] Note that in the American West, water rights adjudications typically take much longer than elsewhere.

Stapilus indicates that there are three factors that make the circumstances now favorable for the "four dams" to be deconstructed[31]:

[30] See Stapilus (2023, November 7).

[31] Ibid.

1. Other US dams have been successfully deconstructed in the last decade.
2. The existing dams need costly repairs if they are to continue in service.
3. Renewable energy—solar, wind, and possibly the new smaller nuclear power—has grown in the Pacific Northwest so that the need for hydropower is diminishing.

In addition, however, we have two other very significant factors that appear to make deconstruction necessary: (4) the algae blooms in the reservoirs resulting from the combination of global warming and agricultural nutrients from the Upper Snake, and (5) the continued salmon population depletions. Restoring the Snake to a flowing river will reduce the water's temperature and help avoid algae blooms. The forces for dam deconstruction and environmental restoration are therefore successfully forming and have begun achieving their aims.

But there is more pending in this Columbia River dam story. Under the *1855 Treaty*, fourteen tribes ceded 11.5 million acres to the federal government's control. In exchange, the tribes were given the right to harvest fish at all their traditional sites, including Celilo Falls—the once prodigious site now covered by the Dalles Dam on the Lower Columbia. On "Indigenous Peoples Day" (Columbus Day, October 12) in 2019, representatives of the Yakima and Lummi Nations met at the former Celilo Falls site and called for the removal of the John Day, Dalles, and Bonneville Dams, all on the Lower Columbia River. Court examinations of the removals are now pending.[32]

Conclusion: Restoring the Columbia River?

Dam removals on the Lower Snake River appear now to be inevitable. The Snake River drains the southern agricultural area of Idaho where plenty of runoff provides fertilizer to the reservoirs of the four dams on the Lower Snake. Global warming of the relatively stagnant reservoir water breeds an intolerable algae problem. It is either clean the agricultural runoff of Southern Idaho, or remove the four dams, or maybe both.

[32] See Flatt (2019, October 14).

The wheat of the Palouse can be shipped by rail, or truck, or perhaps both. A transportation solution will be found, perhaps without Lewiston's involvement. Industry is always changing, always adapting, always finding new ways. The algae problem demands solution. The salmon problem also demands a solution that will be found. This may require that further dam removals and/or rehabilitations also be on the horizon. Why? The crisis with salmon runs is one of severity because of the global warming problem. The "great dams" of the Columbia, almost all constructed during the first half of the twentieth century, are not now looked on as fixed monuments like the nearby mountains. We now know that all dams are temporary and removable structures. They pose issues of environmental justice, and the inevitable issues of nature.

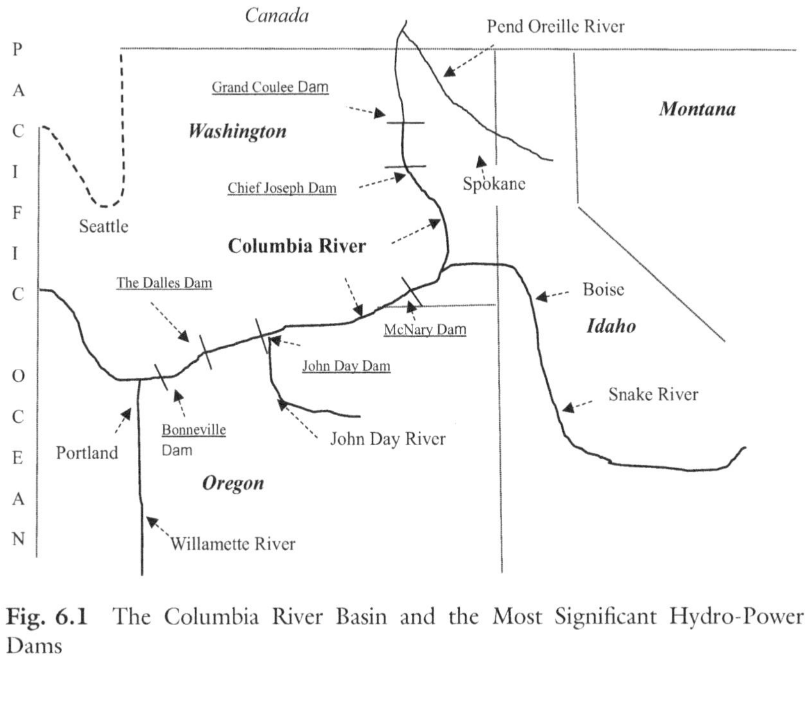

Fig. 6.1 The Columbia River Basin and the Most Significant Hydro-Power Dams

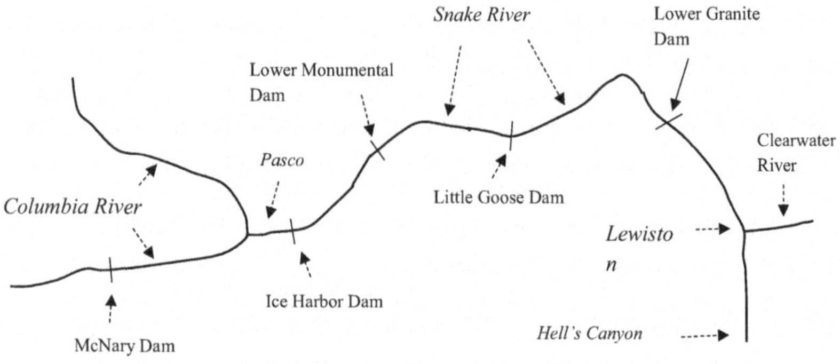

Lake Sacajawea is 32 miles long between Ice Harbor Dam and Lower Monumental Dam.
Lake Herbert West is 28 miles long between Lower Monumental Dam and Little Goose Dam.
Lake Bryan is 37 miles long between Little Goose Dam and Lower Granite Dam.
Lower Granite Lake is 39 miles long between Lower Granite Dam and Lewiston.

Fig. 6.2 Lower Snake River Dams

References

Flatt, Courtney. (2023, October 11). Large Snake River Toxic Algal Bloom hasn't Happened Before. *Northwest Public Broadcasting*. www.nwnewsnetwork.org/environment-and-planning/large-snake-river-toxic-algal-bloom-hasnt-happened-before

Flatt, Courtney. (2021, May 14). One Idea To Remove Snake River Dams May Be Dead In The Water As Inslee And Murray Say They Oppose It. https://www.nwnewsnetwork.org/environment-and-planning/2021-05-14/one-idea-to-remove-snake-river-dams-may-be-dead-in-the-water-as-inslee-and-murray-say-they-oppose-it

Flatt, Courtney (2019, October 14). Northwest Tribes Call For Removal Of Lower Columbia River Dams. *Oregon Public Broadcasting*. www.opb.org/news/article/pacific-northwest-tribes-remove-columbia-river-dams/

Geranios, N. K. (2022, July 12). *White House: To Help Salmon, Dams May Need to be Removed*. Associated Press. https://abcnews.go.com/US/wireStory/white-house-salmon-dams-removed-86666032

Leslie, J. (2019, October 10). On the Northwest's Snake River, the Case for Dam Removal Grows. *Yale Environment 360*, Yale School of the Environment, Yale University.

McClusky, I. (2021, October 16). Removing Condit Dam Spurred Hopes, But It also Begged Questions the White Salmon River Is Slowly Answering. *Oregon Public Broadcasting*.

Michaelsen, R. S. (1995). Dirt in the Courtroom. In D. Chidester & E. T. Linenthal (Eds.), *American Sacred Space*. Indiana University Press.

Reisner, M. (1986, 1987, 1993). *Cadillac Desert: The American West and Its Disappearing Water*. Penguin Books.

Robinson, R. (2021). *Environmental Organizations and Reasoned Discourse*. Palgrave-Macmillan.

Stapilus, R. (2023, November 7). OPINION: A Look at the Fight over the Snake River Dams at Thirtysomething. ridenbaugh.com, www.lmtribune.com/opinion/opinion-a-look-at-the-fight-over-snake-river-dams-at-thirtysomething/article_81fb379c-5058-a87e-1e85ca02bd3b.html

CHAPTER 7

The Low-Head Dams of Industry and the Mahoning River

Our Age of Environmental Reaction[1]

In North America, we are now fully in an era of environmental reaction, one of cleaning up our legacy of more than a century of industrial pollution. In the upper mid-west and northeast sections of North America, local environmental restorations largely involve cleanups of the remnants of these old and abandoned industries. These include our rustbelt brownfields, the derelict abandoned industrial buildings, the dumps of toxic wastes and their associated severe contaminations of our rivers, and the remnants of industrialized dams. These restorations also require the interruption of our current generators of industrial pollution, and also the disruption and prevention of our point and non-point sources of agricultural and suburban-urban toxic runoffs which still flow into our public waters, e.g. road chemicals, fertilizers, pesticides, herbicides, oils, plastics, and the like.

Counteracting these currently generated pollutions involves political challenges that are not present in restoring the poisonous remnants of abandoned industry, but eventually these currently generated problems will necessitate their own remedies. Lessons now learned from our legacy industry cleanups indicate why our current pollution generators must be regulated and stopped. These cleanups are expensive; they

[1] This section partly relies on the "Preface" from Robinson (2023).

R. M. Robinson, *Restoring America's Rivers*,
https://doi.org/10.1007/978-3-031-81758-8_7

require marshalling of scientific expertise. They also require that our environmental advocacy efforts counteract political resistance. Our political environmental emotions that once lurked below the surface, are now erupting. The politics of cleanup now engulf us and act through our local environmental NGOs and governmental advocacy organizations. But the totalities of our political emotions are likely to favor environmental restoration concerns. This is a movement illustrated by the restorations reviewed in this Chapter.

Perhaps during this age of global climate change, it might appear that local restorations should be of secondary importance in the environmental movement. But this notion is contradicted by the emotions and involvements of our local citizen organizations who are ignited by the environmental causes of their locales, and who are devoted to "their" rivers, watersheds, and bays. A national-level observer must now wonder if every waterbody in North America has a citizen advocacy organization that is devoted to its protection, one that lobbies local governments and organizes cleanup efforts. These organizations investigate the sources of their waterbody's pollution; they ponder how to stop the negative externalities plaguing their river. Our local waterbodies are immediately close by, and we know they affect our lives. We are outraged by their cavalier destruction by those who just don't pay attention to their adverse actions, or even worse, by either those who profit from that destruction or who refuse to pay the costs of cleaning their contaminations. Perhaps global climate change should be the top issue in all our minds, but that poses a more remote problem compared to the degradation of the immediate waterbodies that we interact with, that we grew up with, that we have fond memories of experiencing. For this reason, our local advocacy groups "will be heard from!" They manifest the emotional energy of the environmental movement.

My childhood occurred twenty miles north of Boston. My family gatherings were mostly at the rivers, bays and beaches of northeastern Massachusetts and immediate areas. We considered these public waters our family assets, not to the exclusion of other people, but just the opposite. It was along those sites that we and our friends gathered, but all were welcomed and invited. We recommended these public and favorite sites to others. When some of these assets were managed poorly by our "responsible" regional government agencies, that is by the organizations who were created explicitly for their protection, we and our blue-collar society could not be more outraged. Our political force was properly directed, and the required changes occurred. The political message was clear. We might not enjoy the so-called outstanding cultural assets of Symphony Hall, or the outstanding "Museum of Fine Arts," but we did enjoy our family and friends fishing for "stripers" on the Merrimack. We organized clam bakes at Wingaersheek or Nahant Beach, and we watched the "young ones" catch their first bluegills at Lake Quannapowitt. We gathered to cook hamburgers at one of the "by-permit-only" cookout sites in Squannacook Brook State Forest, or we just enjoyed a warm summer afternoon of group-walking along Crane's Beach. We had strong reservation demands for all such activities.[2] We believed that one aspect of our political processes must be to keep those public assets clean because they represented a considerable portion of our societal wealth. Such are the current motivations behind our local environmental restorations and the advocacy organizations that lead these efforts.

This chapter, however, illustrates a different political dynamic than the one I observed in my early years in Eastern Massachusetts. Between 1900 and 1970, in many areas of the US the political dynamic was aimed at creating and maintaining industrial jobs in the midst of a fairly weak rural farming economy. The local river reviewed in the pages below—the Mahoning River—was cavalierly sacrificed by establishing numerous industrialized low-head dams and filling their small reservoirs with the toxic wastes of steel mills. This method of disposing of the toxic byproducts of steelmaking was not required to produce the product, but whatever helped enable the steel mills to stay in the Mahoning River Valley was encouraged, and this included not forcing the steel companies to suffer the burden of paying the costs of cleaning their pollution. Now the big steel mills are gone, but the polluted sediments remain. Our broader society must now bear the costs of cleanup. The local environmental politics dictate this, and the local environmental advocacy organization—in this case the "Friends of the Mahoning River"—has spearheaded the effort.

[2] By "reservation demand" we mean that we want these waterbody resources to exist even when we have no immediate intention to use them. We are willing to expend our tax revenues to maintain them until we do arrange to use them if ever.

The Roles of Local Environmental Organizations

Citizens prefer that their governments establish and maintain environmental quality. Some citizens become outraged and active when they observe severe degradation, especially when it affects them directly. It is therefore normal that our state (or provincial) and local government agencies should be the lead institutions in our local restoration efforts, but non-government advocacy organizations also often provide special motive force.[3] This "special motive force" is an important component in our society's model of local environmental restoration, one that should be kept in mind in order to fully understand a particularly interesting genre of the environmental movement. This essential component provides the "vision of what could be," as introduced and explored previously in Robinson (2021 and 2023).[4] Envisioning what can be accomplished through collective action is the necessary initial ingredient to the restoration process. This "new vision" poses a significant challenge for the environmental experts of our various federal and state agencies, but perhaps it is even a more important challenge for our local environmental advocacy organizations. It is one thing to clean a river or bay of toxic pollution that makes local residents sick; it is another thing to pose a "new vision" for their local river and business efforts, a vision that would alter their economic future, or that is radically opposite their past modality of industry and transportation. The "new vision," if it can be fulfilled, can be frightening to some. This environmental problem is entirely political in its substance. It is the *environmental advocacy organization* (EAO) that can bring this new vision to the greater public. By articulating the knowledge of the experts found in our government agencies—such as the US Fish and Wildlife Service, the US EPA, or the various state environmental agencies—EAOs can lower the fright level. By accomplishing this, the difference between the old modality and the new can be explained to the public so as to bring the visioned restoration to fruition. The "grassroots" political force, therefore, need not offer only a spontaneous and temporary mirage, but it can have a scientifically sound focus that sets in motion an organized lasting political impetus.

[3] Excellent examples of this advocacy include the Chesapeake Bay Foundation, and the Friends of the Columbia River Gorge as reviewed in Robinson (2021).

[4] Ibid.

In this chapter, locally organized river initiatives form the subject. My observations indicate that in some instances the politics of local restoration should not be separated from the politics of the overall environmental movement. But I have previously argued that to be effective, the process for environmental restoration must have certain characteristics of being *fair and reasoned* in order to fully engage our local advocacy organizations.[5] Without these *fair and reasoned* characteristics, the local citizens will likely feel excluded or ineffective, and the process of restoration will faulter. Other chapters illustrate that the politics of the "environmental justice movement" are relevant for this "exclusion and ineffectiveness" issue. But this Chapter shows that the persistence of the effort is also key to inclusion and success. Hence, the "process" is important, and needs to be fair and reasoned or it will likely be ineffective.

The restoration narratives presented in this chapter are actually tales of organized expertise. They are stories about the local political entities who energize the environmental movement to demand local restoration. Some have yet to receive significant federal funding, but they do have significant state and private organizational funding. These narratives concern a heavily abused river with substantial restoration efforts led by dedicated state and local environmental organizations. How else could these restorations occur if they were not supported locally?

As emphasized above, the processes of local restorations vary as to the involvements of the state, and federal agencies. In particular, the involvements of local environmental advocacy organizations also vary. The key to success in all these efforts is, however, the utilization of the expertise of dedicated scientists (biologists, chemists, engineers, and ecologists) who reside in governmental agencies, academia, and NGOs. To a great extent, these scientists even provide the vision of "what could be." We should be aware of these "narratives of vision" since they could potentially provide directions for other restorations. Expressing these narratives is therefore an important task.

[5] This was argued in Robinson (2021).

Poking About the Poisons

Writing compositions about river restorations in New England—as I did in Robinson (2023)—left me with rather mixed emotions since I witnessed many of their degradations. These reviews force remembrances of my boyhood of poking around the brush, dirt, and muck of the streams that flowed into our town's reservoir (Crystal Lake in Wakefield, Massachusetts).[6] This poking around picked up copious amounts of stickers on my clothing and pounds of muck on my shoes. I distinctly recall my first time staring at an abandoned rusty metal barrel hidden by brush at the edge of a pond. The water of that pond flowed through a short brook into the town's municipal reservoir. I was walking at the time with my father and older sister. I was only six years old. My father pondered what poisons were in that barrel that somebody thought had to be hidden and abandoned. But I went back to that pond innumerable times prior to my teenage years, always looking at that barrel rusting away, and always with the same questions.

I also wrote about a similar experience repeated only a few years ago when I walked through some Western New York fields that I was considering for purchase. I walked down a hill, my clothing picking up the familiar stickers, and the dirt and muck along the way. I rather felt at home on this strange but still familiar piece of land. I was heading toward a pond, and a marsh along the edge. On the northern edge of the marsh, behind some juniper brush, stood a dilapidated old semi-trailer, propped up by the deflated remains of tires. I could barely perceive the old dirt road that led to the truck. At its back end, I could see that the trailer enclosed rusted metal barrels, some spilled out on the edge of the marsh. Again, barrels were abandoned by a pond, hidden from any easy view. Scratch around the brush, stickers, muck, and dirt that surround isolated water sites, and what do you find? Such hidden places are often where we abandon our poisons that contaminate our waters. Hidden ponds and streams provide a resource we find convenient for dissipating our toxins. But now consider how much more outrageous it is to observe a company openly and cavalierly polluting one of our much utilized rivers. Such an openly committed sin against all of us means thumbing one's nose at our society; it seems so much worse than a hidden act of defiance.

[6] I previously wrote about the following events in Robinson (2021).

This Chapter reviews some openly committed societal wounds for the purposes of industry, wounds familiar to our twentieth-century industrial history. This chapter documents dissipations of poisons dumped openly into public rivers, but without publicly voiced complaint. Why no complaint? Because at one time in the past we prized these industrial jobs as a force lifting us out of poverty. The cost of severely degrading the environment was not considered overly high when there were suitably available clean environments nearby to enjoy. This is one way that the tragedy of the commons occurs because soon those nearby areas will also be polluted. If degradation of one area is cavalierly allowed, it will likely be allowed in other areas.

As reviewed in this Chapter, these dissipations of poisons into the public's water were committed by steel mills ridding themselves of the costs of controlling their toxic byproducts. But perhaps that behavior is the remnant of an old culture; one that has finally given way to a classier and cleaner culture of restoration. I view the restoration efforts of today's culture as strongly encouraging and deserving of admiration. There is still, however, something to be said for poking around in the brush, and the dirt and muck to find out what is there. Over a lifetime of this sort of experience, I know that one can easily clean the stickers and dirt off. Unfortunately, the poisons remain. Is this the same situation as investigating the dumping of industrial poisons cavalierly into our public waters? Isn't this a purposeful thumbing of one's nose at our public's health and life enjoyment? Today we must bear the costs of environmental restoration and the stress of diligently finding and preventing current degradations.

The restorations we currently pursue do not typically aim at moving back to an original pristineness. But it does mean that at a minimum we must transform an industrial, or agricultural, or urban-suburban polluted environment to an ecologically sound, unpolluted, and sustainable environment. As a society, we aspire that our polluted waters will be restored to a level of cleanliness that will not make people or wildlife sick; that these waters will have a clarity that is aesthetically pleasing; that they will be inviting to native fish and fowl, and other healthy animals. We aspire that the walkways along our water bodies will be inviting, and that they will offer views of sustainable flowers and native vegetation, and not the scenery of poisoned corruption. We aspire that our water bodies will offer unpolluted recreation and provide unobtrusive transportation arteries that are reasonably absent of negative externalities.

As a society, we know that our rivers historically were used to dissipate the industrial, the agricultural, and the sewerage pollutions by sending them to flow downstream to have negative effects on neighbors. We recognize that our ponds, rivers, bays, marshes, wetlands, and estuaries were used as dissipation dumps. Our historical pollutions were costly to our society because of their negative health effects, and because of their deprivations of recreation and aesthetic benefits. But these buildups of past pollutions typically remain and continue to generate ill health, and continue to deprive us of our recreation and aesthetic benefits. These negative externalities persistently linger to impose substantial costs on us even to today. We recognize, however, that restorations are also costly, but in general we recognize that the benefits of restoration far outweigh the costs of cleanup. Our society knows that the only reason for not undertaking these restorations stems from the political protection of narrowly vested commercial interests.[7] Unfortunately, we know that these political protections occasionally plague us. We further recognize that in order to pursue our desired ends and methods for restoration, we aspire that our decisions should be (i) scientifically informed, (ii) inclusive of input from all those affected, (iii) logical in analysis and decision, and (iv) not frustrated by the political influence of those with narrow self-serving commercial or other personal-economic interests. As explained previously in Robinson (2023), these four criteria describe our society's aspirations for a *fair and reasoned* decision process concerning our environmental matters.[8]

The Industrialized Youngstown in Ohio

This Chapter concerns an old but struggling to be restored industrial locale. The area of Youngstown in northeast Ohio is one of America's earliest to industrialize. It was found by John Young who surveyed the area in 1796, and purchased 15,560 acres from the Western Reserve Land Company in 1797. The area has coal and iron ore deposits which led to Youngstown's iron and then steel industrial development. Its first blast

[7] See Robinson (2021, Chapter 10) for a review of one of these "vested interests," the role of Big Sugar in blocking restoration of Florida's "Everglades."

[8] See Chapter 2 or Robinson (2021, Chapter 3) for a full review and examination of the criteria necessary for "fair and reasoned environmental discourse and decisions."

furnace was constructed in 1803.[9] The railroad arrived in the village in 1856, and Youngstown was incorporated as a city in 1867. Starting in the late nineteenth century, and through World War II, Youngstown became one of America's centers of steel production with both Republic and US Steel establishing large production plants in the City.

In the early twentieth century, the Youngstown area had an influx of middle-eastern immigrants (Lebanese, Palestinians, and Syrians). In the early 1920s, Ku Klux Klan (KKK) caused street fighting between the Irish, Italian, and Mid-Eastern ethnics. But the KKK quickly abandoned the area by the 1920s. Spanish speaking Latino immigration in the post-World War II period provided the workers for the steel industry. But the new less expensive technology for recycling steel, plus foreign competition, caused the steel industry to rapidly decline in the 1970s. Between 1960 and 2010, the population of Youngstown declined from 166,689 to 66,982, a decline of sixty percent. According to the 2020 census, the Youngstown population was 60,068, i.e. another 10.3 percent decline compared to the previous decade. In 2023, the racial makeup of Youngstown has 47 percent identifying as "white," 45.2 percent as "African American," and 9.3 percent identifying as "Latino."

The Mahoning Valley

Dams below fifteen feet are called "low-head." Sometimes they are as short as six or even only five feet from their base. They were frequently constructed in the nineteenth and early twentieth centuries for purposes of merely creating a pool of water for either industrial or civic use, that is to make either water extraction and/or waste dissipation easier. Industry, particularly the steel industry, used low-head dams for facilitating both water extraction and waste dissipation. In the process, the steel industry destroyed the life of many rivers and turned the civic lives of surrounding towns into darkened nightmares of polluted water and air. This was the sad case of the Mahoning River that flows through the Youngstown-to-Warren Corridor of northeast Ohio.

The Mahoning River winds through five counties of northeastern Ohio (Portage, Stark, Mahoning, Trumbull, and Columbiana) before it crosses into western Pennsylvania where it joins the Shenango River to form the

[9] See Blue et al. (1995).

Beaver River northwest of Pittsburgh. The latter then flows into the Ohio River. (Figure 7.1 maps these rivers.) This region had nearby coal and iron ore so that starting in 1803, furnaces were constructed to start the iron industry and eventually the steel industry of the Mahoning Valley. Fifteen major steel mills and thirty-five-related smaller plants were eventually constructed along the River to produce a significant portion of the steel of American's nineteenth and twentieth centuries. Numerous low-head dams were constructed to extract the water that cooled the mills' machinery and also the final steel product. The water temperatures of these small reservoirs frequently reached over 100 degrees Fahrenheit and were filled with the toxic chemicals that the steel plants dumped back into the Mahoning, a drinking water source for Youngstown. The poisons settled into the sediments behind the dams. Municipal sewerage and wastewater treatment did not begin in the Valley until the 1950s and 1960s, but subsequent combined-sewer overflows then became common and still exist today in the Summer of 2024.[10]

In 1952, The *US Geological Survey* claimed the Mahoning River was "among the worst polluted streams in the United States."[11] In 1977, the US EPA claimed that the River was not suitable for any recreational purposes. It listed the toxins being dumped into the Mahoning River by the steel mills and related plants:

- 400,000 pounds of suspended solids per day,
- 70,000 pounds of oil and grease per day,
- 9,000 pounds of ammonia nitrates per day,
- 800 pounds of zinc per day,
- 500 pounds of cyanide per day, and
- substantial amounts of other heavy metals, PCBs and PAHs, both of these being long-lived chemical carcinogens.[12]

(See the Appendix to this Chapter for a review of the toxic chemicals commonly found in US industrialized rivers).

As a result, the River's aquatic life was entirely destroyed east of Warren, Ohio. But the *Clean Water Act* of 1972 and its subsequent

[10] See Marino (2022, October 3).

[11] Ibid.

[12] Ibid.

amendments set surface water standards and pollution controls, which along with the changes in the technology of steel making, led to the Valley's steel mill closures. These changes then provided the impetus for the Mahoning River's recovery.

Up through World War II, the Mahoning Valley was second only to the Pittsburg area in steel production. As such the Youngstown to Warren Corridor grew rapidly until the 1970s and 1980s. In 1988, the Ohio EPA labeled the Mahoning River from Leavittsburg to the Pennsylvania border as unfit for any human contact. The propagated slogan became, "Do not touch the water!".

Over the twentieth century, the changes in the Mahoning Valley population reflect those that occurred in Youngstown which peaked in 1930 at 170,000. It stayed stable at slightly below 170,000 until the 1960s when it began a steady decline to its current 60,000 people (approximately). On September 19, 1977, "Youngstown Sheet and Tube Campbell Works"—the last significant plant in the Valley—shut down laying off 4,000 employees.[13] Currently (2024), manufacturing in general is only the third-largest employment sector behind health care and retail trade in the Valley. Nonetheless, as a sector manufacturing is about one-fourth larger as a portion of employment in the Mahoning Valley than it is for the overall US.[14]

The population of Mahoning Valley is now also older as compared to the US. The median age in 2019 was 43.5 years for the Valley, but it was 39.4 years for Ohio, and 38.1 years for the US.[15] The average household size is also decreasing primarily because the average number of children per household is dropping, and correspondingly the number of seniors per household is increasing. Also, the median household income has been increasing but not as rapidly as in the State of Ohio or the overall US.[16] These demographic changes indicate that our broader society has left the old "rust belt" area of Mahoning Valley behind. But as indicated below, the refurbishment of the Valley has also begun.

[13] See Mahoning River Corridor Revitalization Plan at www.mymahoningriver/wp-content/uploads/2021/Mahoning-River-Corridor-Revitalization-Plan.pdf.

[14] Ibid.

[15] Ibid.

[16] Ibid.

Table 7.1 lists the nine low-head dams along the Mahoning River, and Fig. 7.1 shows their locations. Table 7.1 also presents some data concerning the removals of these dams. Note that the Lowellville and Struthers Dams have already been removed. Four others (Center St., Mahoning Dam, Crescent St., and Summit St. Dam) are well along in the removal process in that the appropriate governing bodies have approved their removals, and funding has also been arranged. Two of these nine dams, however, are still "in use" by industry. The poisonous serpent still does its harm.

In 1994, Ohio's EPA tested the water quality at twenty-nine sites along the River. There was no aquatic life found east of Warren. Heavy metals, ammonia, and phosphorous, PCBs, and PAHs were found at toxic levels. (See the Appendix to this Chapter for a review of these toxic chemicals commonly found in industrialized US rivers.) But after further testing in 2013, signs of resurgence in aquatic life were found. The closings of the steel plants and improvements in sewer systems were having some positive effects.

Table 7.1 Mahoning River Low-Head Dam Removals

Dam Name	*Owners*	*Status*	*Budget*
Leavittsburg Dam	Trumbull County Parks	Removal approved by Metro Parks Board	To be determined
Summit St. Dam	City of Warren	Approved by City Council	To be determined
Republic Steel Warren Water Works Dam	BDM Steel Holdings	Currently in use by industry	$3,225.500
Girard Mills Dam	McDonald Industrial Land Corp	Currently in use by industry	To be determined
Crescent St, Dam or US Steel Dam	Briar Hill Slag	City Council approved removal	$2,453,000
Mahoning Ave. Dam	Anthony's on the River	City Council approved removal	$2.619,000
Center St. Dam	City of Youngstown	City Council approved removal	$2,719,000
Struthers Dam	City of Struthers	Removed in July, 2021	$2,325,000
Lowellville Dam	Village of Lowellville	Removed	$2,380,000

The 2001 *US ARMY Corps of Engineers'* "Reconnaissance Study" of the Mahoning River is perhaps the most applicable analysis of the industrial corridor of this waterbody. In that "Study," the *Corps* recognized the Mahoning River as one of the five most severely contaminated US rivers in need of sediment removal and ecological remediation. The *Corps* recommended a goal of returning the River to a "swimmable and fishable" condition in line with the 1972 Clean Water Act and its subsequent amendments. Also, the *Corps'* "Study" recognized the Ohio EPA's 1996 finding that despite the closing of most of the steel mills, the River's ecosystem was slow to improve due to contaminated sediments in the riverbed and on the side banks. These contaminants caused the River's flow to have high concentrations of heavy metals (copper, zinc, lead, chromium, iron, nickel, and cadmium), plus cyanide, ammonium nitrates, and PHAs. (See the Appendix to this Chapter for a review of these toxins.) The concentrations in the Mahoning River were found to be an order of magnitude higher than in comparative control streams. In addition, the concentrations decreased with distance from the legacy steel mill sources. PAHs were also found behind some low-head dams, and the "invertebrate index" of ecological health indicated "poor river quality and severe degradation as compared to the similar reference river," a comparison waterbody that was not influenced by low-head dams.

The *Corps* analyzed five options of potential pursuit. The options ranged from (1) "no action," to (5) "full hydrological dredging" along the contaminated sections. The counties involved have now formed plans to remove the low-head dams and to dredge the identified buildup of contaminated sediment.

Restoring the Mahoning River

The Eastgate Regional Council of Governments represents the thirteen communities along the Ohio section of the Mahoning River. It formed and provides the impetus for "The Mahoning River Corridor Revitalization Plan" which specifies the removal of all nine of the low-head dams in the industrial corridor of the River. Their locations are indicated by Table 7.1 and Fig. 7.1. The removals of these industrialized low-head dams will generate environmental benefits, but the extent of the benefits depends upon other environmental support activities: (1) whether other obstructions (dams or culverts) are removed, (2) whether the riparian areas along the riverbanks are restored, (3) whether toxic sediment is removed, and (4) whether other contaminants such as sewerage or agricultural wastes are prevented from polluting the river.

As listed above, the benefits of removing the Mahoning River's low-head dams depend on restoring other attributes of the River and its streamside environment. The riparian zone includes a river's banks and streamside park-like areas that vary in width and also composition, i.e. as to whether the stream's sides are natural or otherwise developed into impervious surfaces and the like. To be healthy, a river needs a buffer zone, preferably of densely growing native vegetation that creates a natural transition from developed land to the water's edge. Communities along a river generally benefit from these buffer zones.[17] If properly constituted, streamside lands control floods by absorbing water like a sponge, and slowing the flows from impervious grounds (such as roads, parking lots, and rooftops) to the river, the area's main drainage vehicle. The roots of the plants, shrubs, and trees create a porous soil that allows water to infiltrate the ground and be stored. Also, plants transfer humidity to the surrounding atmosphere which reduces the flow to the river.

Riparian zones also include natural wetlands and floodplains, both of which typically need restoration, and both of which can benefit from native vegetation that absorbs and stores large amounts of water. All of this water management helps control flooding. Creating dams, particularly low-head dams, for controlling floods is generally illusory. Severe weather events quickly overcome these control systems, and without the buffer zones, wetlands, and floodplains, the full destructive impacts of significant rain storms are experienced.

During the age of global warming, riparian zones also help cool the water that reaches the river, and also provides shade for the river's edges and the native animals that live in these areas. Native trees such as hickory and sweetgum provide shelter and food to native animals. But plants of various sizes and types provide layers of both vegetation and root systems that create the needed porous soils and anchored stream banks, plus they provide the necessary shade for cooling the river. Native vegetation also provides low-cost maintenance in that dominating invasive vegetation will not require removal. The denser this layered native vegetation the more effective it will be for restoring the river and cooling its waters.

[17] Ibid.

It is important to note that appropriate riparian zones will absorb the type of contaminated runoffs we observe from agriculture, industry, and urban/suburban zones that severely pollute rivers. This runoff otherwise poisons aviary, fish, and other animal life. It feeds algae and stimulates other undesirable infestations.

The above indicates that large riparian zones are desirable, i.e. at least twenty-five feet in width from the river's edge with one-hundred feet being more desirable. Also, the greater the length covered on each side of the river that is covered with an appropriate riparian zone, the healthier the river will be, and the greater its resilience to global warming. The EPA's guidelines for riparian zones also indicate[18]:

- If a wide healthy riparian zone is not possible, then some degree of appropriate buffer between agricultural, or industrial, or urban-suburban developments, or otherwise impervious surfaces and the river's edge is essential.
- "No mow" zones should be established for riverbanks so that thick grasses and natural organic appropriate materials can build up.
- Invasive vegetation should be quickly identified and uprooted.
- Impervious surfaces such as concrete or asphalt should be kept away from the river's edge.
- Pet waste and trash provides obnoxious pollution that should be controlled. To that end, trash receptacles should be convenient.

The Mill Creek Metropark's "Rain Garden Farm" (a streamside park) is an attempt at establishing an appropriate riparian zone. (Mill Creek is indicated on Fig. 7.1.) It is an example of a current landscape architectural design that prevents road and other chemicals from reaching the Mahoning River. It sits below the level of the surrounding area in order to absorb storm water from nearby impervious surfaces. Its vegetation grows on the organic material trapped, and it captures noxious heavy metals that would otherwise pollute the River. In addition, native wildlife is being preserved in the 264-acre Mill Creek Wildlife Sanctuary, a large wetland created in pursuit of the motives reviewed above.

[18] Ibid.

All of these efforts to create riparian zones and wetlands in the Mahoning River Valley, plus the elimination of the point and non-point pollution sources, and the restoration of the River to aquatic health through low-head dam removals, should all be properly viewed as an integrative effort that might ultimately return the Valley to a growing healthy area of desirable location. The *Mahoning River Corridor Revitalization Plan* specifies the dam removals. The Lowellville Dam was the first of the nine dams to be deconstructed. It was the most downstream of the nine and close to the Pennsylvania border. The visible portion of the dam consisted of eight concrete buttresses that pointed downstream and that appeared to have supported a bridge structure. The River flows past these concrete pillars lengthwise. Before removal, they caused navigation hazards for the kayaks, canoes, and other small boats. Lowellville's Mayor sees housing developments along this section of the River after the dam's removal. Ohio EPA's "Water Resource Restoration Sponsor Program" provided the $2.38 million that financed the removal. The same source provided the $2.33 million to remove the Struthers Mill Dam located three miles upstream from the Lowellville Dam. The funding for these removals actually came from a steel company settlement (LTV Corporation) with the Ohio EPA. That settlement will also finance the three dam removals in Youngstown.

Leavittsburg is located about twenty miles upstream from Youngstown on the Mahoning River. It borders the west side of the small City of Warren. In Leavittsburg the local population is opposed to its low-head dam removal. They cite three reasons: (1) the small reservoir is appropriate for their pontoon boats and local recreational fishing, (2) removing the reservoir would expose their untreated sewerage pipes leading from their nearby housing into the water (some houses have Ohio EPA permits for this sewerage disposal), and (3) the reservoir removal means loss of the fire department's secondary water source used in case of emergency. In Leavittsburg, the dam was constructed in the early 1900s for hydroelectric power generation, but it also created a calm reservoir. Along this reservoir, the town built Mahoning Park where canoes were rented, and adjacent cottages were built. Now, local people voice a preference for leaving the dam alone rather than resolving the sewerage and contaminated sediment problems.[19] The park atmosphere is now gone, but people currently still

[19] See Marino (2022, October 3).

float their pontoon boats on the reservoir. Trumbull County Metro Parks owns the Leavittsburg Dam, which is actually two dams constructed for hydropower. Only one dam appears above the water's level. Residents claim that "the water has never been cleaner!"[20] This claim, of course, cannot be true. Signs that indicate opposition to the dam's removal are present and obvious in the reservoir area.

The Metro Parks Board has voted to remove the dam. In January of 2020, Metro Parks obtained $3.2 million from the Ohio EPA to remove the dam. The reservoir's water level is above the pipes from nearby housing that drain raw sewerage directly into the River. The County is under a 2002 EPA consent decree that established a twenty-year plan to eliminate the discharges of raw sewerage into the River through providing sewer service to the town. The sewer project is underway and the County Engineer believes it will be completed prior to the dam's deconstruction.[21]

The arguments in favor of dam removals center on (1) the return of anadromous fish such as the trout and sturgeon which are endangered throughout the upper Midwest and Great Lakes region, (2) the lowering of the river water temperature during our age of global warming, and (3) the associated removal of sediment built up behind the dams by the now defunct steel mills. The first of these arguments is supported by the evidence of success of other dam removals in Ohio and Pennsylvania.[22] The sturgeon is a long-lived fish that is native to Ohio's rivers. It is a popular fish for which the evidence indicates it does recover quickly with dam removals.[23]

With climate change comes more extreme weather events. With stronger flood waters, dams become more vulnerable to being breached, and breaching events pose the most dangerous floods. Among other damages, stream banks are eroded more extensively with these heavy floodwaters. The river without dams has a more natural streambank formation that is likely to be longer lasting. Although building housing close to a reservoir or close downstream from a dam is tempting, it is also

[20] See Marino (2022, October 10).

[21] Ibid.

[22] See the McCoy Dam removal as an example at www.fishandboat/Transact/Angler Boater/AnglerBoater2011/JanuaryFebruary/Document/04spring_ck.pdf.

[23] See *Ohio River Basin Fish Habitat* at http://orbfhp.org/ and also *National Fish Passage Program* at www.fws.gov/pacific/Fisheries/sphabcon/NationalFish passage.html.

dangerous, and low-head dams are the most tempting for making the type of reservoirs that people like to build beside. As a result, dam removals along the Mahoning are still an issue. Some of these appear picturesque, but they are environmentally destructive.

The Eastgate Regional Council of Governments has produced a video that depicts changes to the River in an optimistic vision.[24] Videoed scenes along the river show water flowing over the dams as posing picturesque attractive entities. Numerous kayaks are depicted in colored architectural drawings that appear realistic. This propaganda video depicts a restored and reconstituted industrial corridor that is attractive for kayakers and also for pedestrians walking through streamside parks. The video does not, however, appear to support deconstruction of the dams.

Seven of the original low-head dams remain on the Mahoning River, but four of those have detailed plans for removal. Two of these are in Youngstown (the Mahoning Avenue Dam and the Crescent Street Dam). One of the four is in the City of Struthers, and one is in the City of Warren. All of the costs will be paid by the Ohio EPA's "Water Resource Restoration Sponsor Program" and the Eastgate Council. All entail removal of the contaminated sediments behind the dams.

In Youngstown, the deconstruction of the Mahoning Avenue Dam is projected to cost $2,619,200 of which $1,828,000 is for contaminated sediment removal. The deconstruction of the Crescent Street Dam is projected to cost $2,453,000 of which $1,662,000 is for contaminated sediment removal.

The City of Struthers Dam deconstruction cost $2,324,500 of which $1,162,000 was for removing contaminated sediment. The remains of this dam were particularly ugly and debris catching. The structure once supported a train trestle. This dam once served the "Youngstown Sheet and Tube's Campbell Works" and its coke plant. The costs of this removal was borne by multiple Ohio State agencies.

The Warren Water Works Dam deconstruction is projected to cost $3,225,500. Much of this cost will be for dredging contaminated sediment. Plans for removing the Republic Steel Warren Works Dam, and the Girard Mills Dam, and also the Leavittsburg Dam are still in formation.

Along the Mahoning River corridor, there are two cultures seeking to influence the future. The first manifests those seeking a clean, restored,

[24] See https://eastgatecog.org/projects/environmental/mahoning-river-restoration.

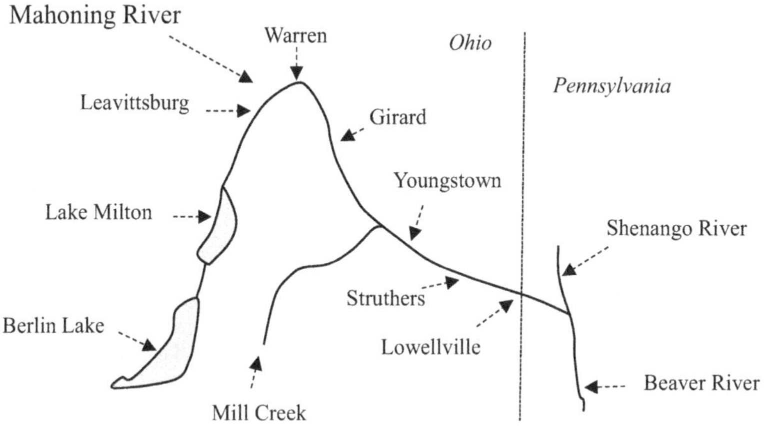

Fig. 7.1 Mahoning River and Low-Head Dams

unimpeded river, a culture best represented by the more youthful kayakers. This group sees the low-head dams as impediments to that cleaner more vigorous future. They do not view the dams as picturesque remains of a "not completely hideous past," but only as impediments. The second cultural group does not manifest the youthful and vigorous interactions with the river, but rather a more "laid back" attitude of preserving at least some of the more picturesque dams. This is an older cultural conflict based on using the small pontoon boats for summer afternoon leisure. The resolution of this cultural conflict is, however, predetermined by the necessity of removing the toxins from sediments below the water levels. Along the Mahoning, the past cavalier poisoning of the River determines the clear options for the future. Either clean the toxic contaminants from the River, or pay the high price of environmental and/or resident sickness.

References

Blue, F. J., William, D. J., William, H. L., & Joan, M. R. (1995). *Mahoning Memories: A History of Youngstown and Mahoning County*. The Donning Company, Viginia Beach, VA. ISBN 0-89865-944-2.

Baselt, R. (2008). *Dispositions of Toxic Drugs and Chemicals in Man*, Biomedical Publications, Foster City, CA.

Marino, J. (2022, October 3). *The Waters of the Mahoning River may be Cleaner, But are They Safe?* Ideastream Public Media.
Marino, J. (2022, October 10). *Upriver on the Mahoning, Trumbull County Residents say Hands Off Our Dam.* Ideastream Public Media.
Robinson, R. (2023). *Environmental Advocacy and Local Restorations.* Palgrave Macmillan.
Robinson, R. (2021). *Environmental Organizations and Reasoned Public Discourse.* Palgrave Macmillan.

Appendix: Toxic River Contaminants[25]

Introduction to the Contaminants of US Rivers

The restorations of US rivers require that a variety of toxic chemical substances be removed or remediated by containment methods. These toxic substances are commonly found in streambeds and/or streamside soils. Poisons such as PCBs or PAHs are carcinogens, and heavy metals are neurotoxins. These substances are usually legacy industrial wastes left behind by abandoned steel mills, or smelter facilities, or pulp and paper facilities.

The list of toxins reviewed here is not inclusive of all the contaminants found in US rivers. The most frequently occurring toxins are, however, included here: PCBs, PAHs, mercury, cadmium, chromium, lead, dioxins, and arsenic. We now live in an age when the poisonous nature of these chemical elements and compounds are known, but for many of these, their toxic nature was also known at the time of their original deposits and contaminations. For example, although the toxic nature of PAHs was unknown until recently, the neuro-health effects of lead poisoning has been known for centuries.

Polychlorinated Biphenyls (PCBs), Their Sources and Effects[26]

Polychlorinated biphenyls (PCBs) are man-made organic chemicals of combined carbon, hydrogen, and chlorine atoms. They have been identified by the EPA as human carcinogens. The arrangement of the chlorine atom in the PCB molecule determines its chemical properties and toxicity.

[25] This appendix is largely taken from Robinson (2023).

[26] See www.epa.gov/pcbs/learn-about-polychlorinated-biphenyls-pcbs.

PCBs have no taste or smell. They were domestically manufactured from 1929 until manufacturing was banned in 1979. Their consistencies range from thin lightly-colored liquids to yellow or black waxy solids. They are non-flammable, and chemically stable with a high boiling point. In particular, they have very good electrical insulating properties. PCBs were used in hundreds of industrial and commercial applications that include:

- electrical, heat transfer and hydraulic equipment,
- platsticizers in paints, plastics and rubber products,
- pigments, dyes and carbonless copy paper, and
- a variety of other industrial applications.

The products that may contain PCBs include:

- transformers and capacitors,
- voltage regulators, electrical switches, and electromagnets,
- hydraulic oil and motor oil,
- fluorescent light ballasts,
- cable insulation,
- thermal insulation,
- adhesives and tapes,
- oil-based paints,
- caulking materials,
- carbonless copy paper,
- floor finishing products.

PCBs enter the air, water, and soils during their manufacture and transportation. They can be released into the environment from poorly maintained hazardous waste sites, illegal dumping, leaks from electrical transformers, or the burning of some wastes in municipal or industrial incinerators. Once in the environment, PCBs do not readily break down. They can travel long distances in the air or water.

A small amount of PCB may dissolve in water, but most stick to particles or soil or sediments that also include organic carbons. Each PCB has a different degradation time with lighter mixtures degrading faster. The heavier PCBs, however, are the ones often found. These are more persistent in the soils and sediments of rivers. Their rate of deterioration is slow.

PCBs in the environment are absorbed by animals and a few plants. Bioaccumulations occur.

in both wild populations and in animals raised for food in backyard operations. The rate of bioaccumulation and therefore concentration of PCBs are impacted by the amount of fat in the animal's body. In general, organisms accumulate higher concentrations if they have higher amounts of body fat since the PCBs are found in that body fat. Once PCBs have entered the body, they are slow to leave it. They typically stay for the life of the animal. Hence, this "bioaccumulation" reaches concentrations in the body thousands of times greater than in the water or soils from which they are absorbed. These high concentrations lead to consumption advisories for fish and waterfowl in the contaminated watershed.

In contrast, most plants do not accumulate PCBs due to their waxy layers which binds the PCBs and prevents them from being absorbed into the plant. Some plants in the squash family do accumulate PCBs from the soil through their roots. Tomatoes also can absorb air-born PCBs through their leaves. Generally, however, most PCBs remain on the surface of fruits and vegetables, often as part of the soil deposited on the plant from rainwater splash.

PCBs have been found to cause a wide variety of adverse health effects including cancers. There are a significant number of non-cancerous effects including impacts to the immune system, reproductive system, nervous system, endocrine system, and other organs. The data indicates that PCBs are probable human carcinogens, and the EPA has classified them as such.

Concerning the reproductive effects of exposures, PCBs were found to reduce the birth weight, conception rate, and the live birth rates of monkeys. PCB exposure was also found to reduce sperm counts. Also, children born to women who worked with PCBs in factories showed decreased birth weight and there was also a significant decrease in gestational age with increasing exposures to PCBs.

The effects of PCBs on nervous system development have also been studied. Newborn monkeys exposed to PCBs show persistent and significant deficits in neurological development including visual recognition, short-term memory, and learning. Studies in humans have suggested effects similar to those observed in monkeys exposed to PCBs including learning deficits. Elevations in blood pressure, serum triglycerides, and serum cholesterol have also been reported with increased levels of PCBs in humans.

Polycyclic Aromatic Hydrocarbons (PAHs), Their Sources and Effects[27]

Driveways and parking lots are common. They are generally either concrete or asphalt, and many of the latter are coated with a black, shiny sealant—usually with the commercial names of "Sealcoat," "Pavement Sealant," or "Driveway Sealer." These are marketed as improving appearance and increasing longevity. These sealers are frequently constituted from *refined coal tar* of the sort usually developed at steel mills. These refinements are significant sources of *polycyclic aromatic hydrocarbons* (PAHs). These hydrocarbons are known carcinogens and cause a variety of environmental damages.

The significant sealcoat products marketed today are either from refined coal-tar or asphalt emulsion. Coal-tar pitch, a known human carcinogen, is a residue of the distillation of crude coal tar, a byproduct of coking coal. It contains approximately 200 PAH compounds. Most coal-tar-based sealcoat consists of 20–35% pitch. Asphalt, however, is a residue from the distillation of crude oil, and is used in asphalt-based sealcoat products. PAH compounds are 1,000 more present in coal-tar-based sealants than in asphalt type sealants.

In the US, coal-tar-based sealants are primarily used in the East; asphalt-based sealants are primarily used in the West. Coal-tar-based sealants are also used in Canada. These geographic differences reflect the location of steel mills which were primarily constructed in the eastern and central US. Steel mills use coke, and as a byproduct they refine coal-tar pitch. Since 2000, PAHs have been found to be increasing in many urban lakes, while other toxins such as lead and PCBs have been decreasing. But in 2003, the City of Austin in Texas found that samples from small tributaries and drainages in largely residential areas have elevated PAH levels. The source was found to be the coal-tar-based particles and dust eroded from the pavements of parking lots. Rainwater washes this dust and particles into the drainage network, and then into streams and lakes.

PAHs compose a large group of organic compounds composed of two or more fused benzene rings that occur in various arrangements. PAHs occur naturally in coal and petroleum products and are also formed by the incomplete combustion of organic matter such as fossil fuels, wood, or tobacco. Some common sources include used motor oil, automobile

[27] See Mahler et al. (2012) for a significant study of these compounds. This appendix is based on their study.

exhaust, industrial emissions, tire particles, and asphalt. Of all possible sources, coal tars and creosote have the highest concentrations of PAHs.

Tires and snowplows combine with severe temperature changes to erode sealed surfaces. Studies find that overall, the annual loss of sealcoat from parking lots averages about 2.4% of the total sealcoat applied. It becomes dust that is carried away with storm runoff, and in many locales, it is washed temporarily into containment ponds and then into rivers and lakes. It causes tumors and other problems in fish. The EPA classifies several PAH compounds as probable human carcinogens. The dust can be breathed in, and PAHs are readily absorbed through the skin. In addition to Austin in Texas, an increasing number of municipalities have banned coal-tar-based sealants. Also, these products are increasingly being dropped from hardware chains.

Mercury Poisoning

Mercury (Hg, atomic number 80) is a naturally occurring chemical element found in rock including coal. It exists as elemental (metallic) mercury, or inorganic mercury compounds, or methylmercury and other organic compounds. Metallic mercury is a shiny silver-white metal (quicksilver) and is liquid at room temperature. It was previously used in thermometers, fluorescent lightbulbs, and some electrical switches. At or above room temperature the liquid metallic mercury can evaporate to become an invisible, odorless gas. Mercury combines with chlorine, sulfur, or other elements to form inorganic salts which can be transported in water or exist in soil. Dust that contains these salts can enter the air from mining. But emissions of metallic or inorganic mercury also come from coal-burning power plants. Inorganic mercury compounds are also being used in soaps and skin creams for skin lightening.

Human exposure to inorganic mercury salts occurs in occupational and environmental settings such as mining, electrical equipment manufacturing, and chemical and metal processing. But microscopic organisms can combine mercury with carbon which converts it from inorganic to organic form. Methylmercury is the most common organic mercury compound found in the environment. It is highly toxic. It usually emerges from coal burning to become airborne. In the US, power plants that burn coal to create electricity are the largest source of emissions; they account for approximately 44% of all man-made mercury emissions.[28] Burning iron ore, coke, and limestone in electric arc furnaces to produce steel are also sources.

Mercury poisoning is a type of metal toxicity. Mercury is a neurotoxin. Symptoms include muscle weakness, poor coordination, numbness in extremities, skin rash and pealing, anxiety, memory difficulties, and/or trouble with speaking, hearing, or seeing. Kidney problems and lower intelligence are some long-term effects. Methylmercury in the blood of developing babies and young children harms their developing nervous systems and affects their cognition abilities.

The most significant source of mercury poisoning is from consuming fish or shellfish that contains methylmercury in their tissues. Note that fish higher up the food ladder are more likely to be contaminated. Breathing the vapors from coal or coke burning from the manufacturing of pig iron or steel are also a more common cause of poisoning.

Birds and mammals that eat fish have more exposures to methylmercury than other animals in the ecosystem. Predators that eat these animals are at risk. Methylmercury is found in eagles and many other birds and animals. This toxin can cause death, but it also causes reduced reproduction capability and slower growth and abnormal behaviors.[29]

Cadmium

Cadmium (element Cd, atomic number 48) is a soft silvery-white metal similar in chemical properties to zinc and mercury. It is found in small quantities in zinc ores, and is a byproduct of zinc production. It has been used as a corrosion resistant plating for steel, and also for various pigments in paint and glass, and for nickel–cadmium batteries. These uses have all found replacements.

Cadmium makes up about 0.1 ppm (parts per million) of the earth's crust. (For comparison, zinc is 65 ppm.) There are no known minable ores for cadmium.

[28] See "2014 National Emissions Inventory, version 2, Technical Support Document," July, 2018, at https://epa.gov/sites/production/files/2018-07/documents/nei2014v2_tsd_05jul2018.pdf.

[29] See *National Park Service*, "Effects of Air Toxics/Mercury on Ecosystems," at www.nps.gov/index.htm.

Cadmium has no known biological functions in humans, other animals, or plants. It is toxic. It is a carcinogen in humans and also causes renal disease, atherosclerosis, hypertension, and cardiovascular disease.[30] Previous to World War II, zinc mining contaminated the Jinzu River in Japan with cadmium, which then accumulated in rice to poison some agricultural workers and consumers, primarily post-menopausal women who otherwise suffered from low iron. The cadmium poisoning caused a painful joint disease. Cadmium can also be absorbed by smoking tobacco, or from consuming some root vegetables (beets or potatoes) or from consumption of crustaceans, mollusks, frog legs, or fungi.

Chromium

The steely-grade, lustrous, hard, and brittle metal element chromium (Cr, atomic number24) is valued for its high corrosive resistance and its hardness, the third hardest element after carbon and boron. Added to steel, chromium produces the alloy stainless steel, a metal that does not tarnish or discolor.

Industrial production of chromium is from chromite ore which is used to produce the iron-chromium alloy—ferrochromium. This is used to produce the alloy stainless steel. But pure chromium metal is produced by heating and leaching chromite to separate it from iron.

Chromium is the twenty-first most abundant element in the earth's crust with an average concentration of 100 ppm. Chromium compounds are found in the environment from the erosion of chromium containing rocks, which are then redistributed by volcanic disruptions.[31] In the US, trivalent chromium is considered an essential nutrient in humans necessary for insulin, sugar, and lipid metabolism. While trivalent chromium is considered non-toxic, hexavalent chromium is a toxic carcinogen. Abandoned chromium production sites often have hexavalent chromium and therefore require environmental cleanup.

The biologically beneficial effects of chromium are debatable.[32] Chromium is accepted by the US National Institutes of Health as a trace

[30] See "ARL: Cadmium Toxicity" at www.aritma.com, and "Cadmium Exposure Can Induce Early Atherosclerotic Changes," in *Wayback Machine, Medinews Direct*, September 7, 2009.

[31] See, Emsley (2001, pp. 495–498).

[32] See Vincent, JB (2013, Chapter 6), and also Maret, Wolfgang (2019, Chapter 9).

element for its effects on insulin, and the metabolism of carbohydrates, fat, and protein.[33] In contrast, hexavalent chromium is highly toxic. Ingestion of hexavalent chromium dissolved in water is linked to stomach tumors. Industrial use in rubber, stainless steel manufacturing, chrome plating, dyes for textiles, tanneries, and other uses are known to contaminate water resources. After hexavalent chromium is ingested, it reaches the blood stream, damages kidneys, the liver, and blood cells through oxidation reactions. Renal and liver failure results. Also, chromate dust is a known carcinogen. It is also particularly toxic to fish because it is easily absorbed through the gills. Because of its role in dyes, paints, and tanning compounds, hexavalent chromium is often found in the soil and groundwater at active or abandoned industrial sites such as old steel mills. These sites need environmental cleanups. Primer paint containing hexavalent chromium is, however, still used in aerospace and automobile finishing.[34]

Lead[35]

The chemical element lead (Pb, atomic number 82) is a dense heavy metal that is soft and malleable with a low melting point. Lead is silvery gray when freshly cut, but in air it tarnishes quickly to a dull gray. Some heavier elements become lead after nuclear decay. When ingested or breathed even in small amounts, lead is toxic especially for small children. It is also toxic for the environment.

At 20 ppm in the Earth's crust, lead is in the upper third of abundance among the elements. This abundance, high density, malleability, and low melting point has resulted in its frequent use for batteries, pipes, bullets and shot, weights, solders, paints, radiation shielding, and leaded gasoline. But lead is a neurotoxin that accumulates in soft tissue and bones in humans. It damages the nervous system and interferes with the function of biological enzymes, causing neurological disorders that range from behavioral problems to brain damage. It also affects general health

[33] See "Chromium," Office of Dietary Supplements, US National Institutes of Health, 2016, at https://ods.od.nih.gov/factsheets/Chromium-HealthProfessional/#h2.

[34] See Baselt, Randall (2008, pp. 305–307).

[35] This section is partly based on several sources that include an excellent wiki review at https://en.wikipedia.org/wiki/lead and www.chemistryexplained.com/elements/L-P/Lead.html and www.dcd.gov/niosh/topics/lead/health.html and www.epa.gov/dwreginfo/lead-and-copper-rule plus the sources as cited below.

through cardiovascular and renal systems. Lead has no confirmed biological role, and there is no confirmed safe level for exposure.[36] A 2009 study indicated that lead "may cause adverse mental health outcomes."[37] Continued exposure may result in bioaccumulation.

Lead is highly poisonous whether inhaled or swallowed. It affects nearly every organ and system in the body. Most ingested lead is absorbed into the blood stream, and airborne levels of lead are immediately dangerous to life. The primary cause of toxicity is due to its interference with the proper functioning of enzymes.[38] Lead can cause severe damage to the brain and kidneys. By mimicking calcium, lead can cross the blood–brain barrier to degrade neurons and interfere with neurotransmission routes and neuron growth.[39] In a child's developing brain, lead interferes with synapse formation in the cerebral cortex. Early exposure causes increased risk of sleep disturbances and excessive daytime drowsiness.[40]

In animals, lead is toxic to the renal system, the cardio system, and the reproduction system. Bioaccumulation occurs up the food chain. Industrial emissions, especially from coal powered plants, cause environmental destruction due to lead. In the mid-1980s, the US environmental regulations reduced lead entering the environment, especially through leaded gasoline and power plants after control devices were required. Blood levels of lead declined after the 1980s.

Dioxins[41]

Dioxins are a group of toxic chlorinated compounds that are *persistent organic pollutants* because they are very stable and last in the environment for lengthy periods. Consequently, they accumulate in the food chain, and once ingested by an animal, they are very slow to leave it. In humans, they have a half-life of seven to nine years. This broad class of polychlorinated compounds have differing toxicity that depends on the position of the chlorine atom in the molecule.

[36] See Bouchard et al. (2009).

[37] Ibid.

[38] US Food and Drug Administration (2015, p. 42).

[39] See Rudolph et al. (2003, p. 369).

[40] See Liu et al. (2015, pp. 1869–1874).

[41] See www.epa.gov/dioxin/learn-about-dioxin.

Dioxins are highly toxic and cause (i) cancer, (ii) problems in reproduction and development, (iii) damage to the immune system, and (iv) interference with hormones. Throughout the world, dioxins are found in the environment where they accumulate in food chains and concentrate in the fatty tissues of animals. According to the EPA, more than ninety percent of human exposure occurs through ingestion of animal fats from meats, dairy products, fish, and shellfish.

Several hundred dioxin chemicals exist. As a collection, they are members of three closely related families:

- polychlorinated dibenzo-p-dioxins (PCDDs),
- polychlorinated dibenzofurans (PCDFs), and
- some polychlorinated biphenyls (PCBs).

Only those PCDDs, PCDFs, and PCBs that have the chlorine atoms in specific positions are toxic.

PCDDs and PCDFs are not created intentionally, but can be produced as a byproduct resulting from human activities such as the backyard burning of trash. Natural processes such as forest fires also produce PCDDs and PCDFs, but PCBs are manufactured products that are no longer produced in the US. Dioxins, however, are a byproduct of waste incinerations or the burning of fuels such as coal. They are also produced by chlorine bleaching of pulp and paper products, and even by cigarette smoke.

Pure dioxin appears as white crystalline needles. In the environment, it is generally attached to soils and dust particles and is therefore not perceived by the eye. Dioxin is not produced or used commercially in the US. But it is a contaminant formed during the production of some chlorinated organic compounds that include herbicides. Over the past thirty years, the EPA and industry have reduced the production of dioxin and its release into the environment. Although the environmental levels of dioxin have decreased in the past thirty years, dioxins are extremely persistent compounds that breakdown very slowly. A large part of the current exposures to dioxins are due to releases that occurred decades previously.

Dioxin can be deposited into drinking water through air emissions from waste incinerations followed by deposits into reservoirs, or into the water through the effluent from chemical factories.

Arsenic

The chemical element arsenic (As, atomic number 33) occurs in various minerals, usually in combination with sulfur and metals, but it also occurs in a pure crystalline form. It has a metallic appearance and has had important applications to industry. It comprises approximately 1.5 ppm of the Earth's crust, the fifty-third most abundant element. Prior to the discovery of DDT, arsenite of lime and arsenate of lead were widely used as insecticides. During the Bronze Age, arsenic was included in bronze to make a harder alloy. The primary use of arsenic is in alloys of lead. For example, it is used in car batteries and ammunition. Arsenic and its compounds, especially the trioxide, are used in the production of pesticides, treated wood products, herbicides, and insecticides. But these applications are declining with the increasing recognition of arsenic's toxicity.

Trace elements of arsenic are essential to the diets of rats, hamsters, goats, chickens, and perhaps other species, but there is no known role in human metabolism.[42] Arsenic poisoning occurs in all these species if quantities are larger than needed. Arsenic contaminated groundwater is a global problem that affects millions of people. The US EPA states that all forms of arsenic pose serious threats to human health. The EPA classifies arsenic as a "Group-A Carcinogen."[43] The US *Agency for Toxic Substances and Disease Registry* ranked arsenic at the top of its "2001 Priority List of Hazardous Substances at Superfund Sites."[44] Most arsenic refinement operations in the US and Europe have closed because of environmental concerns. Arsenic is found in the residual dust from copper, gold, and lead smelters, and is primarily recovered from copper refinement dust.[45]

The toxicity of arsenic to insects, bacteria, and fungi led to its use as a wood preservative.[46] In the 1930s, a process for treating wood with chromated copper arsenate (CCA) began as the most prevalent industrial use of arsenic. Discovery of its wide toxicity, however, led to its being banned

[42] See Anke (1986, pp. 347–372).

[43] See Didyendu and Datta (2007).

[44] See Carelton (2007).

[45] See Brooks (2007).

[46] See Rahman et al. (2004).

in the US and Europe. CCA is still used, however, in other countries.[47] Arsenic was also used in various agricultural insecticides and poisons. For example, lead hydrogen arsenate was a common insecticide for fruit trees until it was discovered to cause brain damage among the sprayers.[48] Other agricultural uses have been phased out except for cotton farming. But arsenic is still used as a feed additive in poultry and swine production to assist in weight gain and to prevent disease.[49]

During the eighteenth, nineteenth, and early twentieth centuries, arsenic compounds were used for treatment of syphilis, which is now treated with antibiotics. Arsenic trioxide has been used for centuries for the treatment of cancer, and it is still used for the treatment of some leukemia. Arsenic is also still used in some particular electronic applications.

Naturally occurring sources of human exposure include volcanic ash, the weathering of minerals and ores, and the mineralization of groundwater. Arsenic is also found in food, water, soil, and air, and as a result, arsenic is absorbed in all plants, but it is more concentrated in leafy vegetables, rice, apple and grape juice, and seafood. An additional route of human exposure is the inhalation of atmospheric gases and dusts.

In the US, arsenic is commonly found in the groundwater of the southwest. But significant amounts are also found in the groundwater in New England, Michigan, Wisconsin, Minnesota, and the Dakotas.[50] Increased levels of skin cancer is associated with exposures to contaminated drinking water even when it is below the standard of 10 ppb. The EPA finds that perhaps twenty percent of private US wells exceed this standard. Bladder cancer has been associated with levels of drinking water below 10 ppb.[51]

References

Anke, M. (1986). Arsenic. In W. Mertz (Ed.), *Trace Elements in Human and Animal Nutrition*. Academic Press.

Bouchard, M. F. et al. (2009). Blood Lead Levels and Major Depressive Disorder, Panic Disorder, and Generalized Anxiety Disorder in US Young Adults. *Archives of General Psychiatry, 66*(12), 1313–1319.

[47] See Mandal and Suzuki (2002).

[48] See Peryea (1998).

[49] See Nachman (2005).

[50] See Welch et al. (2000).

[51] See Chu and Crawford-Brown (2006).

Brooks, W. (2007). *Arsenic: Minerals Yearbook 2007*. United States Geological Survey.

Carelton, J. (2007). *Final Report: Biogeochemistry of Arsenic in Contaminated Soils of Superfund Sites*. EPA, Washington, D.C.

Chu, H. A., & Crawford-Brown, D. J. (2006). Inorganic Arsenic in Drinking Water and Bladder Cancer. *International Journal of the Environmental Restoration and Public Health, 3*(4), 316–322.

Didyendu, S., & Data, R. (2007). *Biogeochemistry of Arsenic in Contaminated Soils of Superfund Sites*. EPA.

Liu, J. et al. (2015). Early Blood Levels and Sleep Disturbance in Preadolescence. *Sleep, 38*(12), 1869–1874.

Mandal, B., & Suzuki, K. T. (2002). Arsenic Around the World: A Review. *Talanta, 58*(1), 201–235.

Nachman, K. et al. (2005). Arsenic: A Roadblock to Potential Animal Waste Management Solutions. *Environmental Health Perspectives, 113*(9), 1123–1124.

Peryea, F. J. (1998). Historical Use of Lead Arsenate. *16 th World Congress of Soil Science*, Montpellier, France.

Rahman, F. A., et al. (2004). Arsenic Availability from Chromated Copper Arsenate (CCA) Treated Wood. *Journal of Environmental Quality, 33*(1), 173–180.

Rudolf, A. M. et al. (2003). Lead. *Rudolf's Pediatrics* (21st edn., p. 369), McGraw-Hill Professional.

United States Food and Drug Administration (2015). Q3D Elemental Impurities Guidance for Industry, US Department of Health and Human Services. p. 41.

Welch, A. et al. (2000). Arsenic in Groundwater of the United States: Occurrence and Geochemistry. *Ground Water, 338*(4), 589–604.

CHAPTER 8

The Very Old and Quaint Early Dams of New England

Introduction: An Old Low-Head Dam in Maine

Camden is a seaside resort town in Maine. It is located on the southwest shore of Penobscot Bay. Its population is only 5,232 according to the 2020 Census, but this population triples during the summer months due to tourists and summer residents. The Megunticook River flows through the center of Camden where it once powered gristmills, carriage factories, and other industrial businesses before the US Civil War. The River then flows into Canden Harbor where during the nineteenth century, its six shipyards launched ten to twelve large sailing vessels annually. Some of these vessels were among the largest under sail for that time.

The Megunticook River is three and one-half miles long. (Fig. 1.1, 1.2, and 1.3 of Chapter 1 maps this area.) It originates at Megunticook Lake and then flows through the remains of former mills and factories, and also over seven low-head dams (ones that are lower than fifteen feet from its base). The seven structures were all constructed during the nineteenth century for and by private interests. Three of these dams are still privately owned, but four (Knowlton Street Dam, Know Mill Dam, Powder Mill Dam, and Montgomery Dam) are now publicly owned. As previously indicated, the Megunticook River flows through the middle of Camden, and many of the old buildings (including the old mill buildings) are built partly on stilt-type supports extending up from and over

R. M. Robinson, *Restoring America's Rivers*,
https://doi.org/10.1007/978-3-031-81758-8_8

the riverbed. The Megunticook also flows completely under some buildings such as the Camden Deli which sits directly over the reservoir behind the Montgomery Dam, and just a few yards from the downstream Dam itself. Among the seven dams, the Montgomery Dam is the closest to Camden Harbor and Penobscot Bay. As with the other dams along the Megunticook, the Montgomery Dam supports a picturesque waterfall. Restaurants overlook the small reservoir behind the dam, and also overlook the waterfall the dam creates. The restaurant owners consider the reservoir and waterfall to be valuable assets. “The waterfall is a treasure. It is a treasure like the harbor, …” claims Tom Rothwell, co-owner of Camden Deli that is located directly over the small reservoir.[1]

In 2019, the town of Camden authorized a “feasibility study” by two engineering/environmental restoration companies. Upon studying the ecological effects (especially the fish passage), flood risk, and repair and maintenance costs, the resulting 2021 “Feasibility Report” recommended removal of the Montgomery Dam. This report depicts and illustrates the future redesigned and undammed Megunticook as flowing down a gentle slope past and around large granite boulders, so that the picturesque view of this flow would in some eyes be even more appealing than the current waterfall. The fish passage of the redesigned river would be enhanced, and without the dam, the possibility of flood damage to the nearby buildings would be mitigated. Dam removal and area restoration would also eliminate the Montgomery Dam’s maintenance and operational costs that are now paid by the Town of Camden.

The 2021 “Feasibility Report” of the Megunticook Watershed restoration, and the Montgomery Dam removal, was funded by the *National Oceanic Atmospheric Administration* (NOAA) and the *National Fish and Wildlife Foundation* (NFWF), the latter being a nonprofit e*nvironmental advocacy organization* (EAO). The opposition to dam removal is organized by *Save the Dam Falls Committee*. “We are happy with the beautiful waterfall the way it is,” said Deli Owner and Committee member Tom Rothwell.[2] The uncertainty posed by the area’s proposed change stimulates the opposition who want a Townwide ballot on the issue of dam removal. A quick vote of “no!’ would tie the hands of the *Select Board*

[1] See www.mainepublic.org/news/2021=08-01/opposition-grows-to-remove-a-200-year-old-dam-that-poses-flood-risk-for-camden.

[2] Ibid.

(the governing body of Camden), and prevent the Montgomery's deconstruction. Opposing this "preserve the dam" movement is the *Restore Megunticook* organization. They want the vote delayed so that citizens have time to become familiar with the proposed vision of the restored area.

The issue of dam removal is, however, not remote nor abstract. The floods of downtown Camden that occurred in January of 2024 were severe. Global climate change now causes stronger storms with greater precipitation amounts. The low-head Montgomery Dam and its reservoir backs much more water into downtown Camden than would occur if the dam were removed. Low-head dams with their waterfalls may be quaint and picturesque, but free-flowing water is also picturesque. Low-head dams, however, can be dangerous during storms, and they do block fish passage and also require significant maintenance costs.

The Public Management of the Megunticook River

Our rivers are generally considered as public property for which water extractions, or any sort of waste dissipation, or building within the river's confines, are highly regulated. Dam constructions are usually only public projects. But the colonial era and nineteenth century were different. Areas like Maine were either wilderness or at least semi-wilderness within which public regulations were lax. "Grandfathered rights" for those who built over the stream in the nineteenth century are apparently still existent in Maine. Nevertheless, the notion of "direct democracy" via a quick ballot measure to prohibit a dam removal appears to be manipulative of the very notion of democratically established legal and regulatory procedure, especially when it is organized and led by those who appear to directly benefit by the prohibition.

Positive externalities will be created by the Megunticook's restoration. The redesigned river will likely be appealing to view; the sound and sight of flowing water will remain. These external amenities will exist for the general public and the local businesses. Negative externalities are certainly generated by the current Montgomery Dam in that fish passage is prevented and flood problems are exacerbated. The "Feasibility Report" indicates that the restoration is clearly necessary, but the onus of change requires that the public officials (Camden's "Select Board") must promote the *vision* of the undammed Megunticook River and its restoration. This is required to overcome the nostalgic desire to retain the old and quaint structures, and to move to an ecologically sustainable river. This is a common general problem for the dam removals that are necessary for restoration of the rivers of the Northeast US. The dams tend to be low-head and in many cases, old and picturesque. These dams and reservoirs recall the childhood adventures of the viewers, as well as a nostalgic past that is in reality distantly past. The *vision of what could be* poses a preferable future, but this *vision* must be promoted. This means that environmental leadership is required, but this has always been the case in the environmental movement.

The Early Industrial Dams of Eastern Massachusetts

Massachusetts is a state of small area but high in population density. It is located on New England's Atlantic Coast, and was one of America's original thirteen colonies. In area, it is the forty-fourth largest of fifty states, but it is the sixteenth largest in population at approximately seven million people. Most of this population (4.4 million) live in Eastern Massachusetts (roughly east of the City of Framingham). This old State exudes a strong sense of colonial history with centuries old buildings, cemeteries, monuments and Revolutionary War sites that often seem omni present. The rivers of Eastern Massachusetts were the sites of the original textile mills of America's Industrial Revolution which began on the Merrimac River in the Cities of Lowell and Lawrence. Eastern Mass was also the location of America's original fishing and whaling industries with Hawthorne's Salem, Kipling's Gloucester, and Melville's New Bedford all located there. America's first environmental movement began at Thoreau's Walden Pond in Concord. It is also the site of America's first iron forge built on the Saugus River, just north of Boston. But now the entire area is densely populated and either urban or suburban.

The City of Boston is New England's largest city. It is a seaport on Massachusetts' central Atlantic Coast. The Merrimac, Parker, Shawsheen, Concord, and Ipswich Rivers drain the area immediately north of Boston. The Charles and Mystic Rivers drain the areas of Boston and Cambridge (located on the south and north banks of the Charles River respectively) and its immediate environs. The Neponset River and numerous smaller rivers drain the area close to the south border of the City. The Taunton River drains southeastern Massachusetts into the Narragansett Bay which is located in the State of Rhode Island directly to the south of Eastern

Mass. (Figs. 8.1 and 8.2 indicate these locations). To use the terminology of the Industrial Revolution, all of these rivers were converted to "working bodies." They were all dammed to provide power to meet industry's demands. But all of these dams have now lost their original purpose, and all pose environmental and safety problems. All have *environmental advocacy organizations* (EAOs) and government agencies seeking either their total removal or reconfiguration. These efforts form an important and necessary component of America's dam deconstruction movement, i.e. deconstruction of decrepit but sometimes scenic impediments to our rivers' restorations. But these dams are also part of Eastern Mass' colonial history and are consequently cumbersome to reconfigure or remove since numerous permits and reviews are required. The remnants of the colonial past must be preserved, or at least managed as though they actually embody the honored history.

From the early seventeenth through the nineteenth centuries, New England's proliferation of low-head dams created the reservoirs that powered mills and dissipated wastes. The Commonwealth of Massachusetts is now home to over 3,000 old dams that serve no purpose. They are artifacts of past industry that turned rivers into ponds of warm polluted water that destroyed fish runs and riverine ecology. They are now public safety hazards.[3]

Small-town Massachusetts often features picturesque waterfalls flowing over historic low-head dams, the remnants of colonial and early industrial heritage. These dams have long-lost their purpose in powering the colonial gristmills, or sawmills, or even the textile mills of the 1830s. But in this area, connections to the colonial and/or early industrial heritage is valued by communities. The ponds formed behind these low dams offer quaint landscapes, an important component of an old culture.

Advocates of dam removals in Eastern Massachusetts wish to restore the runs of millions of herring, shad, and other *diadromous* migratory fish that once penetrated its rivers to spawn. *Diadromous fish* migrate in either direction between salt and freshwater over their life cycle. *Anadromous fish* are born in freshwater, migrate down their river to mature in salt water, and then migrate back up to spawn. These include the herring, shad, steelhead trout, and Atlantic salmon. *Catadromous fish* are born

[3] See "This Spring, Fish Will Migrate Up This Massachusetts River for the First Time in 200 Years," NRDC, April 30,2018, at www.nrdc.org/stories/sspring-fish-will-migrate-massachusetts-river-first-time-200-years.

in salt water, migrate up their river to mature in freshwater, and then migrate back down the river to spawn in salt water. The American eel is *catadromous*. All these catadromous fish were once an important resource used for sustenance, trade, and fertilizer, but dams drastically eliminated these runs.

Removing these old dams would reconnect the ecosystems, and could assist in flood protection and the mitigation of the effects of global warming. Stagnant ponds will reach warmer temperatures than flowing water. The *environmental advocacy organization* (EAO) Ipswich River Watershed Association (IRWA), together with the Massachusetts Division of Marine Fisheries have now worked for decades to restore the Ipswich and Parker Rivers. These restorations are necessary components for the preservation of "The Great Salt Marsh," a spectacular coastal asset immediately north of Cape Ann (shown on Fig. 8.1). To this end, the IRWA seeks to remove the Larkin Mill Dam on the Parker River, and the Ipswich Mills Dam and South Middleton Dam on the Ipswich River. The IRWA also seeks to build appropriate fishways around the Willowdale Dam on the Ipswich and the Howlett Brook Dam on a tributary of the Ipswich. These projects are not easily accomplished because disturbing the past is not done quickly in this area of early colonial America.

Early History of Ipswich River's Dam

The Ipswich River Watershed is north of Cape Ann, i.e. north of Gloucester (shown on Fig. 8.1). It drains one-hundred and fifty-five square miles with a resident population of approximately 160,000 residents, or slightly more than an average of 1,000 people per square mile in a strictly dense suburban setting. This area includes the River, its tributaries, and a considerable marshy acreage. The River extends only thirty-five miles, and empties into Ipswich Bay at the town of Ipswich. The mouth of the Ipswich River partly forms the environmentally sensitive "Great Salt Marsh," an area that state agencies along with the *Ipswich River Watershed Association* (IRWA) are trying to restore and protect.

This watershed includes some of the oldest settlements in New England, i.e. the Puritan settlements that include the oldest "blue blood" families of the original Bay Colony.[4] Names such as Winthrop, Peabody,

[4] The Puritan migration from England occurred between 1620 and 1640. This migration primarily settled in what is now Eastern Massachusetts.

Lodge, Cabot, and Saltonstall all have considerable histories of influence and involvement in Ipswich and Essex Counties, the area of the Ipswich Watershed. These founding family names and descendants are still found in this old area of colonial heritage.

The original Ipswich Mills Dam was the first built in New England, and as such, it was an essential factor in colonial development. The Town of Ipswich (originally called Agawam) was established in 1634 by John Winthrop "the Younger," the son of the Massachusetts Bay Colony's first Governor. The Bay Colony was established in 1630. John the Younger and twelve compatriots sailed into what is now called Ipswich Bay in 1634, and quickly obtained a charter to establish the Town. They perceived that the nearby salt marsh was capable of generating the hay needed for livestock, and that the river was capable of powering a gristmill. This mill was constructed by the son of another Bay Colony founder, Richard Saltonstall, son of Sir Richard Saltonstall, one of the original migrating Puritans who found the Bay Colony with Governor John Winthrop.[5] The mill's dam was built from logs and stone at the Upper Falls located at the "run of the tide" of the Ipswich River, i.e. where the salt and freshwater meet at high tide.[6] This is the essential border that must be traversed by migrating diadromous fish. The Saltonstall family owned this mill until 1729. A sawmill was also quickly constructed on the bank opposite the gristmill in 1634. Other gristmills, sawmills, a fulling mill, dye houses, a bark mill, tanneries, and a hemp mill were all constructed by 1657.[7] With all this rapid colonial development, the area along the River where the mills were concentrated was called "Mill Garden." During the Revolutionary War, the British Colonial Administration closed the Port of Boston after the famous "Tea Party," and the Port of Ipswich consequently grew.

[5] The Saltonstall family included later mayors and governors, and also US senators and representatives. They were significant landowners in Massachusetts, and ten generations attended Harvard College between 1642 and 1959.

[6] The "run of the tide" is the river's upper reach of the high tide.

[7] *Fulling Mills* make woolen cloth. *Bark Mills* convert tree bark into powder for tanning. Hemp Mills convert hemp fiber into cloth or rope. *Gristmills* grind cereals into flour.

In 1827, the mills' dam was rebuilt as a more substantial stone structure.[8] The reconstruction was for the Ipswich Manufacturing Company which produced textiles, the first such production in the Ipswich area. The reconstruction allowed the height of the dam to be adjusted by adding heavy wooden boards at the top so that the stone structure remained low. By the 1830s, Benjamin Hoyt was operating a sawmill powered by this Dam. He also purchased the right to build another sawmill close by, and also the right to lease the waterpower to others. This new sawmill was operated until 1858. In 1868, Amos Lawrence purchased the mill and associated structures and renamed them "The Ipswich Mills Company." He converted the textile cloth manufacturing into making stockings, the largest such manufacturing in the US at the time. This stocking production lasted until 1928. But in 1941, Sylvania purchased the mill buildings and dams to produce World War II-related proximity fuses, plus military and commercial transformers, and tungsten coils. Finally, in 1995, EBSCO Publishing purchased the buildings and the Ipswich Mills Dam and renovated all the facilities.

For the century and more following the construction of the Ipswich Dam, the anadromous fish (shad, the herring called alewives, and salmon) substantially declined. These fish were once valuable resources for the colonists. They were salted or smoked for preservation to provide sustenance and trade. They also fertilized crops. The colonial legislature required fish passages be constructed at all dams, but in 1745, the dam operators were able to pressure the legislature to eliminate the passage requirement. But in 1773, the legislature again required "fishways" of six feet in width, and four feet above the rivers' bottoms. They were required to have gravel and rock on both sides of the bottom, and flat rock sills. With the industrial revolution, however, enforcement of the law was weak, and the herring remained largely depleted. In 1995, fishway passages were again required by the State.[9]

[8] For the history briefly reviewed here, see https://archive.org/stream/ipswichmassac00wisegoog#page/n666/mode/2up, or https://historicipswich.wordpress.com/ipswich-mills-and-factories/.

[9] See Mcquaid (2020, July) for a review of this history at https://historicipswich.net/2023/07/24/alewife-ipswich-river-2/.

The Sad Example of Suburbanization

The deterioration of the environment of the Ipswich River watershed largely occurred because of the post-World War II suburbanization of the Boston area. The population growth of the Town of Ipswich provides an example as presented by Table 8.1. During the decade of the 1950s, the Town's population grew by twenty-four percent; during the decade of the 1960s it grew by another 26 percent. During the three decades between 1970 and 2000, it grew an additional twenty-one percent. During the 1950s and 1960s, the expansion of paved roads was considerable with Route 128 (now Interstate Route 95) and Interstate 93 being completed early in the 1960s. State Route 28 became the epitome of a north–south artery from the inner city of Boston to the suburbanized strip malls that serviced the new real estate developments. These became the major transportation arteries to the North Shore area of Essex and Ipswich Counties; all of these roads and developments penetrated the Ipswich River watershed. The Ipswich River Watershed Association began in 1977 as a natural reaction to try and control what clearly was uncontrolled sprawl.

The various stories that explain our moral motivations for involvement in our *environmental advocacy organizations* are likely to be similar to the personal story of any reader of this volume, and also to my own story. At ten years old, I and my boyhood friend rode bicycles, with fishing rods in hand, north along Route 28 in Reading, Massachusetts. We then diverged onto back roads to reach sections of the Ipswich River to trout fish. (There actually were trout and other desirable fish in the Ipswich in the late 1950s.) We took this route several times each summer from 1957 to 1959.[10] Over the years, we watched the deterioration and degradation of the River until our favorite sections became nothing more than suburban drainage ditches filled with tires, shopping carts, and other

Table 8.1 Town of Ipswich population

Year	*1950*	*1960*	*1970*	*2000*
Population	6,895	8,544	10,750	12,987

[10] My boyhood friend became a civil engineer for the US *Environmental Protection Agency*.

debris. I have traveled many times since to the areas where we once stood, young boys with fishing rods in hand looking for trout. But now, I cannot stand on the banks where we once fished because these formerly special places no longer resemble riverbanks. I can only now stand on the asphalt of private parking lots and other illegally paved surfaces that were once public domain. I can now look at lands that were illegally seized by those who owned property up to the riverbank, but who sought to extend their property-lines beyond the river's marshy borders. Even to today, the seizure of what is, or should be, common property appears to be socially easier than recognizing and collectively pursuing the benefits of common preservation. Perhaps this was a natural reaction for people moving from a crowded city to a roomier suburban paradise. But the consequence is that now the water is too polluted to support the aquatic life of trout. The once natural drainage is now more of a polluted swamp than a free-flowing natural stream, which it once was. But we can all recount similar stories! They motivate us to try to restore and preserve what is left.

The Ipswich River has become a sad example of the tragedy-of-the-commons. It was once a picturesque trout stream. It once fed a marshy coastal area that offered considerable beauty. Sixty years previous, it was still an environmental asset; today it could still be a meandering free-flowing stream that adds considerable sight value to the areas it flows through. It can still be restored so that it will support a healthy *Great Saltwater Marsh.* (Fig. 8.1 indicates this Marsh.) But today it is largely a filled-in quagmire destroyed by colonial development, followed by early industrial development, and then by the real estate development of the last seventy years. But the area's real estate values are below what they could be with an effective restoration. This last seventy years of development happened without the coordinated efforts that would control the environmentally degrading suburban sprawl and its pollution. As a result, the Ipswich River Basin is an example of the "commons phenomena." It occurred despite legal prohibitions as to destruction of the drainage and alteration of the river route, legal prohibitions that were ignored by the local governments responsible for their enforcement. Currently, however, the Ipswich *River Watershed Association* (IRWA) is attempting restoration by purchasing land along the river to protect the ecology of the watershed, by controlling the suburban runoff, by restoring tributary connectivity and drainage, and by restoring the legal riverbanks that have been filled in. Most importantly, however, the IRWA seeks the removal of the dams and culverts that make the Ipswich River stagnant and without its fish runs.

THE EFFORTS OF THE *IPSWICH RIVER WATERSHED ASSOCIATION* (IRWA)

Dam removals are only one necessary component of the preservation and restoration efforts of the IRWA. Its overall main efforts are in ***habitat preservation*** and associated ***land use influence***. For these efforts, the IRWA tries to influence policies and actions among the twenty-nine municipalities within the watershed in order to create and protect their natural areas. The following explanations review four overlapping categories of the IRWA's efforts:

1. Net Zero Policy: This is also called a "Water Neutral Growth Policy." There are thirteen municipalities within the watershed that use Ipswich River water for its municipal supply. "Water neutral growth" requires establishing regulations that reduce water uses for actions such as lawn watering and car washing. But it also requires quick identification and fixing of leakages. Nonetheless, population expansion within the watershed is the obvious major factor influencing water use so that land-use policies that limit population growth within the area is the IRWA's significant target.
2. Habitat Protection: Natural areas need connectivity so that water can flow and wildlife can migrate between them. There are seventy dams and five-hundred culverts and bridges within the watershed, and they break or reduces this connectivity. IRWA seeks to remove and/or modify these physical barriers to improve connectivity within the watershed. For examples, sunken or collapsed culverts need to be rebuilt, or replaced, or widened.
3. Community Resilience Building: As financed by the "Massachusetts Municipal Vulnerability Preparedness Program," the IRWA holds "community workshops" that identifies weather-related vulnerabilities due to climate change factors such as drought, flooding, sustained high heat, extreme storms, and public health problems. The "workshops" explain the state funding sources for the community's infrastructure developments that are intended to resist the effects of climate change. Land-use policies should create and

protect marshes and floodplains for high-flow periods, and develop riparian buffer zones that filters water drainage from impervious areas and that stabilizes riverbanks. (Chapter 7 reviews the design, construction, and resulting benefits of establishing riparian zones.) Explaining these important flood mitigations and buffer zones is part of the IRWA's community outreach efforts.

4. Municipal Services Advocacy: IRWA advocates for enforcement of cost-effective water conservation and stormwater management. In this effort, it advocates for low-cost landscaping that is designed to promote the above reviewed actions.

Within the above categories, the IRWA's advocacy for dam removals, plus the establishment of fishways and culvert remediations pose its more significant and immediately consequential efforts.

Ipswich Mills Dam Removal?

The Ipswich Watershed Association began its analysis and promotion of the Ipswich Dam Removal in 2010. In that year, the Town's "Board of Selectmen" voted to proceed with the analysis of the dam's removal. Grants from the Massachusetts Environmental Preservation Administration (MEPA) financed a pre-feasibility study in 2014 which first investigated the potential costs, methods, and effects of dam removal. At that time, the Massachusetts Office of Coastal Zone Management estimated that 6,900 cubic yards of sediment had accumulated behind the dam and would be released to flow downstream with dam removal. This necessitated analysis of the possible toxicity of the accumulated sediment.

The first sediment sampling was conducted by an engineering consulting company (Horsley Witten Group). The purpose was to investigate if the sediment flows might be toxic to the ecology of downstream areas once the dam was removed. In 2020, a full-feasibility study of dam removal was completed with more sediment sampling and analysis of the effects on downstream wetlands. This feasibility study also planned the final steps necessary for dam removal. As expected, given the benign nature of the upstream manufacturing the sediments were all found to be non-toxic.

Even under these benign circumstances, there were those who preferred that the quaint dam, waterfall, and pond be preserved. For example, the Mill Pond Preservation Association (MPPA) generally opposes deconstructing any of the old dams in Massachusetts. The MPPA questions the benefits of the ecological improvements, and generally favors the scenic value of the existing dams, ponds, and scenic waterfalls. Nonetheless, on October 16, 2023, the MEPA Office issued its permit for the Ipswich Dam removal. A final vote by the Town of Ipswich still remains before removal is permitted, but the sentiment of the Ipswich community supports the Dam's removal which should therefore take place sometime in 2024.

The South Middleton Dam is another colonial era construction that the IRWA seeks to remove from the Ipswich River. It blocks the upper reaches of the Ipswich from the diadromous fish migrations. Owned by the Bostik Company (maker of adhesives and flooring), it is a stone and wood beam dam without a functioning fish passage, and it blocks several upstream tributaries from fish migrations. These include more than 120-acres of important spawning grounds for herring. This dam is classified as one of "significant hazard" by the Massachusetts "Office of Dam Safety" because it has significant wear that must be remedied if the dam is allowed to remain. Due to the liability and expense of reconstruction, the Bostik Company has decided to deconstruct the low dam. The removal process has been lengthy as indicated by the time-descriptions shown below[11]:

- In 2006: Massachusetts Riverways funded a site reconnaissance study for dam removal.
- In 2014: Project Partners (Bostok Inc., IRWA, NOAA, Massachusetts Division of Ecological Restoration) began seeking funding for the removal design and implementation.
- In June of 2014: Design and engineering work began for dam removal.
- In February of 2018: Permitting began for the Ecological Restoration after the dam removal.
- In May 2018: Cultural Resources Survey and Archeological Survey of dam removal was conducted.
- In 2019: Massachusetts Environmental Policy Act (MEPA) Certification issued.

[11] For this time sequence, see Ipswich River Watershed Association at www.ipswichriver.org/south-middleton-dam/.

- In June 2020: Following the Massachusetts Wetlands Protection Act, the Middleton Conservation Commission issued a "notice of intent" to remove the dam.
- In December of 2022: The dam's upstream sediment was sampled for toxicity. The results showed negligible toxicity.

Note that funding was received in 2014 from the National Fish and Wildlife Foundation for the Bostik Dam's removal, but the dam still remains at the start of 2023. It might be removed in 2024.

In addition to the South Middleton Bostik Dam and the Ipswich Mills Dam, the Larkin Dam on the nearby Parker River is also in the process of removal. This was also originally built in 1630 to support a colonial pond and mill. It is now owned by the Town of Newbury, and the IRWA is leading the effort to organize and bring about this deconstruction. There is also both a fish ladder and culvert in need of rebuilding in the Ipswich Watershed, both being currently organized by the IRWA.

Undamming the Meandering Charles River?

The Charles River meanders slowly through a course of eighty miles. Along this course, it flows through twenty-three municipalities in northeastern Massachusetts. It originates at Echo Lake in the Town of Hopkinton, which is only twenty-six miles southwest from the mouth of the Charles River. But since it only drops three-hundred and fifty feet over its eighty miles of length, its flowing current is slow. It bends and abruptly turns in direction numerous times along its course. It has a three-hundred and eight square-mile drainage watershed which includes thirty-three lakes and ponds, thirty-five municipalities, and eighty small tributaries. It flows into Boston Harbor close to the City of Charlestown immediately northeast of the Back Bay section of Boston.[12]

Along the last few miles of its lower downstream, the Charles forms the border that separates the City of Cambridge from the "Back Bay" section the City of Boston; Cambridge is on the north bank and the Back Bay of Boston is on the south. It is important to note that there are numerous parks and public lands bordering both sides of the River along its entire

[12] The "Back Bay" is the area in Boston that borders the Esplanade of the Charles River.

course. Its two most downstream dams are the New Dam, which is close to the mouth of the River, and the Watertown Dam which is only a few miles upstream from the New Dam. The New Dam was reconstructed in 1978 to prevent salt water from flowing into the upstream estuary. Consequently, this dam blocks the migrations of diadromous fish.

There are four large universities along the Charles: Harvard, Brandeis, Boston U., and MIT. The latter institution situated its large quadrangle along the Cambridge side of the Charles. As a result, MIT overlooks the River's reservoir along the Charles Estuary, the banks of which have been designed to be the area's "Esplanade," i.e. a park area for walkers, bicyclists, rowers, sail-boaters, and other recreational boaters. This popular "Charles River Basin" is formed as the reservoir between the downstream New Dam and the upstream Watertown Dam, the City of Watertown being on the north bank of the Charles and to the west of Cambridge. This significant slow-flowing reservoir is a focal point for both cities, and in particular for the students of the significantly prestigious universities located on its banks, i.e. MIT and Harvard, the latter being on both the Boston and Cambridge sides of the River. This Basin is used for popular rowing regattas and small boat sailing. Students of the area's numerous other universities (such as Boston University, Northeastern University, and University of Massachusetts at Boston) also heavily utilize this reservoir. Consequently, transforming the Charles into a faster-flowing river via dam removal would pose a substantial change in the area's character. As a resource, the "Basin Reservoir" between the New Dam and the Watertown Dam would be substantially altered at a significant cost to its present users if the dams were removed.

There are nineteen dams along the eighty miles of the Charles River. Starting in the Colonial Era (1639), forty-three mills were eventually built along this River. In 1814, the first fully-integrated textile mill was constructed on its banks in Waltham.[13] As a result, the River's water has long been severely polluted with sewerage, industrial wastewater, and urban runoff. Toxic chemical plumes, liquids of unnatural colors, abandoned tires, submerged vehicles, and obnoxious smells have been so common that in 1965, a local band's song "Dirty Water" (by the Standells) became an anthem for the Back Bay Boston–Cambridge area.

[13] By "fully integrated," it is meant that the mill makes cotton thread, then cloth, then a final product.

After the 1972 Clean Water Act, the federal, state, and municipal governments began a major campaign to clean up the sewerage and other contaminations flowing into the Charles and then into Boston Harbor. There has been some success in this endeavor, but to date there is still sewerage flowing into the Charles from two of its tributaries (Stony Brook and Muddy River), and persistent problems from "combined sewer overflows" exist along the River's length. These problems become critical after heavy rains. The more severe persistent contaminations, however, stem from the road salts of winter, the lawn runoffs of summer, and the algae blooms endemic to the warming waters of the River's stagnant reservoirs. Global warming poses a substantial threat to the Charles River. Dam removals would help ameliorate the threat by cooling the water and preventing algae blooms.

The Charles River Watershed Association (CRWA) has been a long-term advocate for the environmental restoration of the River. The CRWA indicates that there are three functionless old and crumbling dams along the Charles in need of removal: the Watertown Dam, the Dam in South Natick, and the Eagle Dam in Wrentham. The Watertown Dam stretches 180 feet across the River. It is an aging concrete structure that was originally built to power an early nineteenth-century mill. It is now owned by the Massachusetts Department of Conservation and Recreation. In June of 2021, the Massachusetts Division of Ecological Restoration (DER) issued its feasibility study for removing this dam. The CRWA is now building local support for its deconstruction.

The Eagle Dam in Wrentham was also originally constructed in the early nineteenth century to power a gristmill. It is now without function and is owned by the Town of Wrentham. It is an earthen dam that is classified as "unsafe" by the DER because its spillway was damaged by a flood and not properly repaired. Also, the embankment has been overgrown by trees and other vegetation which weakens the structure. In April of 2021, a feasibility study indicated that the cost of removal would be less than the cost of repair, and that removal would decrease the threat of flooding in downstream communities.

The Charles River makes a wide bend as it flows through South Natick. A point on the bend was selected by Reverend John Eliot to become the first church in Eastern Massachusetts for "Praying Indians." At this "South Natick Meetinghouse," a dam was built in 1720 by John Swain whose family constructed and operated grist mills and sawmills into the nineteenth century. Then the Bigelow family constructed several more water-powered mills at the present dam's site. Up to the 1930s, manufacturing firms (Boston Rubber Cement Company, Boston Bedding Supply, Indian Spinning Mills, and Stewart Manufacturing) located on the dam's reservoir. Starting in the early twentieth century, however, the Charles River became popular in Natick for recreational boating and fishing. The mill pond above the Dam was used for swimming and picnicking. These activities have ceased in recent decades because of the contaminated water. The Dam continues, however, to be a scenic attraction, especially near the "Audubon's Broadmoor Wildlife Sanctuary" located upstream from the Dam.

Floods and ice flows, however, have threatened the Natick Dam's destruction since 1807 when the River's ice breached its wall. The north section of the Dam failed in 1917. In April of 1929, a severe flood breached the Dam and inundated all the downstream properties within 100 yards of both banks. This extensive destruction repeated again in 1932. A new dam design and construction was needed and this began in 1933. This included further development of the park along the reservoir. But in 2023, the ninety-year-old dam again requires either considerable renovation or deconstruction. Global warming means stronger storms, warmer water in reservoirs, and more dangerous algae blooms. Blooms of cyanobacteria (blue-green algae) found in the reservoirs of the Charles are poisonous to the dogs who are walked along the reservoir's park, and also to the humans who boat on the reservoir. Dam removal would also mean cutting sixty trees growing on the earthen embankment of the Dam, and these trees have defenders among local residents. But dam removal would eliminate the maintenance costs, plus the threat of breaching, and also the poisonous algae blooms.[14] Dam removal would also again establish fish passage. Reconstructing the dam, however, would maintain the Town's picturesque landmark for which many residents have nostalgic memories. The Town estimates that the cost of reconstruction would be $2.6 million, but the cost of removal would be $1.5 million.[15]

On November 10, 2022, Natick's "Board of Selectmen" (the Town's governing body) voted four to one to remove the Charles River Dam

[14] See Sudborough, Susannah (22022, July 14) at www.boston.com/news/local-news/2022/07/14/natick-dam-repair-or-remove/.

[15] See Reale (2022, July) at www.wgbh.org/news/local/2022-11-10/natick-will-remove-its-90-year-old-dam-after-contentious-process.

in South Natick.[16] Town officials, however, recognize that it will take years of applying for permits and arranging for engineering studies and contractors before the dam is removed. At the quickest pace, it will be 2026 before the dam is removed.[17]

Concerning the problems of removing dams in Massachusetts, a local commentator observed,

> The Natick dam is like a hundred other dams in Massachusetts, according to a 2011 state audit. They're owned by towns, they're in poor or unsafe condition – and if they fail, they could cause loss of life or major property damage. There have been 67 dam removals in Massachusetts since 1999, accelerating in recent years. Removal remains an uncommon feat, but those dozens of projects offer .... the state's Division of Ecological Restoration a stack of case studies when they consult with towns.[18]

In the Heart of Yankee "Minuteman" Culture: The Shawsheen and Concord Rivers

Two tributaries of the Merrimack flow through the heart of Colonial Minuteman Territory of Eastern Massachusetts. At one time, the Concord and Shawsheen Rivers formed picturesque and marshy meadows as it flowed through the towns immediately north and west of Boston. These were the meadows and streams that spawned our nineteenth-century's "transcendental philosophy of nature movement" of Emerson and Thoreau. They have the potential to be restored to an ecologically sustainable and picturesque series of meadows with a clean meandering river.

The Shawsheen River is twenty-seven miles long. It originates in smaller streams in Lexington, near what is now the old Hanscom Air Force Base on the border of a much older Route 128 (now part of Interstate Route 95). It flows northerly through the towns of Bedford, Billerica, Tewksbury, and Andover before meeting the Merrimack in the

[16] See Beltran and Pescaro (2022, November 10) at www.nbcboston.com/news/local/natick-select-board-votes-4-1-to-remove-aging-charles-river-dam/2890624/.

[17] See "Project: Dam Removal & River Restoration," at https://natickma.gov/2036/Project-Charles-River-Dam-and-Park-Improvement.

[18] See Jeffrey Porter (2022, July 12) at https://www.wgbh.org/news/local-news/2022/07/11/the-fight-to-undam-the-charles-river-comes-to-natick.

old mill-town of Lawrence. Its watershed includes eighty square-miles and twelve towns with a combined population of 250,000. There are trails and parks along the Shawsheen's course. The *Shawsheen River Watershed Association* currently leads its restoration efforts. Because of stormwater runoff and combined-sewer overflows, the *Merrimack River Watershed Council's* monitoring resulted in the Shawsheen currently being classified as "impaired" under the Clean Water Act.[19] The Shawsheen had three significant dams in the Town of Andover: the Balmoral, the Marland Place, and the Ballardvale Dams. The first two of these were originally wool mills, and they were removed in 2017. The latter dam remains.

Andover is a very upper-crust town of 36,363 residents living in 11,305 households according to the 2020 Census. It is highly educated. Phillips Academy, one of the more prestigious preparatory schools (or "prep" schools) in the US, is located immediately south of Andover's town center. (President George W. Bush is a graduate of Phillips.) The median income in 2012 was $118,324.[20] Its population has steadily grown ever since its founding during the first half of the seventeenth-century's "Puritan migration."

Andover was found in 1642 by settlers migrating from the towns of Ipswich and Newbury located to the east. Andover was incorporated as part of the Bay Colony in 1646. In 1753, the Ballardvale Dam and Woolen Mill were constructed. On April 19, 1775, Andover sent three-hundred and fifty "Minutemen" to the battles of Lexington and Concord, and later sent "Minutemen" to the Battle of Bunker Hill, so its colonial identity is soundly established. On the dam's reservoir, a gun powder mill was constructed shortly after, and it produced powder for George Washington's Continental Army. In 1778, Samuel Phillips found Phillips Academy in Andover. Marland Place Dam (or Steven's Street Dam as it is often called) was also constructed in the "colonial era." The Balmoral

[19] See www.merrimack.org/publications/FinalReportShawsheenWatershedStormDrainOutletMappingProject0006MWI.pdf.

[20] As a young student at Central Catholic High School in Lawrence, the author of this volume often would take a bus from Lawrence Center to Andover Center. I would then walk the short distance to Phillips Academy, stick my thumb out and quickly obtain a ride down Rt. 28 to my older sister's house in Reading. As a student at Central Catholic, I always wore a suit coat and tie, and with a load of books under my arms and standing in front of Phillips (a residential Academy), I certainly did not appear to be a threatening figure. I was always quickly offered a ride.

Dam was then constructed as an ornamentation on William Wood's estate. Note that Wood owned the Ballardvale Wool Mill.

In 1838, the owners of the Ballardvale Dam were found liable for flood damage in Andover. In 1888, because of the dams' negative effects on fish runs, their removals were advocated by many of Andover's citizens; but the dams survived their protests. Periodically, however, floods did affect both the dams and nearby property. In 2006, historic flooding occurred on the Shawsheen when homes and roads were inundated. In 2008, the *US Fish & Wildlife Service* (USFWS) and Andover's *Conservation Department* advocated removal of all three dams. By that time, while the Ballardvale Dam was still privately owned, the Marland and Balmoral Dams were municipally owned by the Town of Andover so that their removals became likely.

In 2008, removal of all three dams in Andover became "priorities" for the Massachusetts *Department of Fish and Game* when $1.3 million in funding became available from state and federal sources. Removing these dams would open more than four miles of the Shawsheen to fish runs, and open sixteen acres of spawning habitat. The USFWS estimated that just removing the Marland Dam would reduce flood levels by 3.5 feet. In 2009–2010, sediment studies, plus an engineering feasibility study, and environmental studies were all completed, but removals of the Marland and Balmoral Dams waited until 2017. The Ballardvale Dam, however, still remains.

The Ballardvale Dam is jointly owned by Shawsheen Coating and Converting Company and the Shawsheen River Condos. The Condo's Board does not want removal because it considers the Dam to be a picturesque asset. Required repairs, however, are expensive, i.e. the currently recommended repairs will cost between $280,000 and $460,000. These ongoing repair costs, plus the threat of flood liability will likely force the removal of the final dam on the Shawsheen.

The extraordinarily historic Concord River is formed by the confluence of the Sudbury and Assabet Rivers near the Town of Concord's Center. It flows northerly for 16.3 miles to join the Merrimack River a short distance to the east of Lowell (downstream from Lowell). Like the Shawsheen River, the Concord River is a meandering stream that is generally of slow current, and it flows through picturesque marshy meadows. It has been dammed for more than two-hundred years at the Town of Billerica, approximately eight miles from the Concord's confluence with the Merrimack.

The Towns of Concord and Billerica were both settled by English immigrants during the "Puritan migration" of the mid-1630s. Both towns were incorporated as part of the Bay Colony in the mid-1650s. Billerica was later divided into four separate towns (Billerica, Bedford, Wilmington, and Tewksbury). On April 19, 1775, after the battle of Lexington, "Minutemen" from Billerica joined those in Concord to defend the Old North Bridge over the River from the British troops, and to then drive the British back to Boston. Concord was also the home of Thoreau, Emerson, and Hawthorne. Thoreau's Walden Pond is on the Sudbury River a short distance upstream from Concord's center. In 1849, Thoreau published his first book; it was about a canoe trip with his brother on the Concord and Merrimack Rivers. It is a classic in the body of environmental literature.

In 1711, Charles Osgood dammed the Concord River in Billerica to construct a gristmill for corn.

That dam was reconstructed in 1798 to divert water into the Middlesex Canal. The dam was again rebuilt in 1828. But in 1851, the Talbot Brothers purchased the Dam to power a woolen mill. It has been known as the "Talbot Mill Dam" ever since. Note that both Concord and Billerica are upper-income towns similar to Andover. As with the other dams of this region, Billerica's dam is low-head with a continuous picturesque waterfall.

The Talbot Mills Dam is an old stone structure that has no current purpose other than quaint nostalgia, but it is leaking and dangerous. If removed, it will enable herring, shad, and eels to migrate upstream to over 100-miles of riverbeds and ponds for spawning. Deconstructing this Dam is necessary to restore the Concord River's habitat and ecology. This removal project is led by the Merrimack River Watershed Council, plus federal agencies (NOAA and F&WS) and Massachusetts State agencies (Department of Ecological Restoration and Department of Environmental Preservation). The Dam's owner supports the deconstruction. Feasibility studies were begun in 2014 and indicate that removal is a preferred option. Engineering studies were completed in 2021. All the necessary public hearings have taken place. Funding from state and federal agencies are arranged, but the Dam still awaits deconstruction, which is tentatively planned for either Summer or Fall of 2024.

The Great Taunton Flood of 2005

The City of Taunton in Massachusetts was established in 1637 by members of the Plymouth Colony. Taunton lies approximately thirty miles inland from the original Plymouth Bay Colony settlement on Cape Cod Bay. It is also approximately forty miles south of Boston. The early seventeenth-century colonists established and immigrated to Taunton because it was located on a wide meandering river—the Taunton River—that flows south to a protected estuary and then to the abundant Mount Hope Bay which adjoins Narragansett Bay. Both provide abundant seafood and provide excellent port facilities.

The Taunton River flows southerly from its headwaters at the confluence of the Town River and Matefield River in Bridgewater, Massachusetts. The Taunton River is thirty-seven miles long, and has a large five-hundred and sixty-two square-mile watershed that includes a sizeable natural wetland called Hockomock Swamp (16,800 acres large) which is the largest freshwater wetland in the State. This watershed also includes two-hundred and twenty-one lakes and ponds, and an estimated one-hundred and seventy-three square miles of canoeable river. The Taunton River is probably the flattest river in New England since it drops only twenty feet over its thirty-seven-mile journey. Consequently, the River's tidal estuary reaches as far north as Taunton. The slow flow of this River did not make it amenable to damming. It is therefore the longest New England River without dams. In 2009, The Taunton River was designated a *Wild and Scenic River* by the US National Park Service.[21] This River, however, has eight significant tributaries: (1) Segregansett, (2) Three Mile, (3) Mill, (4) Cotley, (5) Winnetuxet, (6) Nemasket, (7) Assonet, and (8) Quequechan Rivers. These tributaries have more rapid flows and consequently are, or were, dammed.

In 1656, the first successful iron works south of Boston was established on the Three-Mile River.[22] It operated until 1876, and initiated other iron works in Taunton. The Taunton River and its tributaries were used to dissipate wastes from the iron production. But during the nineteenth

[21] See Robinson (2023, Chapter 10, Sections. 4.2 to 4.4) for a review of the requirements for designation as a *Wild and Scenic River*, and also Chapter 11 of this volume.

[22] The Saugus Iron Works was constructed in 1645 on the Saugus River, ten miles north of Boston. It is generally recognized as the first such facility in the colonies.

century, Taunton became known as the "Silver City" when four well-known silverware companies located there: Reed & Barton (established in Taunton in 1824), the Taunton Silverplate Company (established in 1853), F.B. Rogers (established in 1886), and the Poole Silver Company (established in 1892). Again, wastes from metal productions were dissipated in the rivers of this watershed. Other metal products (stoves, textile machinery for nearby mills, and locomotive engines) were all manufactured in Taunton. The Taunton River was also used for shipping the grains from the farms of Eastern and Middle Massachusetts through Narragansett Bay. The population of Taunton was approximately 3,000 in 1790; 25,000 in 1890; 50,000 in 1990; and 60,000 in 2020. Population growth has been consistent and steady in Taunton since its inception in 1637.

After Reed and Barton established its silverware manufacturing business in Taunton in 1824, four dams were constructed on the Mill River all within a mile of downtown Taunton: The Whittenton Mill Pond Dam, the Hopewell Mills Dam, the West Britannia Dam, and the Morey's Bridge Dam. These were all combination wood and earthen dams. By the end of the nineteenth century, the Mill River became lifeless with no fish runs or other functioning ecology. Whether it could ever be restored to some sort of life was questionable.[23]

But in October of 2005, the Whittenton Pond Dam on the Mill River (immediately north of downtown Taunton) threatened to fail after a week of nine inches of rainfall. Because of the threat of flood, two-thousand residents of Taunton were evacuated, and all downtown businesses were ordered closed.[24] A state of emergency was ordered by the City's Mayor. If the dam had failed, the Mill River would have inundated downtown Taunton. As a result, an immediate inspection of all the Commonwealth's high-risk dams were undertaken. Note that the Army Corps of Engineers quickly repaired and converted the Whittenton Dam temporarily into a rock-structure dam.

In the wake of the 2005 flood, the Mill River Restoration Project was organized in 2007. It was led by the *Taunton River Stewardship Council*

[23] See "What a River Means," *The Nature Conservancy*, March 16, 2023, at www.nature.org/en-us/about-us/where-we-work/united-states/massachusetts/mill-river-restoration/.

[24] See "Town Braces for Massive Flood" at www.cbsnews.com/stories/2005/10/18/national/main952230.shtml. CBS News, October 18, 2005.

and the *Mill River Partnership*. The three dams on the Mill River were removed starting with the Hopewell Mills Dam in 2012, then the Whittenton Dam in 2013, and then the West Britannia Dam in 2018. A modern fish passage facility was installed in Morey's Bridge Dam which forms Lake Sabbatia at the head of the River. As a result, thirty miles of additional river habitat has now been opened to fish runs which primarily consist of the approximately 100,000 herring that migrate up the Taunton River to spawn each year, with at least 30,000 traveling up its Mill River tributary. The more than fifteen-year efforts of state agencies and environmental advocacy organizations—such as *The Nature Conservancy*—have resulted in large shad and herring runs and a general Taunton River restoration. Now you can boat from the former site of the Rogers Silver Company on the Mill River, through Taunton Center and then all the way to Narragansett Bay without encountering an obstruction. The Rogers industrial site has now been converted to the *Weir Village Riverfront Park* which offers a boat launch facility.

The Cotley River is an eight-mile long tributary of the Taunton River. It flows northerly to join the latter close to downtown Taunton. The low-head Barstowe Pond Dam was constructed on the Cotley in the 1780s to power a sawmill. It was a fifty-seven-foot long earthen dam with a thirty-foot timber and rock sluice. After the 2005 Taunton flood, the State's *Office of Dam Safety* classified the Barstowe Dam as a "Significant Hazard." The Cotley River Restoration Project is led by the nonprofit "Taunton Development Corporation." The Barstowe Dam removal began in 2010 as a priority project by the State Department of Ecological Restoration. This dam was removed in 2018, but the River's restoration continues.

Deconstructing New Hampshire's Old Mills

New Hampshire also has the remnants of colonial era working-mill dams. They were constructed on most of this State's rivers. A few of their deconstructions are briefly reviewed here.

New Hampshire's "Seacoast Region" stretches along the Atlantic Ocean for only thirteen miles. Salisbury in Massachusetts forms the region's southern border, and the Piscataqua River and Kittery Maine form its northern border. This region's shoreline alternates between rocky and sandy beaches. While the City of Portsmouth is the cultural center of this region, Dover is the area's largest city with a population of 32,741 according to the 2020 Census. It is the oldest permanent settlement in New Hampshire. In 1623, William and Edward Hilton built log homes at the confluence of the Bellamy and Piscataqua Rivers to establish the settlement of Dover (as shown by Fig. 8.3).

The Hilton brothers were fishmongers sent by London's "Company of Laconia" to establish a colony and fishery on the Piscataqua. (The Piscataqua was first explored by the British in 1603 for its potential as a fishery.) Between 1623 and 1633, the area had only grown to three houses. But in 1633, the area of "Hilton's Point" was purchased by Puritans who renamed the settlement "Bristol," but it was renamed Dover in 1637.

The Bellamy River is a fifteen-mile long tributary of the Piscataqua River. It forms at Swain's Lake just nine miles west of Dover. Downriver from the central park in Dover is the site of a dam once used for the former Sawyer Wollen Mills established in 1824 by Alfred Sawyer. This mill site included fifteen buildings on eight and one-half acres. Sawyer became one of the largest woolen mills in the US. In 1889, it was acquired by the American Woolen Company which continued production on this site until 1955. The buildings have since been turned into housing. The old Dam remained until 2018.

Sawyer Mills had two other dams that formed its Bellamy River reservoir. The upper dam actually channeled the River through a narrow pipe of only a few feet in width where it flows under what is now an apartment complex with hundreds of residents. Consequently, a severe flood storm could have resulted in deaths. The lower dam was the first barrier that fish runs encounter along the Piscataqua–Bellamy River combination. In 2018, these dams were removed at a cost of $2 million which was paid by *The Nature Conservancy* (which initiated the removal), NOAA, US Fish and Wildlife Service, and state agencies.

The Ashuelot River is a sixty-four-mile tributary of the Connecticut River that flows through southeast New Hampshire (as shown by Fig. 8.4). It drains the mountainous south-central part of the State, i.e. the Keene–Swanzey area. The remaining buildings of the Homestead Woolen Mills are in the very rural town of Swanzey. When her husband passed away, Sally Brown inherited the Homestead Mill from the Brown family in 2008. The Brown family ran this mill since 1911. Previously, the mill powered a nineteenth-century box-making factory, and in doing so, it blocked fish passage from the Connecticut River. At its peak production, the Homestead Mill employed 500 workers. It prospered using Australian and New Zealand wool for manufacturing textiles for the Manhattan fashion industry, and even blankets for World War II soldiers.

After the decline of New England's textile industry, the Homstead Mill's one-hundred and forty-foot long wooden timber and granite stone dam had no functioning purpose. It either had to be deconstructed or repaired. In 2010, the cost of repair was $1 million. Sally Brown, the owner, selected deconstruction. At the Mill's building, she organized a picnic party to watch the dam's breaching and deconstruction on an August day in 2010. It was the culmination of a twelve-year effort by The *Nature Conservancy* that organized the deconstruction. It should be recognized that the Homestead Dam was significantly picturesque with a continuous waterfall beside a classic New England styled wooden mill building. Nevertheless, it was demolished and the Ashuelot is now an attractive free-flowing river in southeast New Hampshire.

Conclusion: Restoring America's Oldest Industrialized Rivers

The Nature Conservancy reports that more than fourteen thousand dams still block New England streams and rivers.[25] The vast majority are low-head dams originally constructed one or two centuries ago. They were once engineering accomplishments that turned the water wheels that powered sawmills, gristmills, and textile mills. Homes were built along the reservoirs formed by the dams. The views of old mill buildings by waterfalls were attractive and are now nostalgic. But before the 1972 Clean Water Act, New England's rivers were severely polluted. Now the pollutions have been largely cleaned, and the problem is to reconnect the riverine habitats so that the waters can be cooled and ecology reestablished.

The author of this volume grew up in New England, but has now lived in other sections of the Nation for more than fifty years. But I have visited my New England family numerous times in the last five decades. New England has proudly lived in its historical culture. Its past is preserved and apparent everywhere. I feel very at home and comfortable in some of New England's old and familiar industrial areas. But restored rivers

[25] See Jenny Rogers (2017), *Nature Conservancy*, at www.nature.org/en-us/magazine-articles/unleashing-rivers/.

and marshlands are more comfortable, attractive, and certainly healthy. I suspect that sometimes New Englanders cling to the past because they refuse to believe that the future can be cleaner and more prosperous. But that cleaner future can exist. We need to describe this future and illustrate it so that it becomes visceral, and with consequent effort, realized. Note that as of the end of 2024, both the Talbot Mill Dam on the Concord River and the Ipswich Mill Dam of the Ipswich River have been removed despite all the procedural votes required for approving their removal.

Fig. 8.1 Eastern Massachusetts Rivers

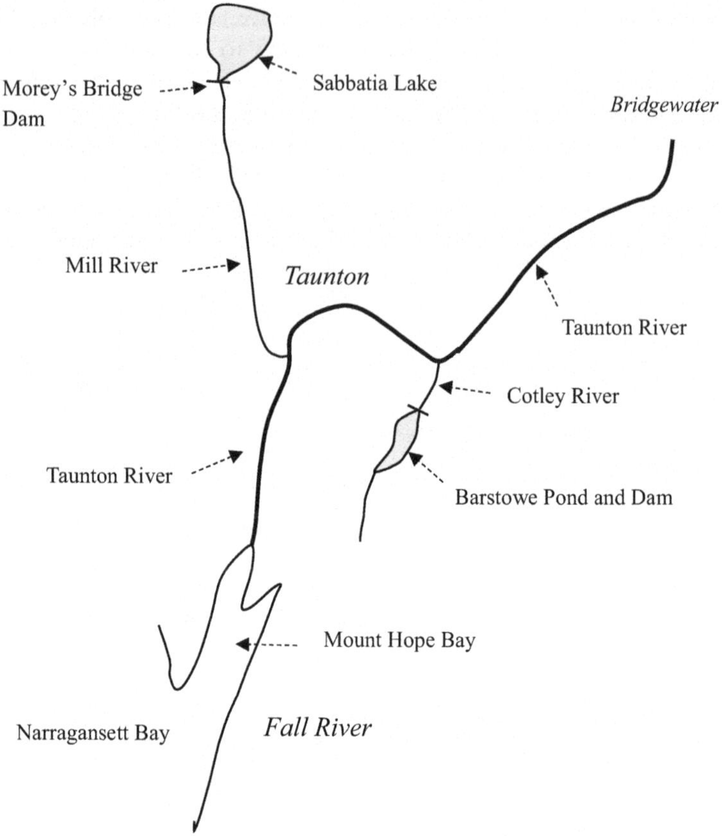

Fig. 8.2 Taunton River of Southeast Massachusetts

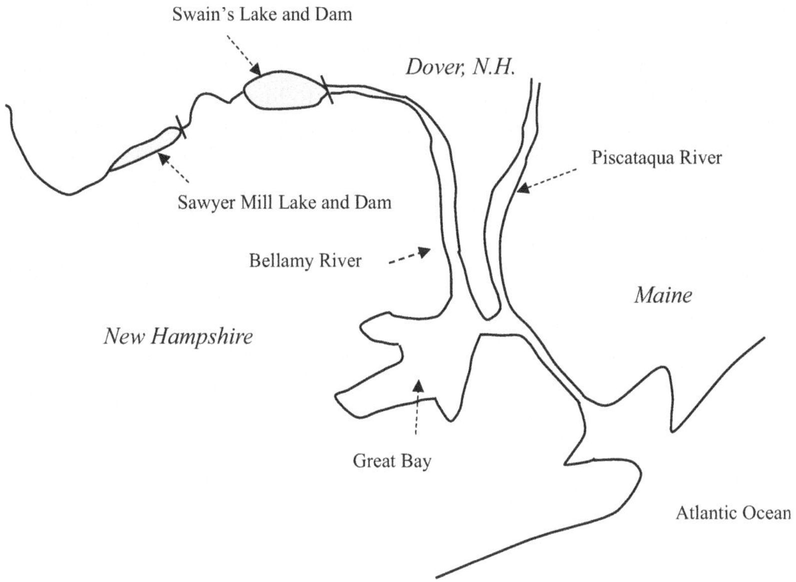

Fig. 8.3 Bellamy River's Dams

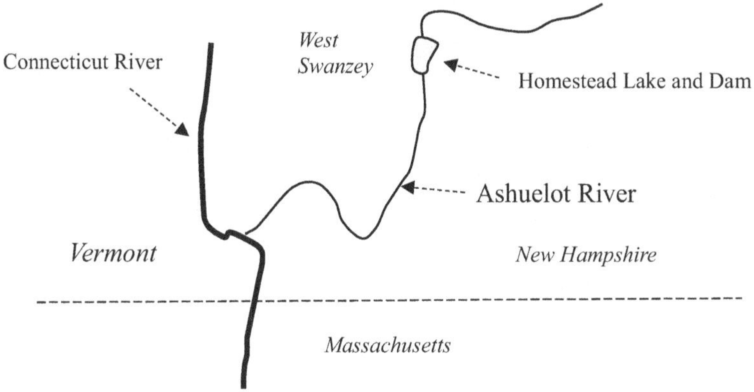

Fig. 8.4 Ashuelot River's Dams

References

Mcquaid, H. (2020, July). "Massachusetts Provincial Law" "An Act to Prevent the Destruction of Alewives on the Ipswich River." *Historic Ipswich,* at https://historicipswich.net/2023/07/24/alewife-ipswich-river-2/

Reale, H. (2022, July). Natick Will Remove Its 90-Year-Old Dam After Contentious Process. *GBH News*.

Robinson, R. (2023). *Environmental Advocacy and Local Restorations*. Springer Nature.

CHAPTER 9

An Old Commonwealth's Industrial and Flood Control Dams

Introduction

Between 1912 and 2021, there were 1,951 dams removed in the US, an average of approximately eighteen per year. But the peak year for removal was 2018 with 111 deconstructions.[1] Generally, in the latter years of this period, fifty to sixty dams were removed per year, which indicates our considerable recent dam removal movement. Over most years of the last two decades, Pennsylvania has led the US in dam removals. It removed 364 dams in the period of 1912 to 2021, more than any other state. Note that as measured in 2021, eighty-five percent of US dams were older than fifty years. It should therefore be expected that since the older northeastern states all have geographies with extensive rivers and tributaries such as are found in Pennsylvania, then the northeast would have more older dams and therefore more dam removals.

The significant watersheds of the Commonwealth of Pennsylvania are formed by the Delaware River, the Susquehanna River, the Allegheny River, the Upper-Ohio River, Monongahela River, and the Potomac

[1] See www.amerucanrivers.org/2023/02/dam-removals-continue-across-the-u-s-in-2022/ and www.americanrivers.org/2022/02/free-rivers-the-state-of-dam-removal-in-the-u-s/.

R. M. Robinson, *Restoring America's Rivers*,
https://doi.org/10.1007/978-3-031-81758-8_9

River.[2] The geography of Pennsylvania's watersheds is one of steep valleys and rivers. This makes them flood prone and therefore its rivers have been extensively dammed for flood control. But most of these dams were earthen-type constructions so that with age, they became particularly vulnerable to breaching during flood events. Eventually these earthen dams become so weakened that without extensive restoration, they fail. In many cases, they are removed rather than restored. But flood control is not the only reason for past dam constructions in the Commonwealth. Even though this area has extensive coal and natural gas available for electric power generation, the steep gradients and narrow valleys of Pennsylvania's many rivers make them appropriate for hydroelectric power. Also, many of its small population centers require dams for municipal water provisions. In addition, the extensive coal and iron ore deposits of this area resulted in numerous dams that provided the reservoirs for cooling blast furnaces and for dissipating industrial wastes. This area, therefore, has had extensive dam constructions in the nineteenth and early twentieth centuries.

Pennsylvania's combination of iron ore and coal attracted industrial populations. Their settlements' locations proliferated the flood control, municipal water, and hydropower dams of the Commonwealth. But after two centuries of heavy industrial development, Pennsylvania's still existing dams are now very old. Many have become weak, leaking and dangerous. The Commonwealth declares these as "hazardous," and attempts to organize the funding required for their removal. For example, many low-head dams were constructed to form reservoirs for sources of potable water. But they lost this purpose. They now appear to be innocuous quaint structures. In actuality, they form dangerous currents that catch and sometimes drown swimmers and other recreators. As explored below, the existence of some of these dangerous dams are not even known to the State's authorities who are responsible for evaluating their safety.

This chapter shows that the dam removal movement is strong in Pennsylvania, and that this is true because the Commonwealth's dam deconstructions are necessary for both fundamental safety reasons and for ecological restorations. Furthermore, it is shown that overall safety and ecological restoration are not independent.

[2] There are four US states that title themselves as "commonwealths:" Massachusetts, Virginia, Pennsylvania, and Kentucky. There is no legal distinction, however, between the term "state" and "commonwealth.".

The Allegheny River

The Allegheny is historically one of America's significant Rivers. It drains Pennsylvania's second-largest watershed. It rises in the north-central mountains of Pennsylvania, only eight miles south of the New York State border (as shown by Fig. 9.1). From its headwaters, it flows northerly across the border into New York State ten miles south of the town of Olean. It then flows west past Salamanca in New York, and then back southerly through Warren, Pennsylvania. It continues southerly for more than a hundred miles to join the Monongahela River in the center of Pittsburgh in Pennsylvania. The confluence of the Allegheny and Monongahela Rivers forms the Ohio River.[3] This strategically key confluence is why the "Steel City" of Pittsburgh exists.

In total length, the Allegheny River is three-hundred and twenty-five miles long. It drains an eleven-thousand five-hundred and eighty square-mile watershed that includes most of Western Pennsylvania and a large portion of southeastern New York State. In 1992, the Allegheny was classified by the US National Park Service as a "Wild and Scenic River." But this watershed has been a significant fossil fuel extraction area for coal, petroleum, and natural gas. North America's first commercial oil well was drilled in Titusville, Pennsylvania, located in the middle of the Allegheny watershed. (It was drilled in 1859 by Edwin Drake.) The industrial use of coal has long polluted this significant River, although this industrial pollution has been largely cleaned in recent decades. The Kinzua Dam, however, was constructed close to the Allegheny's headwaters in the New York border area. It blocks the River six miles upstream from the town of Warren in Pennsylvania. This "big concrete" Dam forms the twenty-four-mile long Allegheny Reservoir which extends up through the New York border to Salamanca. Before the Dam was constructed, the inundated land was part of the Seneca's Reservation. This Reservoir was forced on the Seneca Nation by an "eminent domain taking" in violation of the Nation's *1794 Treaty* with the US. This Dam's construction was motivated by flood control for the downstream cities, especially Pittsburgh. But this purpose poses significant issues of environmental justice for the Senecas as explored below.

[3] The Ohio River flows westerly to the Mississippi forming Ohio's border with West Virginia, Kentucky, and then Indiana with Kentucky.

Prone to Floods

Pittsburgh is a city of more than one and three-quarter million people living in its greater urban area. It is located in the southern portion of the Allegheny River's watershed at the confluence of the Allegheny, Monongahela, and Ohio Rivers. That is where Fort Pitt was established by the British in 1758. Fort Pitt was established to help drive the French from the area during the French and Indian Wars of the 1750s and 1760s. The Fort was named after the British Secretary of State. This was an area rich in coal and iron ore, and during the nineteenth and twentieth centuries, Pittsburgh became one of the largest industrial areas of the US. In 1857, Pittsburgh had 939 factories that employed ten thousand workers who produced $12 million in goods while utilizing four-hundred steam engines.[4] In 1857, its production is reported to have consumed twenty-two million bushels of coal, and to have processed one-hundred and twenty-seven tons of iron ore. In shipping activity, in the late 1850s Pittsburgh was America's largest inland port and stood only behind New York City and New Orleans in tonnage shipped.

In 1859, coke-fired iron smelting was introduced in Pittsburgh where the Allegheny Arsenal and Fort Pitt Foundry became America's most important Civil War armament producing factories.[5] By the Civil War's end, more than half of US steel production and one-third of its glass production came from Pittsburgh. These ratios continued into the early twentieth century. In 1875, the new "Bessemer furnace process" was adopted to meet the large demand of the US for steel rails. This became the age of America's westward railroad expansion. Industrialists Andrew Carnegie, Henry Frick, Andrew Mellon (Banker), Charles Schwab, and George Westinghouse all had Pittsburgh locations. But also during the 1870s, small steel manufacturers began making more specialized finished products—barbed wire being an example—in smaller outlying towns such as nearby Johnstown (fifty miles to the east of Pittsburgh) and Youngstown (which is just across Pennsylvania's nearby border with Ohio).

[4] These statistics are from *Ballou's Pictorial Drawing-Room Companion*, Boston, February, 1857.

[5] Coke is a type of porous gray-coal that when burned with iron ore, it absorbs the oxygen from the ore.

Steel production required two principal location characteristics: (1) abundant river water to cool the output of the furnaces and to dissipate the waste pollutions, and (2) railroads for transporting the coal and iron ore inputs, and also to transport the steel product to markets. By the eve of World War I, Pittsburgh had become a vast railroad hub capable of supplying more than a thousand freight cars per day. For limiting the costs of input transportation, steel production also required that the sources of coal and iron ore be close to the mills. All of these input factors were present in the Western Pennsylvania area, especially around Pittsburgh. The Allegheny, Monongahela, and Ohio Rivers, and their tributaries provided the necessary river water for cooling and waste dissipation. (Chapter 7 reviews similar environmental problems caused by steel production in the area of nearby Youngstown in Ohio.) Various dam constructions, however, were considered necessary for both the functioning of mills and for flood control in the narrow valleys drained by these rivers.

During the late nineteenth century, downstream Ohio River water users suffered the pollution disposal of the Pittsburgh area which did not stop its raw sewerage discharge into rivers until 1939. But Pittsburgh's worst environmental problem was its smoke discharge from its coke-burning steel mills. Its steel production peaked for World War II's armament demands for which ninety-five million tons were manufactured. In 1945, Pittsburgh was called "the Smoky City" since its smog was so thick that its streetlights burned during the day.[6] During the 1970s and 1980s, however, steel production technology changed toward reprocessing recycled steel in electric smelters (mini mills), and foreign imports also posed new competition. As a result, the "big steel" production of Pittsburgh and its surrounding communities rapidly declined. In addition, local coal and iron ore deposits became depleted. Consequently, Pittsburgh entered its "renaissance" era of urban renewal after turning away from coal burning and toward environmental cleanup. It is now listed as one of America's "Most Livable Cities."[7]

The typography of the Allegheny River Basin makes it particularly susceptible to flooding. The Allegheny mountains are part of a series of high ridge lines that parallel Lake Erie's eastern shoreline. This series of

[6] See Lorant (1999).

[7] See Majors (2007) and Winant (2021).

parallel ridges continues into Eastern Pennsylvania. They trap lake-effect storms formed by the west-to-east winds that draw moisture from Lake Erie, and that subsequently deposit the rain or snow on the interior high ridges. During the winter, when the Lake is not covered with ice caused by lengthy cold spells, large amounts of lake-effect snow build up at the higher elevations along the ridge crests. When warmer weather storms arrive in the Spring, large rainfalls combine with considerable snow melt to raise the river and tributary levels in the many narrow and steep-sided valleys. The resulting floods are frequent and severe. The most severe occurred in 1889, 1907, 1936, 1964, and 1996. All of these were winter-spring floods with heavy warm rains that quickly melted accumulated snowpacks in the higher elevations. Dam construction appeared to provide the only solution to these natural disasters.

The Kinzua Dam was first authorized in 1938. Its location is several miles upstream of Warren Pennsylvania on the Allegheny River (as shown by Fig. 9.1). It is a big concrete dam, but it was not constructed until 1965, the year after the March of 1964's unprecedented deluge and severe floods in the Allegheny watershed. But in this watershed, the previous floods caused by dam breakage were particularly severe with destructive effects. These dam breaches and floods must be reviewed prior to our examination of the environmental justice issues associated with the Kinzua Dam. As indicated below, the purpose of the Kinzua Dam was to control the floods along the Allegheny, especially in the Pittsburgh area.

Great Floods and Broken Dams

The "Great Flood of 1889" is often called "the Johnstown Flood." During a deluge of three days in late May of 1889 on a tributary of the Allegheny, the South Fork Dam broke. It was a sixty-nine-foot high earthen dam located on the South Fork of the Little Conemaugh River, only fourteen miles upstream of Johnstown. This small city is located fifty miles east of Pittsburgh. The Conemaugh River is a tributary of the Allegheny, which it joins fifteen miles northeast of Pittsburgh. Lake Conemaugh was the reservoir formed by the South Fork Dam (as indicated by Fig. 9.1).

Johnstown produced specialty steel products. It is located between the steep hills of the narrow Conemaugh Valley, and along the Conemaugh River. Its upstream watershed frequently has large amounts of rain and snowmelt runoff. Consequently, Johnstown was prone to flooding.

The slag wastes from iron furnaces and steel mills were used as landfill for building along the River which further developed it as a residential location and made it even more susceptible to flood devastation.

Although the South Fork Dam was built by the Commonwealth of Pennsylvania, it was purchased by a private group of Pittsburgh's wealthier industrialists who constructed a lodge and used the reservoir as their private lake, the "South Fork Fishing and Hunting Club." In order to construct an access road across the top of the Dam, the Club lowered its height by three feet which reduced the Dam's storage and discharge capacity by half.[8]

On May 28, 1889, six to ten inches of rain fell in twenty-four hours in the Pittsburgh region. Tributaries consequently bore strong currents that were previously unexperienced on these streams. Rail beds were washed away. On the morning of May 31, teams of workers abandoned their attempts to release the high waters of the South Fork Dam by clearing its spillway of debris. At 2:50 PM, the Dam was breached. It took more than one-half hour for the reservoir to empty. On its fourteen-mile downstream path to Johnstown, the flood's strong current carried trees, houses, wagons, and other debris. A seventy-eight-foot high railroad bridge collapsed. At Mineral Point, one mile past the collapsed railroad bridge, no structures were left and all soil above bedrock was washed away although there were thirty families previously residing along its single street. Sixteen people were killed. In East Conemaugh, the next downstream town, a locomotive engine was washed away. In the next town, Woodvale, the Cambria Iron Works had their railroad cars carried downstream. The boilers of the Iron Works exploded, and miles of barbed wire (a specialized product) was carried by the current. In less than an hour of the Dam's collapse, a sixty-foot high wall of water that was traveling at forty miles per hour washed away another railroad bridge in Johnstown.

After the waters receded, 2,209 people were dead and thirty acres in Johnstown were covered with debris that included upstream houses, trees, railroad cars, wagons, steel wire, and other material. The flood extended through Pittsburgh and into other locales. The Johnstown Flood was the worst in the nineteenth-century US. It remains the worst in

[8] See Blakemore (September 28, 2023).

Pennsylvania's history: 1,600 homes were destroyed, $17 million in property damage resulted, four square miles of downtown Johnstown were completely devastated.

An interesting aftermath of the Dam's breaching was its legal effects. The South Fork Hunting and Fishing Club was accused of failing to properly maintain the structure to the point that it was unable to contain any severe storm. In court, the Fishing and Hunting Club argued that the Dam's failure was from a natural disaster—an "Act of God"—and that consequently no legal compensation was required. The Club won its case. But the Flood did subsequently change the law into a "strict, joint, and several liability so that a non-negligent defendant could be held liable for damage caused by the unnatural use of land."[9] Since a dam was an "unnatural use of land," its breaching was not "an Act of God" if the dam was negligently constructed or inadequately maintained. The owners could therefore be liable for the loss of life and property resulting from the breach.

The steep slopes and narrow valleys of the Allegheny Basin make dam construction problematic. The events of May, 1889, should have warned the Commonwealth about the likelihood and dire effects of dam breaching. The lesson did not, however, "stick." In July of 1977, seven dams failed in the area of Johnstown. These were small earthen dams constructed for municipal water use and flood control. The Laurel Run Dam was constructed between 1915 and 1918. (The Dam's location is indicated by Fig. 9.1.) It was an earthen dam with a masonry spillway across Laurel Run near Johnstown. It was forty-feet high and six-hundred and twenty-feet long. But as early as 1943, the spillway was identified by state safety engineers as inadequate to handle large storm events. In 1959, state officials again found that the spillway was less than half the necessary size for maintaining dam integrity, but there was still no remedial action taken by local officials. In 1970, state officials again classified the Laurel Run Dam as a "high hazard" for failure that would result in significant loss of life and property damage.

In July of 1977, a torrential storm dropped 11.8 inches of rain in eight hours. The Laurel Run Dam's reservoir began to overflow the Dam's crest. The Dam was breached in the early morning and a catastrophic flood wave ran through the two-mile valley below the Dam.

[9] See *National Geographic*, "May 31, 1889 CE, Johnstown Flood," published July 10, 2019.

Forty people were killed in the valley. Six other smaller earthen dams were also breached in the Johnstown area by this 1977 storm. Lawsuits were brought against the Greater Johnstown Water Authority (the Dam's owner), the Bethlehem Steel Corporation (the previous owner), and the Pennsylvania Department of Environmental Resources. The Lawsuits were settled out of court.[10]

The Flood Control Act of 1936

On September 16, 1966, the US Army Corps of Engineers completed the Kinzua Dam and closed its gates to begin filling the Allegheny Reservoir. Note that this reservoir is called "Lake Perfidy" by members of the Seneca Nation.[11] The construction of this Dam clearly violated the *1794 Treaty of Canandaigua* between the US and the Seneca Nation. President Washington signed this Treaty which guaranteed the Senecas their Reservation lands along the northern stretch of the Allegheny River. But the Kinzua Dam took 10,000 acres of this promised-in-perpetuity Seneca land which was then purposely flooded to form the reservoir. This "taking by eminent domain" was strongly opposed by the Seneca's. It displaced 600 members of the Nation (130 families), nine Seneca communities, and half of their fertile farming land. This "taking" was appealed in the Federal District Court for the Western District of New York. But in 1957, in *United State v. 21,250 Acres of Land,* the Seneca Nation lost. The Dam's construction began in October of 1960.

The Kinzua Dam's construction was authorized by the US Congress in its *Flood Control Act of 1936*. This *Act* authorized sixteen flood control projects along the Allegheny and Monongahela Rivers. This was in response to the 1936 Pittsburgh flood and the numerous floods partly reviewed above. The worst Pittsburgh flood was the "Great St. Patrick's Day flood" of March, 1936. It also affected wide areas of the Mid-Atlantic US. Prior to March 16, snowpacks had accumulated to substantially exceed normal levels at higher elevations of the Alleghenies. On March 16, warmer than normal temperatures arrived together with torrential

[10] See *Association of State Dam Safety Officials*, "Lessons Learned from Dam Incidents and Failures, Case Study: Laurel Run Dam" at https://damfailures.org/case-study/laurel-run-dam-pennsylvania-1977.

[11] "Perfidy" means *faithless actions, or treacherous, disloyal, deceptive actions, or not acting in good faith.*

rains. Rapid melting of the snow and ice caused record flood levels of twenty-one feet above flood stage on the Allegheny and Monongahela Rivers. The flood levels lasted for four days during which time sufficient damage was caused to close Pittsburgh's steel mills for sixty-thousand workers. Sixty-five percent of the downtown business district was under water. Electric power was down for eight days. This was the flood that led to the call for a flood control dam on the Allegheny upriver from Pittsburgh. It led directly to the *Flood Control Act of 1936* and subsequently the Kinzua Dam.

Numerous responses to the Army Corps of Engineers "Flood Control Plan," however, indicated alternatives to the Kinzua Dam's construction. For example, the Dam might have been located farther downstream and closer to Pittsburgh. But the Kinzua's location of seven miles upriver of Warren, Pennsylvania, also mitigated floods in that small city. (Warren's population was only 9,404 according to the 2020 US Census.) Other alternatives were posed that might have been more cost effective, but they were rejected in favor of the Kinzua Dam and the "taking" of the Seneca Nation's land.[12]

It should be noted that the Kinzua Dam serves more than flood control. It is also a hydroelectric facility that provides 400 megawatts per hour of peak capacity power. It also has a "pump storage" capacity to pump water to an upper-storage reservoir for subsequent generation of electricity when demand peaks. (This energy storage strategy is reviewed below as a feature of some dams along the Susquehanna River.) In addition, the twenty-four-mile Allegheny Reservoir is a popular recreation asset for boating, fishing, and summer beach bathing.

The Rivers and Dams of East Pennsylvania

An *American Heritage River* is a designation applied by the US *Environmental Protection Agency* as created by "Executive Order Number 13061" of President Bill Clinton in 1997. This designation means that various federal agencies will cooperate in disseminating federal funding for purposes of (i) environmental protection, (ii) economic revitalization, and/or (iii) preservation of historic and cultural characteristics. There are fourteen US rivers with the *American Heritage* designation. The criteria

[12] See Schultz (2023).

for this designation were developed by the *Council on Environmental Quality*. The criteria specifies:

- There must be a recognized "distinctive character" to the river, i.e. it has distinctive natural, economic, historic, cultural, or recreational characteristics.
- The river's community has formed a potentially effective action plan that could benefit from federal funding and that addresses the three objectives of (i) environmental protection, (ii) economic revitalization, and/or (iii) preservation of historic and cultural characteristics.
- The *preservation action plan* has strong, broad, and diverse community support as indicated by elected officials, citizen organizations, and businesses.
- The community is willing and capable of forging partnerships and agreements with government agencies and relevant NGOs to implement its action plan.

The *American Heritage River Program* is similar to some other federal programs such as *The Areas of Concern Program* that addresses restoration of industrially polluted locales along North America's "Great Lakes," or the *Urban Waters Federal Partnerships Program* that addresses the restorations of rivers that flow through qualifying urban areas, or the *National Estuary Program* that addresses coastal estuary restoration. These four federal programs are aimed at coordinating the funding grants and efforts of federal agencies with the efforts of local government agencies and *environmental advocacy organizations* to restore specific locales.[13] They all require local planning and control systems in order to receive these grants. Both the Susquehanna River and its tributary, the Lackawanna River, are designated *American Heritage Rivers*. They both flow through the industrial areas of Northeast Pennsylvania.

At four-hundred and forty-four miles, the Susquehanna River is the longest on the East Coast of the US. Its North Branch originates in Cooperstown in New York. Its West Branch rises in Western Pennsylvania and joins the North Branch at Northumberland in Central Pennsylvania (as shown by Fig. 9.2). At 27,500 square miles, its watershed includes

[13] Robinson (2023) reviews the *Areas of Concern Program* in detail. It also reviews the *Urban Waters Federal Partnerships Program* and the *National Estuary Program*.

almost half of the State of Pennsylvania. The Susquehanna empties into Chesapeake Bay in Maryland where it provides half the Bay's freshwater. In 2005, the *environmental advocacy organization* American Rivers called the Susquehanna "America's Most Endangered by Pollution." This River flows through agricultural lands with strongly polluting runoff (animal wastes and fertilizer runoff), plus it suffers from urban and suburban stormwater runoff, and inadequately treated sewerage disposal. All of this runoff and sewerage pollutes Chesapeake Bay, one of America's most important East Coast inlets. Flood control, hydropower generation, and the condition of older dams pose significant issues throughout the Susquehanna Basin.

The Susquehanna River is not only the longest US river east of the Mississippi, it is also one of the world's oldest rivers. Typically, rivers rise in higher elevations where the terrain is steeper. The terrain then becomes flatter as the silt drained by the river accumulates downstream. Consequently, dams are usually located on the upstream steeper terrain rather than along the downstream flatter slopes because a steep gradient allows gravity to accelerate the water falling through electricity generating turbines. The Susquehanna, however, is a very old river with upstream terrain that has been eroded so that it is more of a flat meandering river in its upper reaches, but it has a steeper gradient downstream. Susquehanna's hydroelectric dams are all located in the lower fifty-five miles below Harrisburg in Pennsylvania and above the River's terminal at the Chesapeake Bay. There are four currently functioning hydroelectric dams in this stretch of river:

1. the Conowingo Dam (in Maryland) generates 538 megawatts peak power,
2. the Safe Harbor Dam generates 417 megawatts of peak power,
3. the Holtwood Dam generates 252 megawatts of peak power, and
4. the York Haven Dam generates 135,485 megawatts of peak power.

Four million people reside in the Susquehanna watershed which includes Pennsylvania's State Capitol of Harrisburg. The entire watershed is, however, flood prone. It suffers an average of $150 million in flood damage annually.[14] Storms from North America's Great Lakes sweep across the area, and with the global warming of climate change, these storms carry increasing amounts of moisture. Pennsylvania's *Department of Environmental Protection* (DEP) estimates that the Commonwealth's average temperature was almost six degrees Fahrenheit warmer during the period 1971–2000 as compared to 1900.[15] The previous Great Lakes winter-ice coverage is disappearing so that "lake effect" winter storms are more prevalent and carry more moisture.[16]

Global warming is caused by human made CO_2 and other greenhouse gas emissions such as methane. Pennsylvania is one of the two highest energy producing states in the US. Its industrial and electricity sectors annually produce greenhouse gas emissions that are among the largest in the US.[17] The DEP forecasts an increase of approximately eight percent in the area's precipitation by 2050, and this adds to the flood problem. Pennsylvania's "2021 Climate Action Plan" calls for a twenty-six percent reduction in greenhouse gas emissions by 2025 as compared to 2005. Most of this reduction will come from the electricity generating changes associated with switching from coal to natural gas-powered production. Up to the end of 2023, this "switch" already reduced CO_2 emissions by forty percent since 2005. This progress in emissions reduction can continue through expansion of the alternative energy source of wind power as a substitute for fossil fuels, but the implications of wind power (an intermittent power source) for hydroelectric dam continuance is complex and explored below.

[14] See Hendricks, Karen (2023, March 1), "Climate Change in Pa.: The flood-prone Susquehanna River watershed poses a widespread challenge," *TheBurg*, at https://theburgnews.com/.

[15] Ibid. Note that the State of Pennsylvania is one of four states addressed as a "commonwealth." This is an old Anglo term with no legal distinction between what is a "state" and what is a "commonwealth.".

[16] In the winter, when the Great Lakes are not covered with ice, the west to east winds pick up the moisture from the Lakes and then generate snow on Pennsylvania's and New York's inland areas. Cold temperatures and consequent ice coverage, therefore, reduce winter precipitation and warmer winter temperatures increase it.

[17] See footnote # 14.

Energy Storage

In February of 2024, York Energy Storage LLC applied for *Federal Energy Regulatory Commission* (FERC) approval to construct a hydroelectric dam at Cuffs Run, a tributary of the Susquehanna River in York County, Pennsylvania. (Note that Chapter 1 reviews the FERC regulatory process). This project would be a "storage project" for generating hydroelectric power. The plan is to pump water from Lake Charles, the reservoir behind the nearby Safe Harbor Dam, and store this water into what would be a newly created dammed reservoir (Cuffs Run) with a surface area of one-thousand acres. (These locations are shown by Fig. 9.2.) The new reservoir would be adjacent to Lake Charles. This storage pumping would occur during periods of otherwise low electricity demand. When hydropower is needed, the water in the new reservoir would be released to flow through a new electric generating turbine and return back to Lake Charles. The project's purpose is to create additional energy storage, a sort of battery of electric power. The new reservoir, however, would inundate what is now fifty private properties that include three "preserved farms," and hundreds of forest acres.

The Cuffs Run proponents argue that this storage project will generate clean, renewable energy when peak demands warrant. The project's environmentalist opponents argue that this sort of proposal would once again destroy both an undisturbed and ecologically sound tributary and the valley around the targeted stream, an action similar to the creation of Kinzua Dam. The Lancaster Conservancy (a local *environmental advocacy organization*) also argues that the forest and farmlands are worth preservation.[18]

Ted Evgeniadis is the Riverkeeper for the Lower Susquehanna. He pointed out, "If we continue to put more projects that are going to further impact our river, that's pretty scary because we have a river that needs help. If they have a plan to do a pump storage facility, they can take it somewhere else."[19]

[18] See Deornellas (2004) at https://lancasteronline.com/news/local/2-5-billion-susquehanna-river-hydroelectric-project-soarks-opposition/article_467a8dbe-c763-11ee-8796-9f2a51bcd049.html.

[19] See Matt Klinedinst (2023, December 5), *WPMT Harrisburg*, at www.msn.com/en-us/money/realestate/developer-2-1-billion-chanceford-township-hydroelectric-dam-proposal/ar-AA113qiC. For a brief review of "Riverkeepers," see Robinson (2021), p. 243.

York Energy first proposed this storage project in the 1990s. After FERC's rejection, it proposed the project again in the 2000s, and is now proposing it for a third time. Since the energy needed to pump water uphill to the proposed Cuffs Run reservoir exceeds the energy generated by the water turning the turbine on the return route (at most seventy-five percent of the pumping energy is returned), then from a purely energy analysis the Cuffs Run Project is inefficient. But the economics of energy generation have changed given the advances in the alternative power source of wind turbine technology. Unlike hydropower, wind energy is highly variable; consequentially stored energy capacity is necessary for when the wind turbines are not turning. But pump storage of hydropower could complement an intermittent wind turbine system to allow it to be feasible. Therefore, the interesting question becomes, "With the development of wind-generated power which would intermittently replace the hydropower, why not utilize the existing reservoir capacity to store hydropower as complement to the growing wind-power generation? Why create new reservoir capacity? Is there no unused capacity in Lake Charles?" Since the wind power, it appears, would be replacing hydropower as the primary source of electric generation, then as a result there should be increasingly excess reservoir capacity. This could be used as storage.

But the not so apparent answer is that for years to come, wind power will not be replacing hydropower. Instead it will be replacing electricity generated by carbon-based fuels (coal and natural gas) which are slowly being phased out. Why is the phase out slow? The political power behind West Virginia's and Pennsylvania's coal and natural gas industries is formidable. Electricity generation in Eastern Pennsylvania is still overly reliant on coal-fired power plants which, given climate change, must be replaced by renewable carbon-free sources. In this region, wind power is the preferable alternative to coal, but it needs a battery-type storage to provide power when wind turbines are not turning. Existing hydro power reservoirs could provide that storage, but the cost of further environmental degradation from a new reservoir is difficult for the growing environmental movement to accept. Alternative on-demand complementary renewable possible sources include hydrogen power, new fusion nuclear power, compressed air, new dry battery technology, or combinations of the above.[20] But as long as the hydropower dams along the Susquehanna River exist, and wind power is making fast inroads into the power grid, shouldn't one other source be considered? Using the increasing excess storage capacity of the existing hydro-power reservoirs for the needed energy storage is a viable short-run alternative. The cost of building storage alternatives, however, might be overly high if this would cause permanent environmental destruction.

Wind Power

In 2022, more than seven percent of the globe's electric power was produced by wind-driven turbines.[21] Per unit of energy produced, new wind farms are now more economical to build and maintain than either new coal or natural gas plants.[22] The current major inhibitor of wind turbine expansion is its variable source of intensity. It thereby needs an energy storage system as a complement. Since wind generation is generally stronger in nighttime, combinations of wind and solar can be particularly efficient with solar providing energy during daylight, and wind being the primary source at night.[23] Note that a doubling of wind speed potentially yields an eight-times increase in wind-generated power. For wind farms, however, the optimal distance between the turbine terminals is seven times the rotor diameter so that a considerable area is required. The area must generally be along the top of a narrow ridge since turbulence decreases efficiency if wind turbines are placed in front of each other rather than side-by-side.

A second inhibitor to expansion is that suitable locations for wind generation are often not close to population centers, and therefore lengthy transmission lines are necessary. But this entails the increase in

[20] Note that current significant advances in lithium-ion battery storage are reviewed in Chapter 13. This new technology is rapidly developing for wind power storage in New England.

[21] See "Global Electricity Review 2023," at https://ember-climate.org/insights/research/global-electricity-review-2023/.

[22] See www.lazard.com/perspective/levelized-cost-of-energy-levelized-cost-of-storage-and-levelized-cost-of-hydrogen/.

[23] Ibid.

capital costs to establish the longer lines, and also the loss of delivered power as transmission length is increased. (This is also a significant problem with hydropower particularly in the American Southwest where, for example, the Colorado River's hydropower is generated along Arizona's northern border, but the State's population centers are along its southern border[24].) A third inhibitor is that fossil fuels continue to be subsidized even as the costs of wind technology is currently decreasing. Whereas wind power provided only three and one-half percent of global electric generation in 2015, that portion rose to seven percent in 2021.[25]

Since 2015, much of the cost reduction in wind power generation stems from the wind turbine's improvements in blade design. They are now longer, lighter, and better designed so that power efficiency (the fraction of power generated from the wind energy's potential) has improved. As a result, and because of global warming, wind turbines have been replacing fossil fuel powered electric generation. Hydropower's future primarily remains as a potential complementary storage system, and it might continue in that capacity for decades in the Susquehanna River area. (Note that there have been significant advances in electric battery storage to complement renewable power generation. These advances are reviewed in Chapter 13.)

The Lackawanna River, Its Dams, and an Orange-Red Metallic Goo

The Lackawanna River is a sixty-mile long tributary of the Susquehanna River. It is the Susquehanna River's largest tributary along its northern section. It originates in two branches that flow from glacial ponds and wetlands on the Pocono Mountain Plateau (as indicated by Fig. 9.2). The East Branch of the Lackawanna flows from Lake Lorain, Bone Pond, Independent Lake, and Dunns Pond. The West Branch flows from Fiddle Lake, Lake Lowe, and Lewis Lake. The two branches meet at Stillwater Lake and Dam. The Lackawanna River then flows for forty miles through the anthracite coal area of Northeastern Pennsylvania. This includes flowing through the old industrial City of Scranton. The Lackawanna then connects with the Susquehanna River in West Pittston,

[24] See Chapter 4.

[25] See footnote #20.

which is located southwest of Scranton, but close to its border. The upper reaches of the Lackawanna remain relatively clean and pristine, and it is known as a very desirable brown trout fishing and recreation stream that is relatively close to the population centers of Philadelphia and New York City. This River was classified as an *American Heritage River* in 1997.

Anthracite coal (hard coal) began to be used in what was to become the "Scranton area" in the 1790s. Although prior to 1820, the entire length of the Lackawanna River was pristine, by 1900 the coal and iron industries severely polluted this River from roughly the northern extension of the Scranton urban area through the River's confluence with the Susquehanna. The lowest three miles of the River are particularly and severely polluted from the "Old Forge Bore Hole" which daily dumps more than one-hundred million gallons of mine water into the River.[26]

Between 1820 and 1860, the anthracite coal fields of the Lackawanna Valley were developed by the Delaware and Hudson Company and the Lackawanna Coal and Iron Company (the latter was formed by George and Selden Scranton who found the City of Scranton). In 1822, the Delaware and Hudson Company opened the first anthracite mine and it was located close to the River. The Scranton brothers also discovered iron ore close to the River, and with nearby coal available, they opened iron furnaces at nearby Slocum Hollow. The presence of river water for cooling and waste dissipation, plus the iron ore, coal, and new furnaces left one missing element for a successful iron business, i.e. rail transportation for sending output to market. During this time, the Erie Railroad was building its Buffalo to New York City line, but it needed new rails to be delivered in Binghampton, which is about a third of the way from New York City toward Buffalo. These rails were provided by Lackawanna Coal and Iron. Also, the Lackawanna and Western Railroad was completed in 1852, and the Lehigh Valley Railroad soon followed. The Lackawanna Valley was therefore developing rapidly prior to the Civil War.

By 1900, the railroad, mining, and iron companies had severely degraded the Lackawanna River as an environmental asset. In 1866, the City of Scranton was formed, but by that time the River's water was already unfit for a public potable supply. The expansion of underground mining also polluted local water wells. The City of Scranton dumped

[26] See Abington Regional Wastewater Authority (2024), "Lackawanna River Watershed: An American Heritage River," at https://abingtonwastewater.org/management/lackawanna/lackawanna-river-information.

its raw sewerage into the Lackawanna River until the 1960s. Economic development, however, required that clean potable water be continuously supplied. To meet this need, water supply reservoirs were constructed by damming the Lackawanna upstream at Moosic Mountains. It is important to note that in 1937, the Pennsylvania General Assembly passed a "Clean Streams Act" to prevent pollution of the Commonwealth's rivers, but unfortunately coal companies were exempt from these regulations. This was the era of the common argument that the acidic discharges from the coal mines would kill the bacteria from sewerage, so one pollution canceled the other. But finally in 1966, the Scranton Sewer Authority was formed, and after the 1972 Federal "Clean Water Act" was enacted, the Lackawanna Sanitary Authority was formed in 1975.

A major event in this narrative, however, occurred in January of 1959, i.e. the Knox Mine Disaster. This mine had illegally extended its shafts to under the Lackawanna River. The shafts collapsed and flooded the mine. This started the end of the anthracite era in the Scranton area. The last underground anthracite mine in the Lackawanna Valley shut down in 1966, but thousands of the remaining tunnels filled with groundwater which unfortunately drained into the Lackawanna and Susquehanna Rivers. The large body of groundwater under the Lackawanna Valley is now called the Northern Anthracite Mine Pool. Water percolating through the abandoned mines picked up metallic minerals, manganese, and sulfur. It lacked dissolved oxygen and was highly acidic. When it reaches the River, it coats the River's rock with an *orange-red metallic goo*, a reminder of the mining history of the area.

Major flooding of the Lackawanna Valley occurred in 1902, 1936, 1942, 1955, and 1975. The small flood plains that exist in the Valley have been continually rebuilt after floods. In 1960, the Army Corps of Engineers erected the Stillwater Dam where the East and West Branches of the River join (as indicated by Fig. 9.2). This earthen dam creates Stillwater Reservoir but only when significant flood conditions occur. It is close to dry during other times. The Dam is designed for flood control. It functioned accordingly to control the floods in 1975 and 1993.

One of the Lackawanna Valley's public water supply reservoirs was constructed at Moosic Mountain, about half-way along the sixty-mile River. The O'Conner Dam was built at Eales Preserve, a 2,250-acre area with sixteen miles of trails in Blakely. This Dam was a 128-year-old structure built for providing potable water. The Dam was failing, however, when *The Nature Conservancy* purchased it along with the surrounding area of the Eales Preserve. Up to the time of purchase, the old stacked log and stone Dam was not maintained, and dangerous leaks were prevalent. *The Nature Conservancy* deconstructed the Dam in 2022 and is allowing the Eales Preserve to return to a natural marshy area. The original stream channel has been restored.[27]

Removing Pennsylvania's Low-Head Dams

The typography of Pennsylvania generates numerous rivers with extensive networks of tributaries. This, combined with Pennsylvania's history of aged settlements, coal and iron ore mining, and metal manufacturing, resulted in a high proliferation of dams, mostly constructed during the nineteenth and early twentieth centuries. Most of those were low-head dams of stone masonry or earthen construction, although in the early twentieth century some were constructed or reconstructed from concrete. But generally, these structures have now lost their original purpose and subsequently any usefulness they once had. Pennsylvania is a leader in the annual number of dam removals, but removing the thousands of small dams that currently obstruct numerous small creeks is a chore in need of lengthy time. In 1995, the Commonwealth's regulations for dams were changed to encourage deconstructions once the structures lose their usefulness. The Commonwealth also provides some funding possibilities to facilitate these removals.

This section briefly reviews a few dozen of Pennsylvania's dam deconstructions that have occurred in recent years. Although several hundred removals have occurred in the Commonwealth in the last decade, the sample reviewed here should give the reader an understanding of the differences between Pennsylvania's dam removals and deconstructions occurring elsewhere. The removals here bear a more "small township" flavor and "locally oriented" consensus with clear visions of what can be accomplished by rehabilitating rivers and streams.[28] For example, until recently the Chest Creek Dam once provided the Town of Patton with

[27] See Frank Wilkes Lesnefsky (2022, May 9), "Nature Conservancy removes century-old dam, reservoir in Jessup," at www.thetimes-tribune.com/news/nature-conservancy-removes-century-old-dam-reservoir-in-jessup/article.866d7ba2-0031-51fd-9a97-d3d82b75c602.html.

[28] The author lives in a small town in northwest Pennsylvania close to Lake Erie, and across a narrow street from a steelhead trout stream.

its drinking water. But because of water quality issues, Patton switched to groundwater wells. Water Plant Manager Ken Trinkly asserted, "This is an old plant. They really didn't want water out of the stream (Chest Creek) anymore. There were some problems with the raw water out of the stream. Overall, it was cheaper to go with a well system."[29]

Once the old dam was not being used, the *environmental advocacy organization* American Rivers led the effort to remove the dam. Its Director of River Restoration for Pennsylvania, local resident Lisa Hollingsworth-Segedy, pointed out, "Free flowing waterways are the circulatory system of our land." Dams interfere with fish migrations so that species such as trout will not survive. The silting and stagnant water in the reservoirs behind even the low-head dams raise the water temperature so that cold-water species such as trout cannot survive.

The Oakland Dam was on the North Branch of the Susquehanna River. The seven-hundred and fifty-five-foot long and sixteen-foot-high concrete dam was constructed in 1929 to provide hydroelectric power to a nearby hospital and railroad facility. It provided hydropower until 1990 when its maintenance costs would not justify its continued small power output. In the years after its termination, the dam was allowed to deteriorate into becoming a public hazard especially to boaters. In 2000, the dam developed a breach of over one-hundred feet with protruding steel rebar. Flooding also caused a concentrated forceful current through the breach. In review of this hazard, Dana Rockwell of the environmental advocacy organization "Endless Mountain Heritage Region" said, "There was rebar exposed, broken concrete, and there were trees that got caught up. As for boating, it became a serious hazard."[30] The dam was removed on September 18, 2023. To date (Spring of 2024), it remains the largest dam removal in Pennsylvania history.

This dam removal reconnected two-hundred and fifty miles of aquatic habitat for migrating fish in the upper Susquehanna's watershed. The cost was $450,000 paid for by Federal and State grants that included

[29] See Randy Griffith (March 31, 2011), *The Tribune Democrat*, at https://tribdem.com/news/manager-says-removal-of-chest-creek-dam-will-enhance-quality/article_6699eae2-5367-11e9-b559-7731603ccfef.html.

[30] See Kress, Emily (2023, July 31), "Demolition on deteriorating dam in Susquehanna County," at www.wnep.com/article/news/local/susquehanna-county/deteriorating-dam-being-demolished-susquehanna-river-depot-oakland-american-rivers-endless-mountain-rivers-heritage-region/.

$100,000 from the US Fish and Wildlife Service's "National Fish Passage Program." Removal also remediated the risk of flood damage in the area and reduced the broken dam's hazard to recreation. Note that as reviewed above, global climate change exacerbates the flood damage risk of the Susquehanna. With these risks ameliorated, the Susquehanna Borough Council is planning a park that will encompass the area of the old dam.

In September of 2009, the dam on the East Branch of Saucon Creek in Bethlehem, Pennsylvania, was removed. This Creek is a tributary of the Lehigh River. The "Saucon Creek Watershed Conservation Management Plan" called for the removal of this four-foot high, twenty-foot long obstruction in order to improve fish passage, to cool the water of Saucon Creek, to improve the safety of boaters and other recreators, and to reduce the hazard of flooding. Since the removal, the conditions of Saucon Creek have improved accordingly. The deconstruction of this dam was financed by $69,000 in grants from NOAA's "Community-Based Restoration Program" and the Commonwealth's "Growing Greener Fund."

Tunkhannock Creek is in Blakeslee, Pennsylvania, a short distance northwest of Scranton. It is a tributary of the Susquehanna River. Maple Dam, an old stone masonry structure of three-feet in height and sixty-feet in length, was removed from this Creek in 2017. This removal opened fifteen-miles of river for fish passage, cooled the Creek's water, eliminated a public safety hazard, and reduced the risk of flooding.

Macungle Township is southwest of Allentown, Pennsylvania. Little Lehigh Creek, a tributary of Lehigh River, flows through this township. Millbrook Farms Dam was on this Creek and in this Township. It was an old stone masonry dam, four feet high and one-hundred and fifteen feet long. It was originally built to power a gristmill in 1831. It was removed in 2017 to enable fish passage, to cool the Creek, to eliminate the public safety hazard, and to reduce the flood damage risk.

Three Pennsylvania Game Commission Dams (owned by the "Commission") were removed in 2017. They were earthen dams removed to improve the water quality of downriver wetlands along the Lehigh River (as indicated by Fig. 9.2). As a result, the ecology of the wetlands has improved.

Camp Michaux Lower Dam was on Toms Run in Michaux State Forest, east of Shippensburg and south of Carlisle. It was a four and one-half-foot high and thirty-two-foot long obstruction built in 1928. It was built for recreation, but was largely abandoned. It was removed in 2017 to provide passage for trout.

Dugan Run Dam was on Dugan Run, a tributary of the Susquehanna River northeast of York, Pennsylvania. It was an eight-foot high and eighty-foot long dam removed in January of 2017.

The two Dunbar Dams were on Dunbar Creek, a tributary of Youghiogheny River, south of Pittsburgh. They were twelve-feet high and five-hundred feet long. They provided water for coke ovens at steel mills. They were removed in 2017 to allow the river to cool and its ecology to improve.

Gorson Dam was on a small tributary of the Delaware River in Lehman Township, just north of Stroudsburg. It was an eleven-foot high and five-hundred and sixty-seven-foot long earthen structure built in 1918 for recreation by the Girl Scouts of America. Being in disrepair, it was removed in May of 2017 to enable fish passage.

Bowmans Creek is west of Scranton. In 1911, Mountain Springs Dam was constructed across Bowmans Creek. It was an eighteen-foot high and seven-hundred and twenty-two-foot long concrete dam built for recreation. It was abandoned and in disrepair. It was reconstructed in 2017 to open two miles of stream to fish passage, and to eliminate the navigation hazard for boaters and recreators.

Solomons Creek is immediately southwest of Wilkes Barre. Solomons Creek Dam was a six-foot high and eighty-foot long stone masonry construction across this Creek. At the time of its removal in July of 2017, it was dilapidated and without purpose.

Sunny Brook is Southeast of Reading and northwest of Philadelphia. Tyler Hill Dam, a thirteen-foot high and one-hundred and sixty-foot long stone masonry dam was constructed across this Brook in 1820 for irrigation purposes. It was subsequentially used for recreation, but it was deconstructed in 2017.

Wildcat Run is a tributary of the Susquehanna River southeast of Harrisburg. Wildcat Run Dam was s fourteen-foot high and one-hundred and twenty-foot long dam across this tributary. It was removed in January of 2017.

Bushkill Creek is a tributary of the Delaware River, close to the confluence of the Lehigh and Delaware Rivers in Easton. To improve water quality and aquatic habitat, four low-head dams were removed from this Creek in 2019.

Conococheague Creek is in Greene Township, south of Shippensburg. In the Fall of 2019, two dams that had formed Scotland Pond on the Conococheague were removed. Four other dams had previously been removed from this Creek, and one other was removed subsequent to the Scotland Pond Dam removals. This opened fifty-seven miles of the Conococheague. When a final dam is removed, all eighty miles of this creek will flow free of obstructions. The removals were financed by grants from American Rivers, the National Fish and Wildlife Foundation, and from the Commonwealth of Pennsylvania.

Eckenrode Mills Dam was on Chest Creek, about half-way between Pittsburgh and State College. It was a stone masonry dam, five feet high and one-hundred feet long, built in 1830. It was obsolete and removed in 2017 to reconnect the stream and habitat for trout and amphibians. The removal effort was led by American Rivers as part of its efforts to rehabilitate the West Branch of the Susquehanna's watershed from its coal mining legacy. Also as part of this effort, the Garmantown Mill Dam and Patton Dam were removed in 2019. The Cambia County Conservation District, Trout Unlimited, and the West Branch Susquehanna Restoration Coalition have expended more than $30 million in their effort to clean the mine drainage and coal refuse in the Chest Creek area. The three dams removed (Eckenrode Mills, Garmantown Mill, and Patton Dam) open sixty-three miles of West Branch Susquehanna River headwater streams to both fish and amphibian passage, all as part of the overall rehabilitation effort of this watershed.

Darby Creek is a tributary of the Delaware River near Philadelphia in Southeast, Pennsylvania (as indicated by Fig. 9.3). This is in the area where in 2004, the "Athos 1" tanker spilled oil that fouled miles of the Delaware and its tributaries including Darby Creek. A $1.6 million effort to restore Darby Creek was funded by the damage settlement and the Commonwealth's "Growing Green Program," plus the National Fish and Wildlife Foundation. As part of this restoration effort, three dams (the Darby Borough Dam, the Hoffman Park Dam, and the Kent Park Dam) were removed. In addition, abandoned railroad trestles were removed, and an eight-hundred-foot section of the Creek plus ten acres of creekbank were restored. The barrier removals allowed resident fish to be reestablished, and passage for American shad, hickory shad, blueback shad, alewife, river herring, American eel, and other migrating fish. The free-flowing creek alleviates flooding and enhances safety for recreators. The restored vegetation improves benthos and prevents erosion. After the removals, juvenile blueback herring were quickly observed returning to the Creek to spawn.[31]

In the Spring of 2021, an eleven-year-old boy was swimming with a friend on Codorus Creek in York County (as indicated by Fig. 9.2). The boy was swept under by the circular and turbulent current of the low-head dam that created the inviting small reservoir. The boy drowned. Low-head dams are dangerous for swimmers, people fishing, and kayakers because they create these turbulent currents. They are so dangerous that since 1998, Pennsylvania law requires that owners of these dams post warning signs to swimmers and recreators. These dams are known as "drowning machines" because of their danger. A 2019 law strengthened the warning sign requirement, but since the Codorus Creek Dam was unlisted on any official list, the owners were therefore not forced into compliance. The Dam was not monitored for the signage by any Pennsylvania authority.[32]

The Jackson Township is ten miles southwest of York in Pennsylvania, and Codorus Creek flows through the Borough of Spring Grove in this Township (Note that another Jackson Township is close to Pittsburgh). This Creek and area is part of the "Indian Rock Dam Flood Risk Management Project" of the US Army Corps of Engineers (ACE). The section of Codorus Creek that contains this above-mentioned unlisted low-head dam is owned by the ACE. This federal agency, however, had no record of the dam's presence. The dam is not on the "US Inventory of Dams" that is kept and maintained by ACE. Pennsylvania's state agencies are now trying to compose a complete list of its dams, and to also identify those that should be deconstructed.[33]

The City of York in Eastern Pennsylvania has long planned a "Codorus Greenway" along Codorus Creek. The urban area of York needs to have its Creek cleaned of the considerable amount of debris within its streambanks. It also needs access points installed on both sides of the Creek, and most important, it needs to deconstruct the Bascule Gate Dam, which

[31] See Laura Craig, PhD (November 20, 2014), *NOAA Blog*, at https://response.restoration.noaa.gov/about/media/after-opening-pennsylvania-creek-fish-watching-recovery-follow.html.

[32] See Harri Leigh (2021, April 1), "Low head dam in Codorus Creek where 11-year-old drowned could be removed," *Fox43*, at www.fox43.com/article/news/local/york-county/low-head-dam-drowning-removal/521-6438dee9-4678-99cd-20b857b3afd2.

[33] Ibid.

lies within the City. Since 2016, the Dam's gate has been stuck in the open position so that it gathers debris, increases the flood damage hazard, and poses considerable drowning risk. Since, 2016, the Dam's operating mechanism and housing have further deteriorated. Silas Chamberlin, an Officer of the "York County Economic Alliance," indicated that, "It's rewarding to see actual work going on in the Codorus Creek, removing a 40-year-old dam that has brought good things in terms of recreation but has posed a safety risk."[34]

In 2018, ACE told the City of York that the Bascule Gate Dam was noncompliant with flood control requirements, and therefore the City needs to remove the dam. Its deconstruction will be part of the "Codorus Greenway Project," which is planned for completion in April of 2024. Silas Chamberlin indicates that, "The Codorus Creek Greenway is all about reconnecting neighborhoods to the Creek itself with new access points, making it attractive and sending a signal that we want to invest in our natural places. We also want to invest in our neighborhoods and their quality of life."[35]

Not all dam changing activities in Pennsylvania are deconstructions. Occasionally, a reconstruction of an existing dam is undertaken. For example, Philipsburg Borough is immediately west of State College in Central Pennsylvania. Cold Stream and Cold Stream Dam are on the east border of Philipsburg. This is an earthen dam for which the impoundment is popular for fishing, and the park around this reservoir is used for recreation of various sorts. In 2009, this dam was evaluated as unsafe by the Commonwealth's *Department of Environmental Protection* (DEP) because its spillway had inadequate capacity. Note that this is not an unusual situation particularly with the increase in atmospheric water content due to global warming since the result of this is that storms have greater precipitation. The Borough owns the Dam and the park around it, so rather than remove the dam, it decided to reconstruct the spillway and to restore the earthen embankment.

The dam and its seven-acre reservoir was originally created in the nineteenth century to power a sawmill and service an ice house for the area

[34] See Leah Hall (2023, December 4), "Demolition of York Dam begins, part of Codorus Greenway Project, *Fox43*, at www.fox43.com/article/news/local/york-county/demolition-york-dam-0rdered-cororus-greenway-project/531-b25b8aea-2bab-48ca-8b04-2727636ac673.

[35] Ibid.

(Ice was cut from the reservoir during the winter season and stored in the icehouse for summer use). The Dam was reconstructed in 1936 after flooding destroyed some of the earthen structure. But subsequent maintenance was sporadic.[36] Nevertheless, during the decades of the 1970s and 1980s, the reservoir was popular for swimming and fishing. At the insistence of the DEP, the Dam was reconstructed in 2016 with a $2.2 million "Community and Economic Development" grant, and a $.2 million line of credit to the Borough.

The "*Mon*"

The Monongahela River (often locally called "The Mon") flows one-hundred and thirty miles from the north-central coal country of West Virginia to Pittsburgh in southwest Pennsylvania (as shown by Fig. 9.1). It joins with the Allegheny River at Point Park in the center of Pittsburgh to form the Ohio River, which along its lengthy course creates the borders between Ohio and West Virginia, Ohio and Kentucky, and also Indiana with Kentucky. The Ohio then drains into the Mississippi River near Paducah in Kentucky. These three Rivers (Monongahela, Ohio, and Mississippi) have historically been major US transportation arteries for barge traffic, especially for the coal shipped along the Monongahela to Pittsburgh. Some of this coal is shipped farther downstream via the Ohio River.[37]

The Monongahela is formed by the confluence of the West Fork River and the Tygart Valley River at Fairmont in north-central West Virginia. It then flows through close-by Morgantown, just ten miles south of the Pennsylvania border. The Monongahela then generally flows northerly along its entire length to Pittsburgh. Along its route, it drains a seven-thousand three-hundred and forty square-mile watershed which in West Virginia is considerably mountainous, and constitutes "underground coal mine country." Over its entire length—including its West Fork and Tygart

[36] See G. Kerry Webster (2017, March 16), "Cold Stream Dam project on Philipsburg nears completion," *The Centre County Gazette*, at www.statecollege.com/centre-county-gazette/cold-stream-dam-project-on-philipsburg-nears-completion/.

[37] As a branch of the US Marine Highway System, the Monongahela River is designated "M-79" by the US Maritime Administration, a division of the US Department of Transportation. See.
www.maritime.dot.gov/grants/marine-highways/marine-highway.

Valley feeding tributaries—it falls more than three-thousand feet. But from the confluence of the Monongahela's two feeder streams, it only falls a relatively flat two-hundred and eighty feet to Pittsburgh's Point Park where it joins with the Allegheny to form the Ohio River. Through utilizing a series of locks and dams, the Monongahela is made navigable along its entire one-hundred and thirty-mile main stem length. It has historically been a transportation artery primarily for coal barges. Along the River's length, the Pittsburgh District Army Corps of Engineers has created, restored, and maintained a series of nine "locks and dams" that facilitate the Mon's barge traffic along with some degree of recreational boating.

In 1817, the Commonwealth of Pennsylvania authorized the Monongahela Navigation Company to construct sixteen dams with navigation locks to primarily facilitate the transportation of coal between what was to become West Virginia (not yet a state) and Pittsburgh.[38] Note that in the US, the early nineteenth century was the age of private turnpikes, bridges, and canals so that "private navigation companies" of this sort were not unusual.[39] During the nineteenth century and up through World War II, "the Mon" was heavily used by the steel mills and other industries of Pittsburgh (such as concrete manufacturing) to obtain its needed coal. Despite the closing of many of the mills in the 1980s and 1990s, the Monongahela was recently still an important waterway for industry. For example, US Steel's "Mon Valley Works"—the Edgar Thompson Plant for basic steel manufacture, the Irvin Plant for steel finishing, and the Clairton Plant for coke production—operated through 2021. (Much of this production has, however, changed with new sustainable technology, and also has shifted to Arkansas.) As a result, coal barges were recently common on "the Mon."

During the early eighteenth century, coal barges were rectangular wooden boats of one-hundred feet in length, twenty feet in width, and six feet in depth. These wooden boats carried three-hundred and thirty tons of coal and required a river depth of seven feet. They had no power source but were carried by the river's current. After the downstream trip, these wooden barges were dismantled and sold as lumber. With the eighteenth

[38] See Bloom, Suzi (2020, October 1), "Southwestern Pennsylvania's Iconic Barges," *Rivers of Steel*, at https://riversofsteel.com/swpa-iconic-barges/. Note that West Virginia did not become a State until 1863.

[39] See Hession and Sardy (1969, pp. 87–89).

century's advent of steam engines, paddle-wheelers pushed the barges and could return them upstream once emptied. Shipyards in the Pittsburgh area produced more than 3,000 of these steamboats between 1811 and 1888.[40]

In 1824, the US Congress authorized the Army Corps of Engineers to remove snags and obstructions along these navigable rivers, and to construct dams and dikes to concentrate and deepen flows to enable navigation. After the Civil War, in 1869 the "Rivers and Harbors Act" authorized the construction of dams and locks, the first being completed in 1885. Currently modern barges on the Mono are pushed by tugboats with combined crews of eleven members. They move as many as fifteen barges at once with a capacity equivalent to one-thousand semi-tractor trailers. The Corps of Engineers maintains the depth of the River to at least nine feet. In 2023, the Port of Pittsburgh was the fourth busiest inland port in the US. Approximately one-thousand barges pass through the "Mon's" nine dams and locks per year.[41]

The US Coast Guard regulates boat traffic along the coastal and inland waterways of the US. This includes the Monongahela, Allegheny, and Ohio Rivers. As of 2019, the Coast Guard has been proposing to establish a "regulated navigation area" (RNA) in the "Pitt Pool," which forms a twenty-four-mile navigational water pool around the City of Pittsburgh. The RNA might well be necessary because of the volume of both commercial (chains of barges) and recreational traffic within the "Pitt Pool." In particular, the volume of recreational boats have substantially increased due to the various entertainment events centered on the "Point Park" area. This is now a popular venue for recreational boaters. The Port of Pittsburgh contains twenty-seven marinas and twenty-one public boat launch sites in the Greater Pittsburgh area. The Pitt Pool alone has twelve marinas and three public boat launches. As a result, over the previous three years (2021–2023) an average of 7,860 recreational boats annually used the three locks along the "Pitt Pool." Recreational boaters, however, are often ignorant of the right-of-way regulations so that the commercial vessels must halt their transit rather than proceed into what they consider to be unsafe and chaotic conditions. It should be noted that the Point Park area contains multiple entertainment venues near the riverfront.

[40] Ibid.

[41] Ibid.

These include a National Football League stadium, a Major League Baseball stadium, a casino, a United Soccer League stadium, an amphitheater, a state park, a science center, museums, a convention center, monuments, and restaurants. All of this encourages travel to the Pittsburgh area by recreational boats who then loiter close to Point Park. The recreational boaters generally anchor or drift in the "Pool" making channels congested for commercial traffic. The Coast Guard's proposed regulations would prohibit persons and vessels from loitering, anchoring, stopping, mooring, remaining, or drifting more than one-hundred feet from any riverbank in the RNA, i.e. the "Pitt Pool area." It would also prohibit impeding the safe passage of another vessel to any launching facility or port facility within the RNA.[42] These regulations would be promulgated and enforced by the Coast Guard in support of the commercial traffic.

West Virginia University's "Water Research Institute," along with the University of Pittsburgh and Carnegie Mellon University, have monitored the water quality of the Monongahela and its tributaries. In particular, they have monitored the water inundation of approximately 1,200 abandoned mines in the Monongahela River's watershed. They found consistent increases in the water levels in these abandoned mines which now contain billions of toxic gallons polluted by sulfuric acid and other contaminants from the coal refuse. If this is eventually released into the Monongahela's tributaries of Dunkard Creek or Ten Mile Creek then this polluted discharge would be catastrophic for the Monongahela's fish and other aquatic life.[43] Although the "Mon" was once considered dead, after the provisions of the 1972 *Clean Water Act* were implemented, over seventy species of fish and aquatic life have increased in population, species such as smallmouth bass. The aquatic life of the Monongahela has been rebounding and will likely continue to do so provided the contamination is contained. This contamination now also includes pollution from the Marcellus Shale gas drilling that is endemic to the area. Over the last half-century, even the aquatic species that were particularly sensitive to pollution have rebounded.[44]

[42] See USCG-2019–0118 at https://www.regulations.gov/docket/USCG-2019-0118 and also www.regulations.gov/document?D=USCG-2019-0118-0002.

[43] See Bob Ventorini and Matt Kinsey (2009), "Monitoring Fishes of the Monongahela River," *Three Rivers Fisheries*, at https://pfbc.pa.gov/images/reports/2011bio/8x04_01mon.html.

[44] See Bob Ventorini (2010), "Monongahela River Biological Monitoring Study," *Three Rivers Fisheries*, at https://pfbc.pa.gov/images/reports/2011bio/8x04_01mon.html.

An essential question concerns the extent that the "Mon" will continue to serve the needs of Pittsburgh industry at the expense of the recreational needs of the population. The Port of Pittsburgh is governed by a private "Commission" formed in 1992. This consists of fifteen members who represent various industry-related organizations. The Commission's task is to promote the business development of its associated inland waterways and transportation system. Mary Ann Bucci is the Commission's "Executive Director." She explains, "Pittsburgh is exceptional in that it is an inland river port without a publicly owned or operated port facility. All of the terminal facilities are privately owned and operated."[45] But the demand for coal is rapidly declining, and the condition of the Port's infrastructure is declining with it. The Monongahela's locks must be periodically restored. There are seventeen lock systems in the Port of Pittsburgh's district, and nine of these are on the Monongahela. Once restored, a lock system is expected to last fifty years. For example, the Braddock Locks were built in 1953, and needed to be restored in 2004. The Opekiska Locks were restored in 1964, but in 2012 they were closed to weekend recreational traffic. In 2015, however, the Opekiska Locks were reopened for summer weekend recreational traffic. The Elizabeth Locks were restored in 1980 but are now being permanently removed along with the Elizabeth Dam. This removal will begin in 2024, and will restore thirty miles of unimpeded navigable waterway. Once completed, the recreational net benefits derived are estimated at $200 million for the region.

Summary Conclusion: From Industry to Restoration

The Commonwealth of Pennsylvania poses a most significant example of polluted remnants from old heavy industry amid a geography of extensive complexes of rivers and watersheds. The Commonwealth leads the US in dam removals, but these were generally small (low head) dams on small streams originally constructed for flood control, for the creation of potable water reservoirs, and for industrial uses. These structures lost

[45] See Heater Ervin (2018, April 2), "The Port of Pittsburgh: a Vast River Traffic Region," *The Waterways Journal Weekly*, at www.waterwaysjournal.net/2018/04/02/the-port-of-pittsburgh-a-vast-river-traffic-region/.

their original purpose. They became derelict dams that have been deconstructed at significantly high rates. But Pennsylvania's dams were also constructed for navigation and hydropower. These larger structures are slowly being redesigned, and in some cases removed.

The geography of Pennsylvania creates its flood control and river navigation problems. Ecological restoration and recreation create the complexities for the Commonwealth's planning and management of its navigation and flood control solutions. Derelict dams are being removed, but these are the easy tasks. The problems of flood control and navigation are slowly being addressed, and this will continue. It appears that it is the very nature and history of the Allegheny, Susquehanna, and Monongahela Watersheds that makes progress in restoration slow but inevitable.

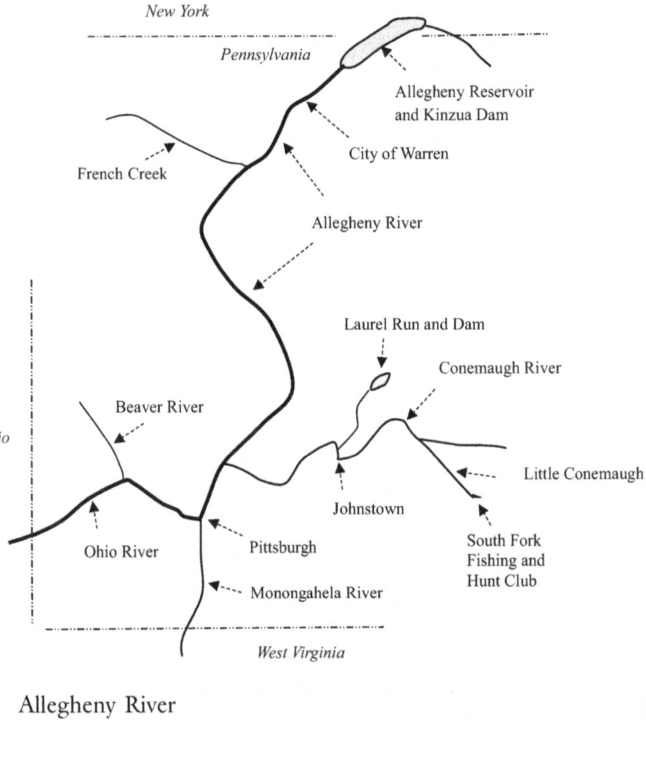

Fig. 9.1 Allegheny River

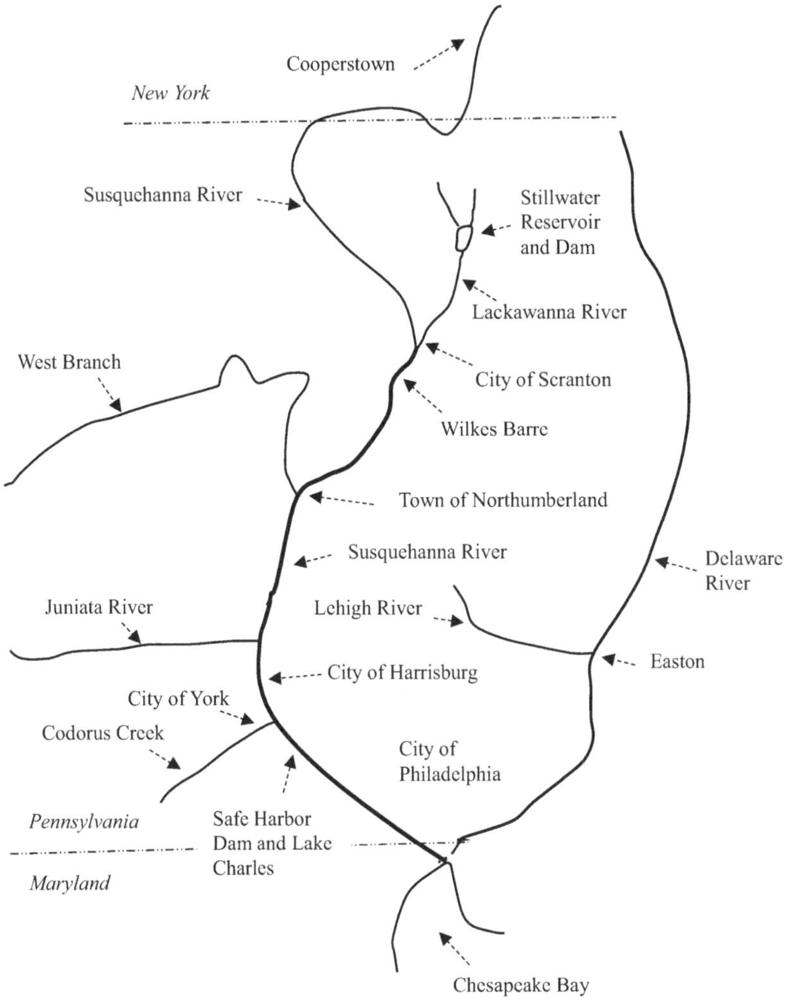

Fig. 9.2 Susquehanna River

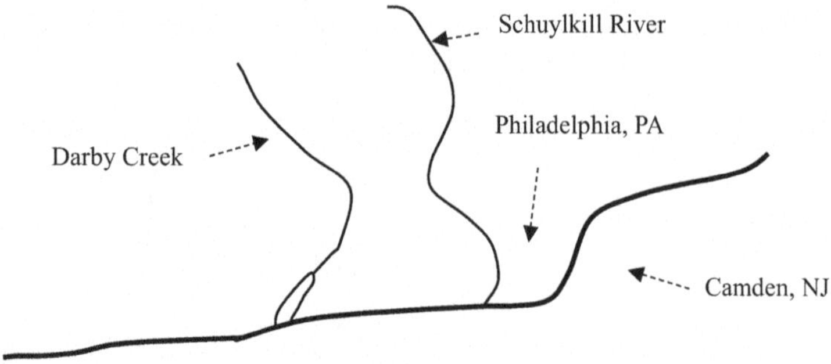

Fig. 9.3 Darby Creek in Philadelphia

References

Blakemore, E. (2023, September 28). How America's Most Powerful Men Caused America's Deadliest Flood. *History*, A&E Television Networks, see www.history.com/news/how-americas-most-powerful-men-caused-americas-deadliest-flood

Hession, C., & Sardy, H. (1969. *Ascent to Affluence, A History of American Economic Development*. Allyn and Bacon.

Ingham, J. (1991). Iron and Steel in the Pittsburgh Region: The Domain of Small Business. *Business and Economic History, 20*, 107–116.

Lorant, S. (1999). *Pittsburgh, The Story of an American City*, 5th ed. Derrydale Press, ISBN 0-9674103-0-4.

Majors, D. (2007, April 26). Pittsburgh Rated "Most Livable" Once Again. *Pittsburgh Post0Gazette*. www.post-gazette.com/pg/07116/781162-53.stm

Robinson, R. (2023). *Environmental Advocacy and Local Restorations*. Springer Nature.

Robinson, R. (2021). *Environmental Organizations and Reasoned Discourse*. Springer Nature.

Schultz, G. (2023, May 23). Public Outcry, a Broken Treaty, and the Controversial Construction of the Kinzua Dam. *National Archives,* Posted in "American Indian History."

Winant, G. (2021). *The Next Shift: The Fall of Industry and the Rise of Health Care in Rust Belt America*. Harvard University Press.

CHAPTER 10

Dam Safety Standards, Rehabilitations, and Failures

Introduction: Federal Dams and Safety

This book is primarily about the dam removal movement, but this movement is entwined with the dam reconstruction and safety progress that the US is struggling to properly organize and finance. The reconstructions generally include attempts at establishing modern fish passage facilities that might enable migratory fish (anadromous species such as shad, salmon and steelhead trout) to reach their spawning grounds, but they also include redesigned spillways and other safety improvements. Dam removals would, of course, eliminate the fish passage problem, but flood control via a patchwork of dams, and hydropower are now so engrained in our national river management systems that complete elimination of dams is not currently on the political horizon. From the standpoint of environmental restoration, we are now pursuing "second best" solutions which currently allow for some dam removals and reconstructions, and we may be doing this for generations to come.

In this vein, dam safety standards for both construction and maintenance have been a national concern for only the last fifty years (since the early 1970s). Unfortunately, the US had to experience some major dam failures before finally recognizing that an effort toward establishing national standards was especially warranted. The record indicates that the older state-level inspection requirements would not prevent repetitive disastrous failures. But the problem was not just the domain of states;

R. M. Robinson, *Restoring America's Rivers*,
https://doi.org/10.1007/978-3-031-81758-8_10

federal agencies were also cavalier in establishing and applying their own safety standards.

Various federal agencies construct and operate dams. The most significant of these agencies are the Army Corps of Engineers and the Bureau of Reclamation. (See Chapter 4 for the Bureau's role in the Colorado River Storage Project.) The Bureau particularly suffered through some disastrous failures before accepting some necessary lessons concerning dam design and construction. These lessons are reviewed here. But before this examination, it is necessary to review some of the motivations for the Bureau's attitude of proliferating dams in support of irrigated agriculture in the midwestern and western states. This "attitude" has much to do with the Bureau's aggressive leader for the decade of the 1960s, Floyd Dominy.[1] Chapter 4 reviews Dominy's aggressive leadership style during his tenure as Director of the Bureau during the 1960s. It should suffice here to point out that the Bureau's aggressive expansion of dams during this period did not emphasize safety as a high priority. Big western agriculture needed irrigation, and as a result federal dams proliferated to provide this necessity. The sites for these dams were consequently extended to locations previously thought unsafe such as the Teton Dam's location reviewed below.

As expressed by Marc Reisner, Southern Idaho's Snake River Plain is an otherwise "useless place made rich" by irrigation (Reisner, 1993, p. 384). This is an old volcanic plain that similar to the achievements of the Central Arizona Project and California's Central Valley (see Chapters 4 and 5), was completely transformed into large, irrigated plots that generate considerable wealth. The problem with this volcanic granular soil is that it is overly porous. Water tends to drain through the soil quickly so that annual watering of more than ten feet per year is required. This is an inordinate amount of water. To ensure adequate irrigation, a Teton River Dam for the eastern section of the Snake River plane was studied as a possibility during the 1920's and the 1930s. It was eventually constructed as an earthen dam in the early 1970s. (Fig. 10.1 indicates the River's location.)

[1] Marc Reisner (1986) wrotes much about Dominy and the Bureau that was not complimentary of either. His title of *Cadillac Desert* reflects his judgment of the foolishness of attempting to convert the US western desert lands into an agricultural paradise and national breadbasket.

The immediate impetus behind the planning for the Teton Dam was the occurrence of a drought in 1961 followed by a flood in 1962. Following these events, the local Irrigation District found it easy to organize support for the dam in the substantially homogeneous Mormon-agricultural area of Southern Idaho. Crop production (potatoes and sugar beets) had been growing steadily (even during the drought) given the abundantly available groundwater. But a big federal project to be paid for by US taxpayers was nonetheless attractive as a subsidy to the Southern Idaho agricultural interests. One barrier to obtaining this subsidized water, however, was 1969's *National Environmental Policy Act* (NEPA) which required an *environmental impact statement* (EIS) for any new project such as a dam.

In 1969–1970, the reactions to the environmental degradation of the post-World War II economic expansion coalesced and led to establishing the ambitious *National Environmental Policy Act* (NEPA) which is generally recognized as the first major environmental law in the US. It requires every federal agency to assess the environmental effects of any proposed action prior to its undertaking. Consequently, the Bureau's view of "dominating nature" was inconsistent with the 1970s environmental movement that created NEPA.

NEPA establishes a national policy applicable for all federal agencies "to use all practicable means and measures, including financial and technical assistance, in a manner calculated to foster and promote the general welfare, to create and maintain conditions under which man and nature can exist in productive harmony, and fulfill the social, economic, and other requirements of present and future generations of Americans." (See 42 U.S.C. 4331(a)). To further this rather broad and bold motivation, NEPA requires (1) a detailed EIS that analyzes the short-term and long-term effects of the proposed actions by federal agencies, plus (2) the identification of any adverse effects that cannot be avoided, and (3) an analysis of alternatives to the proposed actions. (See 42 U.S.C. 4332(2)(c).)

To ensure that federal agencies will cooperate, NEPA also established the *Council on Environmental Policy* (CEP) within the Office of the President. The *Council* was to oversee NEPA's implementation through interpreting environmental regulations and providing guidance to the Executive Branch of the US Federal Government. The purpose of the CEP's creation was to provide the primary impetus to the federal government's environmental policies as manifested in the President's various

"Executive Orders" pertaining to such important matters as (i) floodplain management, (ii) wetlands protection, (iii) coral reef protection, (iv) invasive species elimination, (v) migratory bird protection, (vi) wildfire risk management, (vii) forest and rangeland protection, and (viii) climate change remediation as examples of some recent "Orders."

But in 1970–1971, the requirements of the newly required EIS were yet to be established. The Bureau's EIS for the proposed Teton Dam was only fourteen bureaucratically designed pages. This cavalier brevity provided meaningless analysis since the Bureau assumed that the EIS was just a bureaucratic form of little consequence. NEPA's requirements were yet to "have teeth." But the Teton Dam's vacuous EIS drew the attention of the "Idaho Environmental Council," a group that included scientists from the nearby Department of Energy's "Nuclear Testing Station" in Idaho Falls. Their analysis of the EIS revealed that the Bureau's cost–benefit calculations were not objectively established. They were heavily biased in favor of producing a high benefit-to-cost ratio. A more realistic analysis, however, indicated that the expected benefits would be less than one-half of the direct costs of providing the irrigation, and this ignored the most essential fact that a beautiful and ecologically valuable river (the Teton River) would be destroyed by the dam. This appeared to totally contradict the intent of NEPA. The US taxpayers would be heavily subsidizing the potato and sugar beet production of Southern Idaho and would receive in return less than half as much in value in the form of the produce grown, i.e. the costs would be more than twice the benefits. Although nationally the politics were shifting toward environmental protections, this was not, however, the case in Southern Idaho.

Given the accurate but adverse benefit-to-cost calculations, why was the Teton Dam constructed? The political support for this dam was organized by the Idaho Falls local newspaper, the *Post-Register*. This local influential institution believed it had significant roles to pursue in bringing as many federal dollars as possible to subsidize the area's agriculture while also campaigning against "big government" and the Nation's "creeping socialism."[2] (The contradiction in these pursuits should be obvious). The *Post Register* propagated the politically attractive conception that the Teton Dam was really a flood control project (which should have meant it would be constructed by the Corps of Engineers, who had

[2] See Reisner (1993, p. 388).

jurisdiction over flood control projects, rather than the Bureau of Reclamation), and also that the "Idaho Environmental Council" was just a committee of "extreme environmentalists."[3] The *Post Register* rejected any rational cost–benefit approach for evaluating any federal dam or irrigation project. Any sort of rational analysis that indicated a dam was not warranted should be simply ignored. It might not be worthwhile from the standpoint of the overall taxpayer, but it would be worthwhile from the standpoint of local subsidized farming interests.

Under the Post Register's political leadership, Idaho's Congressional representatives were strongly in support of constructing the Teton Dam. Their support had become so hardened that Jerry Jayne, the President of the "Idaho Environmental Council," and who was also a scientist with the Department of Energy, stated "I can talk to the loggers; I can talk to the ranchers; I can talk to the mining companies; but I can say nothing to the irrigation farmers. They're not reasonable. They don't listen. They're true believers. They're like communists, only in reverse."[4]

The canyon where the dam was to be constructed, however, was not appropriate for such an earthen structure. The eastern Snake Plain is all basaltic rock, the remnants of volcanic activity. Yellowstone and its volcanic remnants are near to the east, and the Cascade Mountains—also volcanic—are to the west. Professor Robert Curry, who instructed geology at the University of Montana, was critical of this site for an earthen dam. Curry was a consultant for the considerable mining interest of the area. The Idaho Environmental Council also consulted with Curry concerning the proposed Teton Dam. Curry recalled that, "The stuff they were going to build the dam on – all those ash flows and rhyolitic rock – may look solid to you, but it's really a veneer, ... It's brittle; it's cracked. It could peel off .... They were going to scrape away the worst of it and then say that they were anchoring the dam in bedrock. But it isn't really what most geologists would call bedrock. The dam was not going to have a true bedrock foundation. It was such an obvious lousy site to a trained geologist, it makes you wonder what happens to human judgment inside a bureaucracy."[5]

[3] Ibid.

[4] Ibid, p. 389.

[5] Ibid, p. 390. *Rhyolitic rock* is very rich in silica. It is generally brittle and cracks easily as with obsidian. Unlike granite-type rock, it is not an appropriate bedrock for heavy structures such as dams.

In an internal memorandum, the Bureau's geologists recognized that within the previous five years, five earthquakes occurred within thirty miles of the planned site of the Teton Dam. Official Bureau documents in support of the dam, however, did not recognize this danger.

But other geological examinations in the canyon where the dam was to be located revealed two significant additional concerns: (1) water tests revealed extraordinary levels of ground absorption, and (2) fissures in the rock walls were very large, even cavernous. Both of these characteristics questioned the integrity of the dam since they would promote significant leakage that would cause the dam to weaken. Bureau officials argued that "grouting" (a process of filling in fissures in the dam's structure and contact walls) would stabilize the Dam and its abutments. During construction, however, the volume of grouting was more than twice that expected. But this was still not adequate. Although the structure appeared to be only close to completion on October 3, 1975, nevertheless the Dam's gates were closed at that time and filling the reservoir began.

But the winter of 1975–1976 saw unusually large snowfall in the Teton Mountains. The Spring melt brought larger than safe flows for filling the Dam. For an earthen dam, the Bureau's standards for a safe fill rate was one foot per day. This rate would allow problems to be quickly discovered and fixed. But on March 3, 1976, the fill rate doubled to two feet per day. The reason for this was that the Dam's spillways and outlet were not yet completed, and as a consequence the gates remained closed. Subsequently, by Mid-May of 1976 the rate of fill reached four feet per day. On June 3, leaks were perceived on the abutments, and the flow rate was recognized as uncontrollable. The earthen Teton Dam was washed way in minutes. A twenty-foot high wall of water rushed down Teton Canyon until it reached the Snake River.

Eleven people died in the subsequent flood. Four thousand homes were either destroyed or badly damaged. The topsoil was permanently washed away in the Teton River Valley. The Bureau's reputation for safe dam construction was also left destroyed.

The experience of the Teton Dam's failure reveals four defects in the Bureau of Reclamation's procedures related to the project's planning and construction:

1. This dam was pursued because of political pressure, not because it was warranted by rational cost–benefit analysis.

2. In evaluating the dam's feasibility, the essential geological information and characteristics were cavalierly ignored.
3. The Bureau's own standards for completion-before-filling, and the necessity of a slow fill rate, were ignored because the spillway and gates were not yet operable.
4. Destruction of one of the Nation's more beautiful rivers was undertaken in 1975 without a serious environmental impact analysis. This occurred despite 1969's prior passage of the *National Environmental Policy Act* (NEPA).

At this time, the ability to enforce a national dam safety policy appeared problematic. Apparently, the politics of dam safety had yet to ripen. A further series of tragic dam failures was needed to awaken serious concerns. By 1977, these tragedies occurred. Attempts at developing such a policy are examined next.

Developing the Politics of Safety

One of America's worst and most politically unsettling dam failures occurred in West Virginia in the early 1970s. On February 26, 1972, a series of coal slurry impoundment dams burst on the Middle Fork of Buffalo Creek causing a flood of thick black slurry water to rush through sixteen coal mining communities (see Fig. 10.2). The rushing flood killed one-hundred and twenty-five people. An estimated one-hundred and thirty-two million gallons of black waste solution crested at thirty feet while rampaging down and through the coal towns of Buffalo Creek Hollow. Five thousand people (generally of mining families) lived in the sixteen towns of the Hollow. Besides those killed, eleven-hundred and twenty were severely injured; five-hundred and fifty houses were destroyed; over four thousand people were left homeless.

Between 1962 and 1972, three impoundments were built on the Middle Fork of Buffalo Creek. "Dam 1" was built farthest downstream, and after it filled with the pumped slurry waste from an open strip mine, "Dam 2" was constructed several hundred yards upstream. The process was repeated for the farthest upstream "Dam 3." But the three dams were not constructed according to any engineering safety standards. Dams 2 and 3 were constructed on coarse mine waste rather than bedrock. Although the trio of impoundments were inspected by mine safety officials, these inspectors were not trained in the engineering of dams and consequently not capable of evaluating dam safety. The impoundments were inadequate for restraining the several inches of rain that fell on the area immediately prior to the Saturday morning of February, 26, when fifty-acre feet of new rain filled the pool of Dam 3. The pool level rose to less than three feet of the impoundment's crest when Dam 3 began to crack. Nevertheless, no evacuation order was given because no trained authority was on hand. Dam 3 failed at 8:00 AM. Dam 2 and Dam 1 failed within minutes after Dam 3. The approximately four-hundred-acre feet of slurry water moved through the Hollow at twenty feet per second. It took only a few minutes to destroy the first three miles of towns.

West Virginia's Governor Arch Moore Jr. quickly appointed an ad hoc commission to investigate the causes of the disaster. But the appointees consisted of nine men with ties to the coal industry. The commission's purpose appeared to be to whitewash the responsibility of both Pittston Coal (the owner of the strip mine and slurry impoundments) and the State's safety regulators. In response to this white-washing, a second commission was formed—The Citizen's Commission—with the intention and results of finding both Pittston Coal and the State's regulators as negligent. A more serious investigation of dam safety issues was therefore warranted. Norman Williams, the Deputy Director of West Virginia's Department of Natural Resources, called for making strip mining of coal illegal in the State. He claimed that the negative externalities involved with strip mining would eliminate this type of coal extraction if regulations forced these costs to be internalized by the mining companies.[6] The efficacy of strip mining, and its cruel externalities dumped on a public resource such as Buffalo Creek, poses one important set of issues entwined with this disaster, but public safety associated with dam constructions pose others.

The State of West Virginia sued Pittston Coal for $100 million in disaster relief damages. In 1977, three days before he left office, Governor Moore accepted a settlement of $1 million from Pittston.

As the politics of dam safety evolved, it was not the State of West Virginia that addressed the issues, it was the federal government. Because of the Buffalo Creek Dam failure disaster, in July of 1972, the US Senate held hearings on dam safety. Senator Henry Jackson opened these hearings by saying, "These occurrences (pointing to Buffalo Creek and other

[6] See Montrie (2003).

dam failures) point up the grim fact that there are innumerable dams in existence throughout the United States which are under no effective public control to insure that they were competently designed and constructed initially, and that they are being adequately inspected and maintained, or that the design remains adequate under current hydrological conditions." These hearings led to the passage of Public Law 92-367, which created the National Dam Inspection Program, and gave the Corps of Engineers (ACE) the authority to begin an inventory and inspection of all the Nation's dams. Nevertheless, the inventory list was slow to develop and safety standards remained inadequate until another dam failure occurred in 1977, i.e. the Kelly Barnes Dam disaster in Georgia.

Toccoa Falls has an impressive one-hundred and eighty-six-foot drop on Toccoa Creek in far northeast Georgia (see Fig. 10.3). The Kelly Barnes Dam was located on Toccoa Creek where it impounded the forty-acre Kelly Barnes Lake. This Dam was four-hundred feet in length with a crest of twenty feet in width. It was located about one-half-mile upstream of Toccoa Falls Bible College. The Falls were located between Kelly Barnes Dam and the Bible College. In the early morning of November 6, 1977, after five days of rain (seven inches of rain total) Kelly Barnes Dam failed at 1:30 AM. An estimated one-hundred and seventy-six million gallons of floodwater washed downstream through the Bible College killing thirty-nine people on its Campus.

The Kelly Barnes Dam had evolved from at first being a wooden spruce structure to being converted to a rock-crib structure, and then finally to an earthen dam. The rock-crib structure was built in 1899 to power a small hydroelectric generator. But in 1937 Toccoa Falls Bible Institute constructed an earthen dam over the rock-crib structure. It had a low outlet with a ceramic intake on the upstream side of the dam. After World War II, in order to increase the power capacity, the height of the Dam was raised to forty-two feet above bedrock. But in 1957, the "Institute" evolved into a "College" with greater power needs. Its power source was consequently changed from using the hydropower dam to using an external source from the electric grid. The maintenance of the impounded reservoir, however, was ignored, although it continued to be used for recreation. In effect, the dam was abandoned.

If we apply today's standards (see below) to the Kelly Barnes Dam as it existed in 1977, it would be classified as "high hazard" because of the at-risk populace on the campus below it, and it would also be classified as in "unsatisfactory condition" because of its blocked outlet and spillway that resulted from its complete lack of maintenance. Trees and brush had grown to completely cover and weaken the earthen structure. Over the previous twenty years, maintenance was nonexistent, and inspections were performed only by local firefighters who were untrained in dam safety evaluations.

The day after the tragic dam disaster, Governor Busbee of Georgia asked President Jimmy Carter for technical assistance in assessing the engineering, hydrological, and other safety issues related to the failure. This was immediately granted by President Carter who assigned the US Army Corps of Engineers (ACE) the responsibility for the analysis. President Carter, however, also directed the Corps to inspect dams nationwide. A three-year investigation revealed deficiencies in the majority of non-federal dams. Follow-up investigations by the National Academy of Engineers and the Federal Emergency Management Agency (FEMA) revealed that the Nation's state-oriented dam safety regulations and inspections were largely inadequate. This was the revelation that finally brought about a more organized approach to dam safety standards.

In response to these investigations, the Association of State Dam Safety Officials (ASDSO) was organized in 1983. They organized National applicable regulatory standards as reviewed below. Prior to their review, it is revealing that ASDSO found the following relevant facts about US dams at that time[7]:

- There are more than 88,600 non-federal dams in the US.[8]
- The number of "high-hazard potential" dams (see below for this definition) where loss of life is probable should the dams fail has increased almost twenty percent over the previous ten years to more than 16,000.
- Seventy percent of US dams are regulated by states; five percent are regulated by the federal government; and the remaining twenty-five percent are unregulated.
- Sixty-five percent of dams are privately owned; thirty-one percent are government owned; the rest have unknown ownership.

[7] The ASDSO "standards" and other information is presented in "The Cost of Rehabilitating Dams in the United States," Association of State Dam Safety Officials, 2023, at Info@DamSafety,Org.

[8] Federal dams are those owned by federal agencies such as the Bureau of Reclamation, or ACE.

- Many risky dams can only be rehabilitated through loans or grants from either governments or NGOs. Half of the states have programs for grants or loans to rehabilitate risky dams. Federal grants have been increasing for these rehabilitations (as reviewed below).

ASDSO has now formed various classifications for inspected dams:

- "High Hazard Potential" means that if the dam fails, then there is a reasonable probability of loss of human life.
- "Significant Hazard Potential" means that in case of failure, there is no reasonable probability of loss of human life, but there is a reasonable probability of economic loss, or environmental loss, or disruption of society's valuable infrastructure such as bridges and roads.
- "Low hazard potential" means that in case of failure, there is no probable loss of human life, or probable significant economic loss, or environmental loss, or social disruption.

Given these classifications of potential effects, ASDSO also suggests the following classifications for the safety conditions of inspected dams:

- "Satisfactory" means that there are no safety deficiencies existing in the dam. Acceptable performance is expected under all likely or even extreme hydrological and/or seismic conditions and risks.
- "Fair" means there are no dam safety deficiencies under normal conditions, but more extreme hydrological and/or seismic conditions could pose safety issues.
- "Poor" means that there are dam safety deficiencies under conditions which might realistically occur. Remedial action is therefore necessary.
- "Unsatisfactory" means a dam safety issue is recognized that requires immediate or emergency action.

A dam that is classified as both "high hazard" and in "unsatisfactory or poor condition" would therefore be in immediate need of remediation. A dam classified as both "low hazard" and in "satisfactory condition" would have no priority for remediation. Mixtures of "fair" or "poor" condition with "significant hazard potential" would have lower but still some priority for remedial actions. In 2023, the ASDSO estimated that repairing the nation's non-federal dams would cost $157.5 billion, and that $34.1 billion is needed just for the "high-hazard" classified dams.

The challenges to the safety of US dams stem from the age of the Nation's dam infrastructure, plus endemic original but still existing design flaws, and the changing new hydrological factors associated with global warming. These safety challenges are manifested in numerous events of recent years. A few of these are reviewed here. For example, the 2017 "Oroville Dam emergency" and the 2019 "Spencer Dam failure" illustrate the effects of unanticipated extreme stress on current structures. These two events are reviewed next.

Although Southern California is desert, the Sierra Mountains of Northern California are rain forest. In recent years, the global warming phenomena appear to be creating intense "atmospheric river storms" that form over the Pacific Ocean and that then transport large amounts of precipitation to the Sierra's rain forest. The dam structures of this area are consequently stressed for controlling the resulting flood waters. The events of February, 2017, involved the Oroville Dam on California's Feather River which clearly manifested some of these new hydrological problems.

In late 1955 and early 1956, floods on the Feather River reached record levels up to that date. Sixty-four deaths and $200 million in property damage resulted from these floods. The State of California constructed the Oroville Dam on the Feather River (seventy miles north of Sacramento) to control these floods. (See Fig. 5.2 in Chapter 5 for the location of this Dam.) The Dam was constructed by the California Department of Water Resources (DWR) as the tallest earthen US dam. It is located upstream and east of the City of Oroville for which it provides flood control and hydroelectric power. The Dam's reservoir, Lake Oroville, is the second largest in California.

The Oroville Dam is one of the large subprojects that created the overall California State Water Project. Construction began in 1961 and was completed in 1968. These projects distribute the flow of the Sacramento and San Juaquin Rivers into California's Central Valley and also the California Aqueduct for the purpose of agricultural irrigation and municipal use.

The rainy winter of 2016–2017 was California's wettest for the previous hundred years. The flows from the Feather River into Oroville Lake were so heavy that in January, the spillway on the Dam was opened to lower the Lake's level, and it again was opened for drainage in early February. On February 7, however, engineers noticed a large section of the concrete spillway and its foundation had eroded. The damage was too severe to repair without opening the secondary emergency spillway and closing the damaged primary spillway. The flows were so substantial, however, that this considerably eroded the secondary earthen spillway so that its releases had to be halted. The erosion caused by the secondary spillway's releases were so extensive that the integrity of the entire Dam became questionable. Officials therefore had to release more flow from the primary spillway which further eroded that spillway and its foundation. Because of continued excessive flows and the high degree of erosion, 188,000 downstream residents had to be evacuated on February 12.

Subsequent investigations indicated that the concrete and rebar construction of the primary spillway was defective and therefore inadequate. In addition, the backup secondary spillway was poorly designed with respect to the positioning of the flow. One must keep in mind that these were the characteristics of a major dam of "high-hazard" that was constructed in the 1960s. In addition, inadequate prior inspections were found to be a significant cause of the emergency. Furthermore, with the increased precipitation resulting from global warming, plus the original faulty dam design, plus faulty construction, and inadequate inspections, the current efforts of the ASDSO certainly appear well-timed and needed. A "high hazard" dam should not be constructed with design flaws, nor with inadequate inspections during construction.

But in addition to the Oroville problems, the events of the Spencer Dam also provide lessons for safety supervisors. Spencer Dam was on the Niobrara River in north-central Nebraska, five miles southeast of the town of Spencer and thirty-nine miles upstream of the Niobrara's confluence with the Missouri River (see Fig. 12.1 in Chapter 12). Spencer Dam was constructed in 1927. It was thirty-seven hundred feet long and twenty-six feet high. The combined length of just the powerhouse and spillway section was four-hundred and four feet. It had nine gates for releasing flows. The rest of the dam was earthen embankment. The Dam's reservoir was 16,847 acre-feet in capacity. In 1935, the Dam was partially breached by an ice flow that blocked the gates from opening. This ice problem was endemic to this Dam.

The Niobrara River is a tributary of the Missouri River. Rivers in the Missouri River Valley carry significantly large amounts of silt. The silt carried by the Niobrara required "sluicing" by opening the gates for two weeks per year to remove the built-up sediment. The Spencer Dam was constructed for flood control and hydropower, but the hydropower was terminated in 2015 due to the increasingly uneconomical costs of generating electricity in the exceedingly silty river.

Although the Niobrara River had become a significant source of irrigation, its environmental conservation had become more important, and the intention (specified in a plan) was to have the nine spillway gates removed so that the River could flow more freely. Its reservoir was no longer needed. But in 2019, the gates had yet to be removed. In addition, the Dam's management had no records of the previous ice problems that plagued the Dam; it appeared to have no operating knowledge of ice preventing the gates from opening if they were left in place. Consequently, the dangers associated with not being able to open the gates during winter floods were not recognized by the Dam's operators because knowledgeable and informed supervision was not present. In effect, the Spencer Dam had been abandoned in a similar way to the Kelly Barnes dam reviewed above. On March 14, 2019, after a severe storm across the Northern Plains, heavy ice broke up on the River and blocked the opening of the gates. The Dam was breached. The downstream flood washed out bridges and roads. It destroyed some commercial buildings, a farmhouse, and also killed a farmer.

The implicit danger of large dams, especially when they are abandoned, needs to be recognized. Such is a key purpose of the ASDSO. Institutional knowledge of the effects of ice damage on the spillways of dams, and the dangers this poses, requires the existence and effective functioning of an association such as the ASDSO. Further problems are posed by the "extreme" conditions of climate change, and these conditions demand revised inspection standards for all "high hazard" dams that might be affected. Some further evidence of this conclusion is presented below through a review of dam safety problems in the States of New York and Vermont.

Dam Inspections in New York

The State of New York has seventeen watersheds that span fairly divergent geographies. Their geographies vary from the Lower Hudson Estuary of the significant metropolitan area of New York City, to the suburban robust river area of the Upper Hudson and the State Capital at Albany, to the sizeable Adirondacks Mountains of the north-central resort area of the State, to the series of ridges that parallel the coast of Lake Erie and that form the Finger Lakes of west-central New York. In addition, the Allegheny, Susquehanna, and Delaware Rivers all originate in New York, but they drain its southern border regions into Pennsylvania.

The State of New York has more than seven-thousand dams. They range from one foot to two-hundred and thirteen feet in height. The State regulates these through its Department of Environmental Conservation (DEC). It applies a "high hazard" rating (see the above ASDSO classifications) to those dams for which failure is expected to result in loss of human life. But besides the "what if" rating, the condition of these dams can also be classified as "unsafe," meaning it is highly likely to fail; or "unsound," meaning it has such serious deficiencies that its safety is questionable. The State also has a "deficiently maintained" classification that indicates its deficiencies are in need of correction for safety reasons.

Of New York's approximately seven-thousand Dams, the DEC is attempting to inspect nineteen hundred and twenty. These dams have an average age of eighty-five years, and many are over one-hundred years old. New York attempts to inspect all "high hazard dams" at least every two years. As of March of 2024, the DEC lists ninety-three "high hazard dams" that are also classified as either in "poor" or "unsatisfactory" condition. In August of 2024, nine of these dams split $766,000 in restoration funding from FEMA. (Reviews of federal rehabilitation funding are presented below.)

Despite a system of classifications and inspections, New York has suffered some dam failures that should be reviewed here. One of the more significant breaches occurred to the Hadlock Pond Dam in Fort Ann, a small community approximately fifty miles south of Lake Champlain. This was an old earthen dam first constructed in 1896, but reconstructed only a short time prior to its 2005 failure. Hadlock Pond was two-hundred and twenty acres that were only used for recreation. Six permanent homes and fourteen camp-type dwellings were located on its shore. The Dam and Pond were town owned; this was considered private ownership which required only privately arranged inspections according to NYS's safety requirements. This dam was repeatedly classified as in "poor condition." It was not well maintained.

On July 2 of 2005, the Hadlock Dam was breached. The flood destroyed seven homes and washed out local roads and State Route 149. An investigation report composed by an appointed independent engineering company revealed that an internal erosion burst a sixty-foot wide hole in the four-hundred and fifty foot long earthen dam. The spaces left in the earthen structure during its construction, plus buried frozen soil, likely caused the failure. An ongoing proper inspection during construction might have prevented these defects.

The Dams of the Lake Champlain Watershed

Lake Champlain separates northeastern New York State from northwestern Vermont. It is a freshwater lake of hour-hundred and ninety square miles—the thirteenth-largest lake in the US. The City of Plattsburg in New York is on the Lake's western coast; the City of Burlington in Vermont is on the eastern coast. Lake Champlain drains north via the Richelieu River which empties into the Saint Lawrence Seaway. The Champlain Canal connects the Lake to the Hudson River.[9] The Lake is one-hundred and seven miles long, but only fourteen miles across at its widest section. It lies between the Green Mountains of Vermont and the Adirondack Mountains of New York where it drains large sections of both.

The La Platte, Lamoille, Missisquoi, Poultney, and Winooski Rivers, plus Lewis Creek, Otter Creek, and Little Otter Creek all flow into Lake Champlain from Vermont. The Ausable, Little Ausable, Bouquet, Great Chazy, Little Chazy, La Chute, Salmon and Saranac Rivers, plus Putnam Creek, all flow into Lake Champlain from its New York side. The geography of Lake Champlain is of a shallow lake (forty-feet at its maximum depth) that resembles a widened river with numerous tributaries. As such, it suffers from two endemic problems: (1) phosphorus and other nutrients flow into the lake from the farms and municipalities along the Lake's shores and tributaries, and as a result (2) toxic algae blooms plague these waters every summer, which is a problem especially exacerbated by global warming.

To maintain the aquatic life of the Lake, in 1996 the "Lake Champlain Basin Program" was organized by the Governors of Vermont and New York, plus the US EPA and officials of Quebec. Its goal has been

[9] The Champlain Canal is a sixty-mile connection of the southern tip of Lake Champlain with the Hudson River, which it joins at Schenectady in New York. It was constructed simultaneously with the Erie Canal, and completed in 1823. It was created to primarily serve the agricultural and wood product interests of the Lake Champlain Basin. Its current primary use is by recreational boaters. See Whitford (1906) for a history of this canal.

to reduce phosphorus and other toxic contaminants from flowing into the Lake. It also seeks to control nuisance invasive species. This "Program" has not been entirely successful, particularly in the Missisquoi Bay of the northeast corner of the Lake. The problem consists of the blue-green algae blooms (cyanobacteria). Agricultural-waste runoff is the major source of the problem, but the treated wastes of the municipal and industrial sources and suburban runoff are also significant.

Nevertheless, the Lake's warming problem is also exacerbated by the numerous dams along its tributaries. The reservoirs formed behind these dams are pools of relatively stagnant water easily heated during the increasingly warmer summers. Besides adding to the warming problem, the dams also block fish passage to spawning areas and consequently interrupt the ecology of the entire Basin. Dam removals have therefore become a contemporary priority on the rivers and creeks feeding into Lake Champlain. The Saranac River is one of the more significant of these feeding sources. It has ten sizeable dams, two of which are currently being removed. Figure 10.4a indicates the locations of the Saranac's ten dams.

The Saranac River originates in the Adirondack Mountains of north-central New York. The Saranac has more than three dozen source lakes and ponds north of upper Saranac Lake (see Fig. 10.4a and 10.4b). It flows northeasterly for eighty-one miles while dropping fifteen hundred feet over its course. It enters Lake Champlain at Plattsburgh in New York. Because of its navigable rapids, the Saranac is popular among kayakers and canoers.

As part of the 1990 *Federal Energy Regulatory Commission's* (FERC) relicensing requirements, a new fish ladder was demanded for the upstream hydro-power facility of Treadwell Mills Dam. (Chapter 1 reviews the FERC licensing requirements.) The Dam's owner, Patriot Hydro, was also required to remove the remnants of two of its downstream derelict structures: Indian Rapids and Fredenburgh Falls Dams. But if the farthest downstream barrier—the Imperial Mills Dam—is also not removed, then these upstream changes will have no impact on restoring the once prodigious salmon runs along the Saranac.

In the early 1800s, Dewitt Clinton, a naturalist and at that time the future Governor of New York, commented on the extraordinary salmon runs on the Saranac River near where it empties into Lake Champlain.[10]

[10] See Dewitt Clinton's (1822), "Letters on the Natural History and Internal Resources of the State of New York," E, Bliss & E. White, New York, NY.

But in 1796, Zephaniah Platt—the founder of Plattsburgh—constructed the River's first dam three miles upstream from Lake Champlain at the site of what was to become the Imperial Mills Dam. In 1800, New York established a law to protect salmon spawning runs by requiring fish passages for any and all dams. The Imperial Mills Dam, however, did not have a fish passage. New York State sued Platt to restore the salmon runs. But in 1819, Platt won the case on the grounds that the dam was entirely on private property. As a result, the barrier remained, and the Atlantic salmon ceased migrating to the upstream spawning grounds. After two-hundred and thirty years, however, the Lake Champlain Basin Program now seeks to restore salmon and other fish to the upper Saranac. Consequently, the NYS Department of Environmental Conservation (DEC) is now installing a modern fish ladder at the Imperial Mills Dam. Its planned completion is 2026. As a result, combined with the removal of both the Indian Rapids Dam and the Fredenburgh Falls Dam, plus a recently installed fish ladder at Treadwell Mills Dam, for the first time since 1796 the upper Saranac River might soon be opened to salmon and other migrating fish.[11]

Some fishermen and other naturalists are consequently pleased by these restoration prospects; but not all are pleased. Trout Unlimited, the *environmental advocacy organization* (EAO), is arguing that the total removal of the Imperial Mills Dam is necessary. It argues that since 2022, New York State has classified this dam as "high hazard" and "unsound." It follows, as Trout Unlimited and others argue, that the Dam is unacceptably dangerous to the City of Plattsburgh, and that it is also a barrier to fish migration despite the installation of a new fish ladder.

At a minimum, therefore, the Imperial Mills Dam will have substantial safety upgrades over the next few years. This is also true of the upstream Bartlett Carry Dam. In addition, prior to 2027, four of the eight hydroelectric dams on the Saranac—Lake Flower, Franklin Falls, Union Falls, and Treadwell Mills—are all up for FERC relicensing review. Four other Saranac River dams have been relicensed since 2000. But the upcoming relicensing examinations do pose possibilities for substantial changes in the Saranac's ecology. Waiting for safety improvements to be approved and financed, especially at the Imperial Mills Dam, is not within the spirit

[11] It should be noted that since the 1970s, the US Fish and Wildlife Service have been stocking some land-locked salmon from Lake Sebago into Lake Champlain.

of the ASDSO standards. In Particular, it is asking the citizens of Plattsburg to endure the danger of a disastrous dam breach over whatever time is required to bring this dam up to safety code.

Quassaick Creek is a tributary of the Hudson River. It flows through the small city of Newburgh fifty miles north of New York City[12] (Fig. 10.5 depicts its location). This Creek once had dams and reservoirs that powered the Stroock Felt Mill, plus a woolen mill, a paper mill, a candle factory, and a sawmill, all operating during the nineteenth century. The Stoock Felt Mill produced carriage blankets, army blankets and uniforms for World War I. The dams have all been derelict since the early twentieth century. The Riverkeepers is leading the movement to remove these dams in order to restore the once abundant river herring and other fish, and to eliminate the flood threat and other dangers posed by the dams' remnants. During a 2022 field exploration by Tracy Tullis, along with the Riverkeepers' ecologist George Jackman, the cracked remains of the Walsh Mill Dam were exposed after some hacking away at the brush. This derelict dam once powered a sawmill.[13] The Walsh Dam remains in danger of crumbling during a sizeable storm. A low-income housing project—The Mullins Courtyard—was built on the Quassaick Creek floodplain. It was inundated by the Quassaick's flood waters during Hurricane Ida in 2021. A collapse of the upstream dams due to another calamitous storm would destroy this housing complex. But the Walsh Dam remains.

The Glenmore Reservoir is created by Glenmore Dam on Florence Creek in Annsville, New York (see Fig. 10.6). This Dam is more than one-hundred years old and has deteriorated to the point of needing approximately $12 million in repairs. It supplies water to the small city of Oneida (the Dam's owner) and several other water districts in the area. It has been in need of concrete repairs since at least 2006, but the Dam's spillway now needs reconstruction, and consequently the DEC's rating is "unsound." If the Dam fails, the City of Oneida would be liable for the reconstruction and the resulting flood damage. In addition, the DEC has ordered the Dam's Administrators to release more flow during the summer season to support fish survival. The City has authorized the $12 million expenditures for reconstructing the Dam in the Spring of

[12] See Foderaro (2020, January 20).

[13] See Tullis (2022).

2024. The costs of necessary repairs, however, might be higher than the currently authorized amounts.

The initial report on the Dam's needed repairs was composed in 2006 when they were estimated to be $4 million. Repairs, however, were delayed until repeated "unsound" ratings were applied by the DEC. Now (Spring of 2024) the expenditures on repairs necessary to warrant a "sound" rating," are $12 million, but they are projected to quickly grow to $14 million if repairs are further delayed.

Adirondack's Dams

At 6.1 million acres, New York State's Adirondack Park is the largest such entity in the lower forty-eight states.[14] It is located in north-central New York State (north of the State Capital of Albany). It was established as a state park in 1892 for "the free use of all the people for their health and pleasure," and also for watershed protection.[15] Fifty-two percent of this Park's land is privately owned, which leaves only forty-two percent publicly owned in what is titled the "Park's Forest Preserve." All of the 6.1 million acres, however, are regulated by New York State's Adirondack Agency whether public or private.

The Adirondack Park has forty-six distinctly high peaks, two-hundred thousand miles of streams, and twenty-eight hundred ponds and lakes. One-hundred and five towns are within its borders, and its permanent residents number a hundred and thirty thousand with two-hundred thousand seasonal residents. This wilderness oriented Park is a product of the nineteenth-century romantic movement in culture and the arts. With the advent of America's nineteenth-century's transcendental movement, initiated by naturalists Dewit Clinton, Henry David Thoreau, Ralph Waldo Emerson, Theodore Roosevelt, and John Muir, and also with John Tyler Headley's *Adirondack, or Life in the Woods* (1849), the Adirondack Mountains became a destination for a recreational contact with nature for New York City's business people and families. But the area was also a target for logging interests. These competing interests resulted in restrictions on logging in 1885. In 1892, the Park was created. In 1912, New

[14] The National Forests in America's West, and also its designated Wilderness Areas are larger, but Adirondack Park is the largest area designated as a "park."

[15] See "Annual Report of the Forest Commission, 1892 at www.archive.org.

York State included all private land within the Park's border as subject to its tight land-use restrictions by the State's Agencies. Overall, more than five-hundred dams have been constructed within Adirondack Park; their average age is ninety years.

Within the Adirondack Park, there are twenty-seven dams rated as "unsound" by the State's DEC, ten of these are also "high hazard" and seventeen are "intermediate hazard"[16] (These statistics are for May of 2023.) As of 2022, half of the fifty "intermediate hazard" dams in the Park have not submitted the necessary engineering inspections required by the DEC. Two of these "high" and "intermediate hazard" dams (shown with underlines) are reviewed below (see Fig. 10.7).

Rainbow Lake Dam is near the town of Indian Lake in the Adirondacks. It was constructed in 1929 to create a small recreational reservoir to serve as a summer vacation retreat. In 1979, the Army Corps of Engineers classified this Dam as unsafe and in need of emergency repairs. It exhibited substantial seepage, overgrown banks, and fissures in its concrete spillway.

In 1981, Jason Roberts Development, a Delaware Corporation, purchased Rainbow Lake Dam. This corporation was owned by a religious nonprofit with a Brooklyn address, but it was managed by Jason Roberts Operating Corporation, a New York State registered company. The owner of this operating company was Gary Spindler, a New York City businessman who paid the taxes on the Indian Lake property and who also corresponded with the DEC. In 2013, the DEC ordered the "operating company" to submit an emergency action plan to either restore or remove the Dam. But the Operating Company ignored the order and skipped the subsequent judicial hearing. The DEC then referred the case to the New York State Attorney General's Office which negotiated a consent decree with associated work permits for fixing the Dam. The Attorney General's Office also established a deadline of April, 2016, for repairing the dam. But this deadline was also ignored by the Dam's owners. When the Attorney General's Office sued in 2018, it referred to the Dam's ownership as a "mess of unauthorized and dissolved shell corporations."[17] Spindler argued that the religious ownership and Delaware

[16] See "List of Dam Removals in New York" at https://en.wikipedia.org/w/index.php?title=List_of_dam_removals_in_New_York&oldid=1217944093.

[17] See Matson (2023, May 5).

registration shielded the owners from liability. New York State argued that the "defendants have created a series of corporations in a vain attempt to shield themselves from liability."[18] Spindler and the other owners admitted liability and Rainbow Lake Dam was finally removed forty-three years after the Army Corps of Engineers' first warning concerning the Dam's danger. No maintenance was performed during this lengthy interim period.

Garnet Lake was once Mill Creek Pond, one of a series of ponds on Mill Creek in the Adirondacks. It is a 306-acre reservoir initially formed behind a timber crib dam constructed in 1853 to serve the logging industry of the area. This area was cleared of timber prior to the turn of the century. This small area has the magnificent scenery of Blue Mountain to the immediate east, and Ross Mountain to the north.

In 1905, Frank Maxam purchased eighty-nine acres with a half-mile frontage along Mill Creek Pond. He constructed a commercial lodge for hunters and fishermen, plus he changed the dam into a more substantial rock-crib structure with a hydro power generator. He also expanded the pond into its current 306-acre size by flooding the marshy area where Mill Creek flowed into the reservoir. He then changed the name of the pond to Garnet Lake which was considered a more commercial name.

By the 1920s, Garnet Lake had become a popular recreational destination. Seventy-five percent of the lake shore was public lands, but the private homes were constructed along the Lake's shore and some of these were year-round residences. Garnet Lake Conservation Association (GLCA) was established in 1956 by the area's homeowners for the purpose of maintaining the Lake and Dam. The GLCA, along with the Township of Johnsburg (the town that encompassed the Lake), reconstructed the dam into a concrete and stone structure with a spillway of three culverts that spanned twenty-five feet. The dam was classified as "high hazard" because of the downstream small city of Johnsburg.

In 1981, the Army Corps of Engineers indicated that the spillway's culverts were too small for managing a major storm. The Corps indicated that the dam needed remedial work and a larger spillway. The GLCA added the emergency spillway on the side of the dam at a cost of $41,000. In 2015, the GLCA submitted an Emergency Action Report to the DEC, which in turn responded in 2018 with a dam rating of "unsound to fair."

[18] Ibid.

A remedial engineering design has been formed that will replace the dam with a larger spillway. The GLCA is waiting for the DEC to review and approve the plans.

The Garnet Lake Dam illustrates that the process of dam safety inspections and remediations is functioning, although very slowly. Remediations are occurring in the Adirondacks although at a slower pace than optimal.

Sacandaga River is a sixty-four-mile tributary of the Hudson River (see Fig. 10.7). The Sacandaga drains the approximately one-thousand-square-mile southeastern quadrant of Adirondack Park. It flows into the Hudson at Hadley, New York, where the Conklingville Dam creates the forty-two square-mile Great Sacandaga Lake. This reservoir is entirely confined within the Park. The origin of the Sacandaga River is Lake Pleasant in the central Park region.

In response to periodic flooding, the ninety-five feet high Conklingville Dam was constructed in 1930. In particular, the "Great Flood of 1913" inundated Albany and Troy, and other Hudson riverside towns. As was then typical of these floods, the Great Flood caused a typhus epidemic in the central Hudson area. The Conklingville Dam and other flood control measures largely resolved the periodic flooding problem, but the ninety-three-year-old structure currently needs significant remediation that include:

- foundation repairs,
- replacement of deteriorated concrete, and
- replacement of outlet valves.

Given the importance of this large dam for flood control in the Hudson River Valley, the State of New York has now allocated $20 million for the above repairs starting in 2023.

The Village of Indian Lake is in the center of the Adirondack Park. The earthen-stone structure of Indian Lake Stone Dam is three miles to the south of the Village. This 47-foot high, 550-foot long structure impounds the Indian Lake Reservoir, a 4,225-acre body that extends into south-central Adirondack Park. The first version of this dam was built in 1850 for the purpose of controlling the flow used for moving the logs harvested in this area. This dam was reconstructed in 1898 for flood control and recreation. It is operated by the Hudson River—Black River

Regulating District. It is a “high hazard” dam. It was last renovated in 1987.

Over this Dam’s history, leakage through its masonry walls has been a problem, and regrouting has periodically occurred. Regrouting is scheduled to be part of the Dam’s two-year reconstruction begun in October of 2023. New gates will be installed, a secondary spillway will be constructed, and a new canoe and kayak launch created. The costs of these repairs have been increasing rapidly and are conditionally estimated as $11 million for all repairs to be completed over the two-year construction period.[19]

But there are other “high hazard” and “unsound” dams (HHU dams) in the Adirondacks in need of emergency remediations. The Towns of Dannemora, Bolton, Indian Lake, and Chester each own one of these HHU dams.[20] For over a decade (up to 2023), the small City of Dannemora has avoided the DEC required annual updates to its Chazy Lake Dam, presumably because it is waiting for state or federal financial assistance. (Note the review below of sources of federal rehabilitation assistance.) Similarly, Loon Lake Dam—which is northeast of Saranac Lake—is waiting for $450,000 to repair its spillway. Also, the Town of Indian Lake is waiting to repair its Lake Adirondack Dam, and there are other Adirondack Dams waiting for financial assistance prior to making emergency repairs.

But these are dams in the relatively less populated region of a state park environment. What about the much more populated area of the prosperous suburbs immediately north of New York City, such as in Westchester County? There the 126-year-old “high hazard” Lake Isle Masonry Dam is being repaired. Lake Isle is a 62-acre body surrounded by numerous single-family homes and townhouse condominiums. A dam failure here would flood Pelham, Mount Vernon, New Rochelle, and Eastchester. Multiple sources of funding—state, federal, and municipal sources—will finance the $11 million reconstruction effort beginning in 2023.

Vermont’s dams are similar to New York’s and have similar safety problems. This similarity is explored next.

[19] See Craig (2023).

[20] See Matson (2023, Nay 13).

The Safety of Vermont's Dams

In July of 2023, Vermont's Capitol of Montpelier suffered its worst flood. Montpelier is the smallest state capital in the US. It is a quaint New England town with an old Main Street that spans the bottom of a central hill. At the top of that hill, Vermont's Capitol building sits and overlooks this small city. On July 9 through July 12, 2023, the lowest segment of Main Street was flooded up to the level of the first-floor windows (four feet) of the street's old red brick buildings. The flood waters came from the North Branch of the Winooski River.

A short distance north of Montpelier, the Wrightsville Dam holds back a reservoir that is 1.75 miles long and 0.2 miles wide. The Wrightsville Dam and Reservoir are on the North Branch of the Winooski River which flows into Montpelier (see Fig. 10.8). The torrential rains of July 9 through July 12 brought the level of the reservoir close to the Dam's maximum capacity, but the dam held. This was not the case throughout Vermont during this storm. But the system of dam safety inspections has solicited criticism from the State's Auditor as reviewed below. The flooding of 2023, and breaching of the other dams in the State, has now led to questioning the safety of all of Vermont's dams.

State officials recognize that more than a thousand dams exist in Vermont. They range from the large flood control and hydroelectric structures to the small remains of derelict dams. The oldest dates from the Revolutionary War era, and the median average age is seventy-seven years. Currently, Vermont's Department of Environmental Conservation's (DEC) "Dam Safety Program" oversees approximately four-hundred dams, and Vermont's Public Utility Commission supervises twenty hydro-electric dams. Federal agencies supervise dozens more. Approximately one-third of the dams inspected under Vermont's Dam Safety Program are classified as in "poor condition."[21]

A 2022 Vermont State Auditor's "Report" concerning the DEC's "Dam Safety program" indicated a series of defects[22]:

- Vermont's DEC estimates that there are at least a thousand existing Vermont dams that are not listed or recognized.

[21] See D'Auria (2023).

[22] See "Report of the Vermont State Auditor," Report No. 22-02, (February 11, 2022), "Department of Environmental Conservation's Dam safety Program."

- A 2019 report by the American Society of Civil Engineers assigned a grade of only "C" to the State's dams because so many were in poor deteriorating condition.
- Ten of the dams classified as in "poor condition" have been so rated for at least eighteen years. The DEC made recommendations to dam owners to improve dams in "poor condition," but it did not provide timeframes for implementing those recommendations. It lacked enforcement authority to require dam owners to implement those recommendations. It only followed-up to discover if its recommendations were pursued by checking at its next inspection, which was typically more than five years later.
- The DEC often did not complete their inspections reports, and it also did not provide its reports to dam owners in a timely manner.
- The DEC has not been inspecting all dams within the required timeframe (two years) and has not recorded all inspections in their inventory, i.e. it did not maintain a centralized inventory that included its "condition" ratings of inspected dams, and its "hazard potential" ratings.
- Insufficient staffing has been a persistent problem at the DEC.
- The Vermont's Public Utility Commission (PUC) and the Vermont's DEC have not been using the same hazard potential definitions, and the PUC has not been inspecting as frequently as the DEC.

Chester is a small city in central Vermont (see Fig. 10.9). The Chester Reservoir and Dam are northwest of this town. They were constructed on a tributary of the Williams River. Vermont's DEC rated this dam as being in "poor condition" in 2006, and also of being a "significant hazard" since it is upstream of Chester. The flood waters of July, 2023, topped the crest of this dam. This Dam barely held but is was severely eroded. It currently needs a major reconstruction. During this flood emergency, three small Vermont dams did fail (Hands Mill Dam, Clark Sawmill Dam, and Lyons Dam), but fortunately no injuries were reported.

Removing small dams would help mitigate floods, but the process for removal in Vermont is usually long and complex. Whether they are owned by local municipalities or other private entities, dam removals need the permission of their owners. The problem is that many small dams have uncertain ownership, or have long been completely abandoned (no owners). But when consent for removal is obtained, and the required feasibility studies conducted, other reviews are generally required before deconstruction begins. Michael Braun, the Executive Director of the Friends of the Winooski River points out, "You need a stream alteration permit, or if it is a bigger dam, you need a dam order from the Safety Division. We have to go through historic preservation review, and lots of other regulatory review."[23]

In Vermont, dams are being removed at a rate of approximately five per year. Given the lengthy process for removal, it will take decades to deconstruct Vermont's deteriorating decrepit dams. The defects in Vermont's reviews and process of removal or rehabilitation are apparent.

Federal Programs for Local Dam Safety

Through the "National Dam Safety State Assistance Grant Program," federal funding has been available for dam safety actions although in small amounts, and then only for "high hazard potential" dams. For the fiscal year 2024, federal grant funding of $211 million is available for state rehabilitation actions. The "National Dam Safety Program" is a partnership of federal, state, and other stakeholders. It provides assistance to state, local, and nonprofit organizations for planning, redesigning, and rehabilitating non-federal "high hazard" rated dams.[24]

The National Dam Inspection Act of 1972 directed the Army Corps of Engineers (ACE) to inventory America's dams with the purpose of pursuing safety examinations and improvements. (See the section titled "Developing the Politics of Safety" presented above for the origins of this 1972 Congressional Act). The *National Inventory of Dams* (NID) originally included 42,000 entries in 1975, but it now has 92,275 dams. The NID was reauthorized as part of the "Water Resources Reform and Development Act of 2018." This NID data base has seventy field entries for each dam including location, size, type, inspection dates and classification, regulatory information, and inundation maps. The latter shows the sequence of flooding areas in case of dam breaching. Classifications of "high hazard" or "significant hazard potential" are indicated. The NID includes all those dams that meet minimum height and reservoir

[23] See D'Auria (2023).

[24] See www.fema.gov/emergency-managers/risk-management/dam-safety.

size standards.[25] Users include emergency managers, safety professionals, infrastructure owners, community leaders, business owners, and all those residents who seek to understand the relations between dams and their communities.[26]

"The National Performance of Dams Program" is maintained by the Department of Civil & Environmental Engineering of Stanford University. Their 2018 review and analysis of "Dam Failures in the US" examines the effects and causes of American dam failures.[27] In addition, FEMA maintains a list and analysis of case studies of dam failures at its "Lite Case Studies Summary."[28]

Conclusion: Safety and Its Funding

The proliferation of old dams in the States of New York and Vermont illustrate the problems encountered throughout America's eastern watersheds, and to a lesser extent throughout its western areas. In particular, our dams were constructed to serve narrow interests (business and/or municipal interests) and with little thought to their ecological impacts. Hacking through the original state of nature and bending it to the will of some entrepreneurial pioneer for achieving economic success provided our motives for those initial dam constructions. The story of Zephaniah Platt's 1796 dam construction on the Saranac River (reviewed above) is an appropriate illustration. That first structure was eventually rebuilt into the Imperial Mills Dam that served the economic functioning of early nineteenth-century Plattsburg, and that still exists to block the Saranac for fish migrations beyond the first three miles from Lake Champlain.

That old dam is once again being refurbished, this time to better serve the Saranac's ecology. But its engineering safety is also being questioned given its obvious "high hazard" threat to its downstream City of Plattsburg. These two issues—safety and ecology—were likely not of significant

[25] See www.fema.gov/emergency-managers/risk-management/dam-society/national-inventory-dams.

[26] For examples of the "inundation maps, see https://nid.sec.usace.army.mil/#/dam-videos/nid-videos.

[27] See https://npdp.stanford.edu/sites/default/files/reports/npdp_dam_failure_summary_compilation_v1_2018.pdf.

[28] See www.fema.gov/sites/default/files/documents/fema-dss-wise-lite-case-studies-summary.pdf.

concern to Zephaniah Platt in 1796. Nevertheless, his dam endured. But what about the coal slurry dams of Buffalo Creek in West Virginia, or the Teton River Dam and its ill-conceived location in eastern Idaho? These were twentieth-century structures of cavalier designs. Both failed with tragic results. Neither safety nor ecology were of significant concern to their builders; only crass narrow economic interest motivated their construction.

For some other dams reviewed above, such as the Spencer Dam on Nebraska's Niobrara River, or the Kelly Barnes Dam on Toccoa Creek in Georgia, maintenance posed the critical issues that led to failure and tragedy. A lack of attention from responsible officials caused these tragic failures.

America's time of confronting its "original state of nature" is now a century and a half past, and its political structure for dealing with our society's natural confrontations, such as controlling our watersheds, has been established. We now have our organized ASDSO classifications for dams, and state agencies for inspections. But these are shown above to be understaffed. Their inspections are too infrequent and do not produce consistent enforcement. We do have federal assistance programs for remediations of local dam safety problems, but funding is shy of adequacy for an effective national program.

Senator Henry "Scoop" Jackson's 1972 Congressional speech referred to some of the dam failures reviewed above. Scoop Jackson said, "These occurrences point out the grim fact that there are innumerable dams in existence throughout the United States which are under no effective public control to insure that they were competently designed and constructed initially, and that they are being adequately inspected and maintained, or that the design remains adequate under current hydrological conditions." His speech helped lead to the creation of our "National Dam Inspection Program," and gave the Corps of Engineers the authority to begin to inventory all of America's dams and to inspect its federal dams. But fifty years later we now know that we are shy of achieving a satisfactory result. Our dam removal and remediation movement might refocus a bit to bring safety to the forefront of its public consciousness raising. Safety improvements need greater federal funding. For example, as reviewed above, Lake Isle Dam in Westchester County is being refurbished with some federal funding to prevent a tragic failure. But our Adirondack dams are equally in need and equally dangerous according to their "high hazard" ratings. They have been waiting for some federal assistance for either reconstruction or removal financing. Establishing a queue for reconstructing highly hazardous and poor conditioned dams is not an optimal policy.

We now have our National Inventory of Dams (NID) well established and near completion. We have over 92,000 dams classified. We know those dams that are classified as "high hazard" or "significant hazard." We know the conditions of those structures. We have our inundation maps formed for anticipating the consequences of dam failures. We have an industry of engineering companies ready to reconstruct or remove the offending structures. We have the funding mechanisms in place, but we need Congress and the states to recognize the immediacy and priority of needs and therefore funding.

Fig. 10.1 Idaho's Teton River

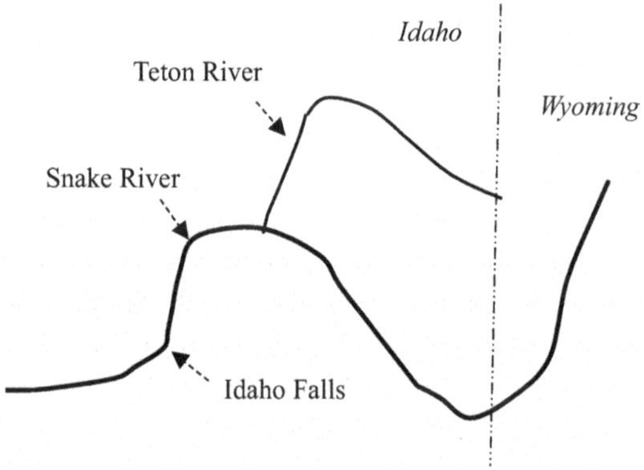

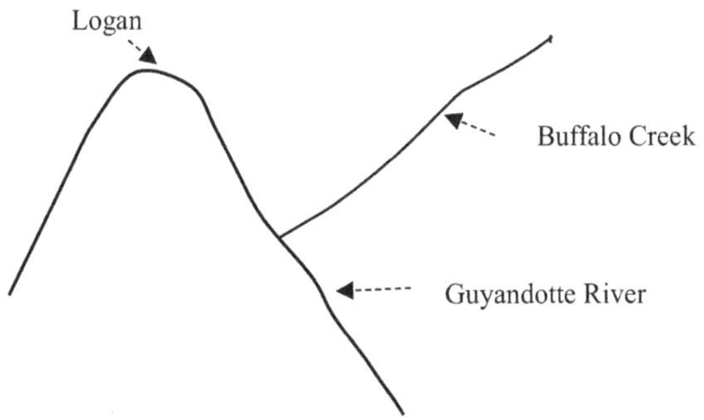

Fig. 10.2 S.W. West Virginia's Buffalo Creek

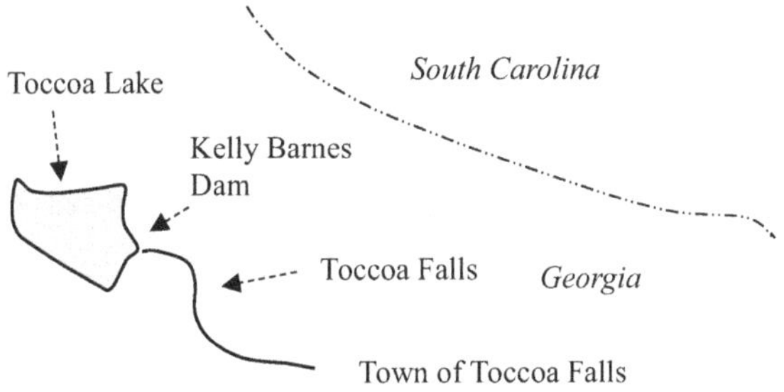

Fig. 10.3 Georgia's Kelly Barnes Dam

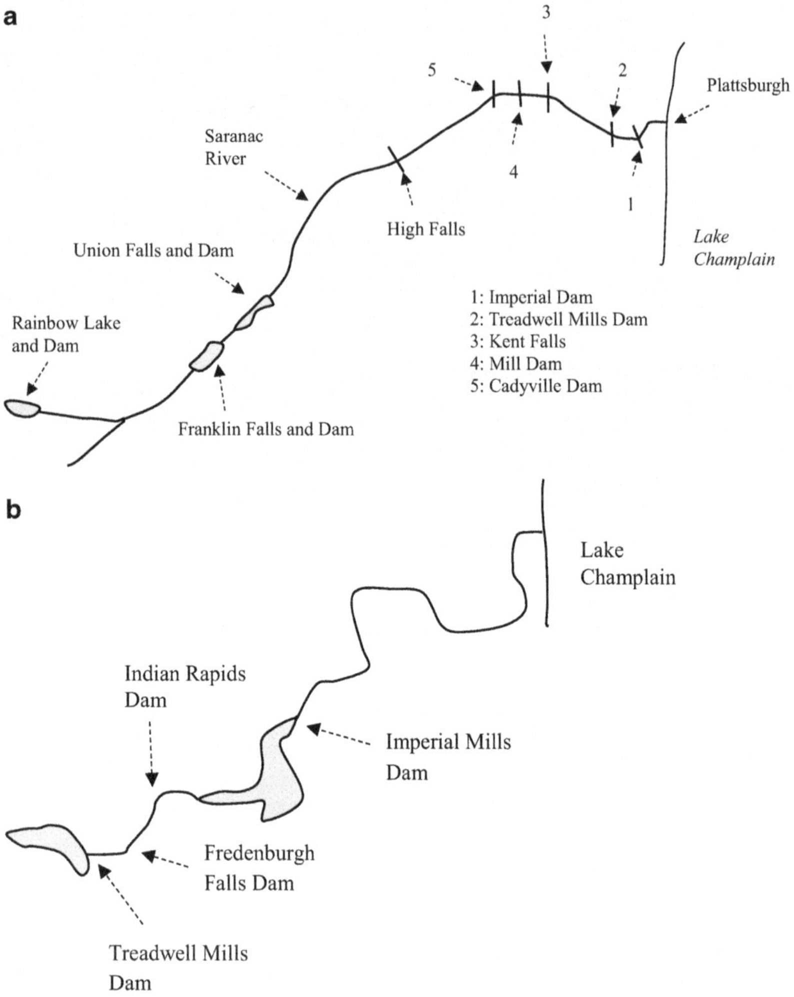

Fig. 10.4 (a) Saranac River Dams (b) Plattsburgh's dams

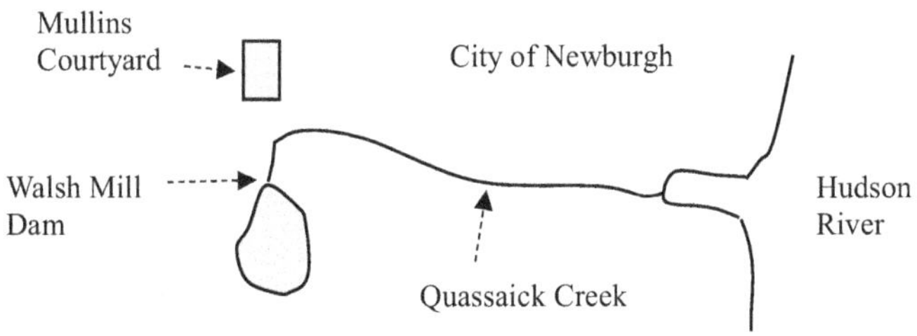

Fig. 10.5 Quassaick Creek

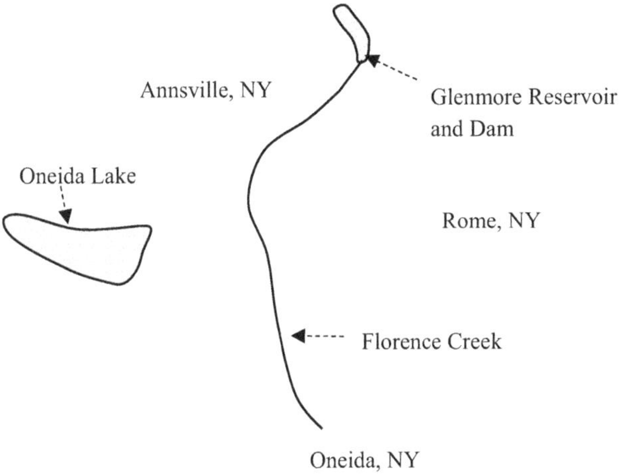

Fig. 10.6 Glenmore Reservoir and Dam

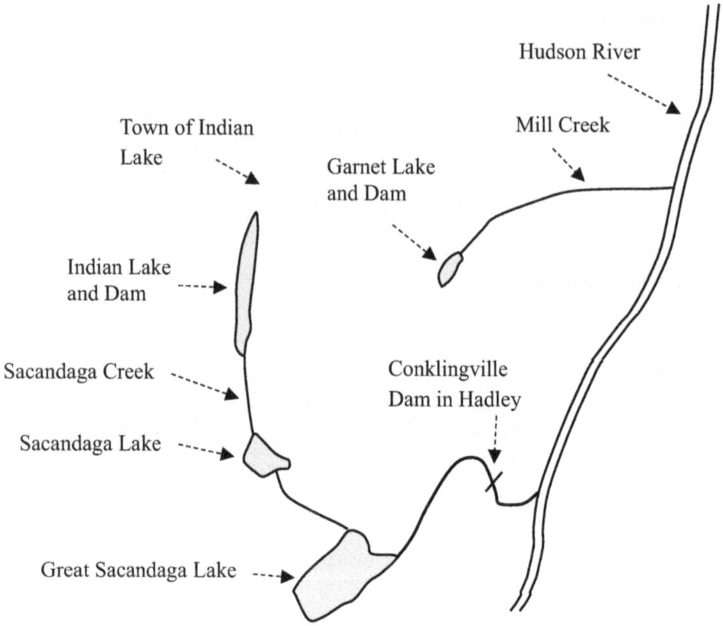

Fig. 10.7 Some Adirondack Park's Dams

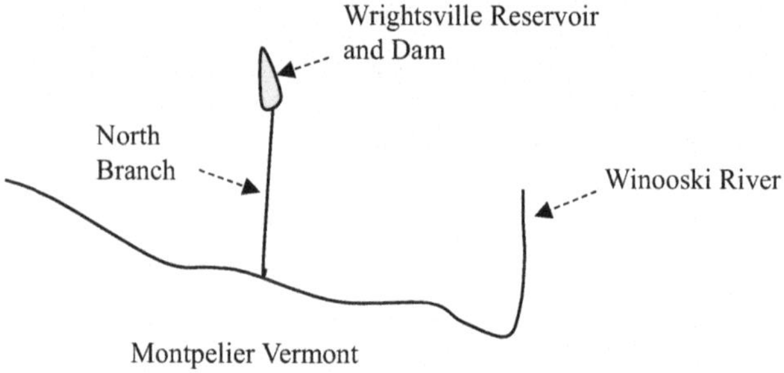

Fig. 10.8 Vermont's Wrightsville Dam

Fig. 10.9 Vermont's Chester Dam

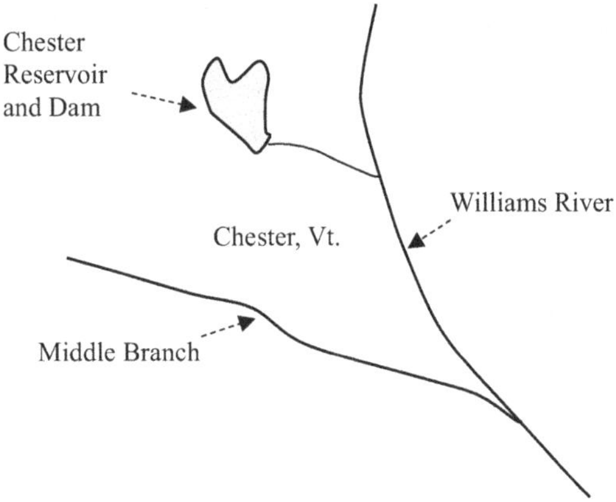

References

Craig, G. (2023, August 26). *Indian Lake Stone Dam repairs to begin in October*. www.adirondackexplorer.org/stories/indian-lake-stone-dam-repairs-to-begin-in-october

Foderaro, L. (2020, January 20). It's Fish vs. Dams, and the Dams Are Winning. *New York Times*. www.nytimes.com/2020/01/20/nyregion/its-fish-vs-dams-and-the-dams-are winning.html

Matson, Z. (2023, May 13). In the Adirondacks, Town Dam Owners Watch and Wait. *Adirondack Explorer*. https://adirondackestg.wpenginepowered.com/stories/in-the-adirondacks-town-dam-owners-watch-and-wait

Matson, Z. (2023, May 5). Nearly a Decade after DEC Commissioner Ordered Rainbow Lake Dam Near Indian Lake Removed or Repaired, the Structure Was Taken Down Last Year. *Adirondack Explorer*. https://adirondackestg.wpenginepowered.com/stories/dec-vs-jason-roberts-rainbow-lake-dam

Montrie, C. (2003). *To Save the Land and People: A History of Opposition to Surface Coal Mining in Appalachia*. University of North Carolina Press. https://archive.org/details/tosavelandandpeople00mont

Pisani, D. J. (2002). Floyd E. Dominy. www.waterhistory.org/histories/dominy/

Pritchard, C. (2023, August 5). Glenmore Dam in Need of Repairs. *Rome Sentinel*. www.romesentnel/news/glenmore-dam-in-need-of-repairs/article_39767c92-32d8-11ee-8371-a3d9bd1d0a0a.html

Reisner, M. (1993). *Cadillac Desert: The American West and Its Disappearing Water*. Penguin Books.

Tullis, T. (2022, November 4). *Ghosts of Capitalism: The Push to Dismantle America's Decrepit Dams*. www.pressreader.com/usa/the-guardian-usa/20221105/281487870303611

Whitford, N. E. (1906). *The Champlain Canal*. https://web.archive.org/web/20100628005458

CHAPTER 11

Answering "The Call of the Wild" in a Kayak

Introduction: Whitewater and Its Call

This chapter presents a tale of two contrasting American whitewater rivers. One lies in a northeastern rustbelt area that is "being restored." The other is a stereotypical western river of cowboy lore. Both attract whitewater paddlers in significantly large numbers. Both have popular scenic hiking trails along their banks. Both are supported by a successful and necessary undamming movement that is motivated to restore these rivers to being amenable to natural recreation.

Whitewater enjoyments are inherently exhilarating because of their nature and locations. That is why they initially attract participants. Whitewater rafting and kayaking have significantly expanded in the last half-century. They bring new participants to experience rivers in need of restoration and preservation. This fact motivates this chapter's review of the movement for restoring and preserving two accessible rivers that are distinctly very different in character, but still have strong similarities in attracting restoration advocates.

Our images, stories and legends of *The Great American West* all project into our cultural philosophy. Many of our American authors—John Muir, Jack London, Theodore Roosevelt, Zane Grey, Louis L'Amour, Dee Brown, and Ken Kesey to name a few—bind these images together to compose our view of rugged natural experiences that add to the canon of what it means to be American, and what it means to experience its bold

R. M. Robinson, *Restoring America's Rivers*,
https://doi.org/10.1007/978-3-031-81758-8_11

settings.[1] In contrast, the American East is picturesque but with gentle vistas. The leisure culture of America's East is perceived as mostly passive relaxation on a sandy beach, or on a quaint island like Martha's Vinyard. But American lore also includes Thoreau's paddling trips described in *A Week on the Concord and Merrimack Rivers* (1849),[2] and also in *The Maine Woods* (1858).[3] The latter describes an excursion in an eighteen-foot canoe on the great Moosehead Lake of central Maine. Our American lore also includes the greatest of all raft trips, the quintessential American account of a smooth but nonetheless hazardous float on the spine of American philosophy, the Mississippi River in Twain's *The Adventures of Huckleberry Finn* (1885).[4] (Mark Twain was knowledgeable about the hazardous Mississippi waters.) We also have James Dickey's novel *Deliverance* upon which the movie of that name was based.[5] It is a fictional whitewater canoe trip down a fictional northwest Georgia river, but those sorts of rugged rivers do exist in the hills of Tennessee, Georgia, and North Carolina. Summer whitewater kayaking is common on these rivers.

But America's West is a place of broad vistas and adventurous pursuits that particularly challenge one physically. Typically, one does not passively enjoy the West; one is drawn to actively participate in its great vistas. These participations elicit physical efforts that can substantially test one's grit and tap one's energy. To experience America's West means to interact with its glaciated mountains, its high purple-sage desert, its cathedrals of gigantic Ponderosa Pine, Douglas Fir, Redwood, and Sequoia. But for many it also means riding the whitewater flowing through scenery of steep and twisting colorful canyons where elk, mule deer, bobcat, mountain lion, pika, mountain goats, and big horn rams reside, and where steelhead and salmon battle the river's current.

[1] See Roosevelt's *Ranch Life and the Hunting Trail* (1888) and *Hunting Grisly and Other Sketches* (1893). Also see Muir's *My First Summer in the Sierra* (1911) and *Steep Trails* (1918). Also see Dee Brown's *Bury My Heart at Wounded Knee* (1970), Jack London's *Valley of the Moon* (1913), and also Ken Kesey's *Sometimes a Great Notion* (1964).

[2] Originally self-published in 1849. Reprinted by Princeton University Press, 1980.

[3] First published in *The Atlantic Monthly* in 1858. Reprinted by Princeton University Press, 1972.

[4] Reprinted by the University of California Press, 2003.

[5] Published in 1970 by Houghton Mifflin Co, Boston, MA.

The American West is much more mountainous than the East, and consequently whitewater abounds in the West. Whitewater provides one of the key popular ways to interact with Western rivers since it flows inherently through majestic western scenery, most frequently through rugged picturesque canyons. These steep flows are capable of carrying one along a uniquely fast and physical ride. Whitewater can take one on a life affirming adventure down the iconic Salmon River of Idaho (the "river of no return") or the Rogue River of southern Oregon (the setting of western stories by Zane Gray and Jack London). These are great whitewater rivers to experience; they pose an American challenge for our more recent generations who do not appreciate floating on the unnaturally placid reservoirs found behind dams.

The factors that motivate the movement for restoring America's eastern and western rivers include the public's demands for recreational experiences such as whitewater interactions. Kayaking is a relatively new form of this whitewater experience. It is largely a product of the 1970s with its new technology for producing less expensive and durable polyethylene kayaks, a new and radical modification of the old-birch bark and animal-bladder canoes of Native Americans. Do not underestimate the degree of influence that kayaking has had on the grass-root politics of undamming America. This influence has been considerable since kayaking facilitated a much expanded access to America's natural river settings, and therefore helped to stimulate a movement for preservation. Significant numbers have sought this access in recent decades. Their numbers have contributed to the politics of river restoration.

Wild and Scenic Rivers

A river is a natural *common property resource*, i.e. a "commons." As such, rivers are liable to suffer from the "tragedy of the commons," that is the phenomena of common property resources being degraded. Avoiding this "tragedy" always involves collective management.[6] "The Wild and Scenic Rivers Act" of 1968 sought to establish a National Policy of using local collective control for maintaining, and in some cases restoring, the undeveloped nature of many of America's rivers. This Act sought to accomplish this by encouraging local communities to collectively manage their rivers

[6] For a review of this phenomena, see Robinson (2021, Chapter 10).

so as to avoid these "tragedies." Section 1 (b) of this Congressional Act specifies:

> It is hereby declared to be the policy of the United States that certain selected rivers of the Nation which, with their immediate environments, possess outstandingly remarkable scenic, recreational, geologic, fish and wildlife, historic, cultural, or other similar values, shall be preserved in free-flowing condition, and that they and their immediate environments shall be protected for the benefit and enjoyment of present and future generations. The Congress declares that the established national policy of dam and other construction at appropriate sections of the rivers of the United States needs to be complemented by a policy that would preserve other selected rivers or sections thereof in their free-flowing condition to protect the water quality of such rivers and to fulfill other vital national conservation purposes.

Note the use of the wording "selected rivers or sections thereof." The intention is to allow for sections of rivers to be designated "wild or scenic." The local management considerations are further addressed in Section 2 (a) of the Act:

> The national wild and scenic rivers system shall comprise the rivers .... that are designated as wild, scenic or recreational rivers by or pursuant to an act of the legislature .... or political subdivision of the States .... through which they flow, that are to be permanently administered as wild, scenic or recreational rivers by an agency or political subdivision of the State .... that are found by the Secretary of the Interior, upon application of the Governor of the State .... to meet the criteria established in this Act ....

This means that to achieve the status of "Wild and Scenic," a river must first:

1. be designated as a "wild, or scenic, or recreational river" by its resident state,
2. be managed by an agency of that state,
3. be "free flowing" and possess one or more "outstandingly remarkable values" (ORVs to be explained below),
4. have its free-flowing character, water quality, and ORVs be permanently protected.

Under this Act, it is not the federal government that preserves and maintains these rivers, but state and local governments that declare the river as "wild and scenic" and that institute the permanent land-use policies that maintain it. The definition of what is meant by a "wild and scenic river" is also specified in Section 2 (b):

(1) "Wild river areas" are those rivers or sections of rivers that are free of impoundments and are generally inaccessible except by trail, with watersheds or shorelines essentially primitive and waters unpolluted. These areas present vestiges of primitive America.
(2) "Scenic river areas" are those rivers or watersheds still largely primitive and with shorelines largely undeveloped, but accessible in places by roads.
(3) "Recreational river areas" are those rivers or sections of rivers that are readily accessible by road or railroad, that may have some development along their shorelines, and that may have undergone some impoundment or diversion in the past.

Under this Act, sections of a river might be classified as a "wild river," and sections classified as a "scenic river," and sections classified as a "recreational river," and consequently the "Wild and Scenic" designation can be applied provided the land-use policy of these areas are judged by the National Park Service as sufficient to maintain these designations. The Act specifies that it is the National Park Service that will judge the efficacy of the local land-use policy, and that consequently might apply the "Wild and Scenic" designation to the river. To achieve this designation, the US National Park Service (NPS) must evaluate the river according to the four criteria listed above.

There were eight initial "Wild and Scenic Rivers" specified by the 1968 Act. They include:

(1) Middle Fork of the Clearwater River in Idaho,
(2) Eleven Point River in Missouri,
(3) Feather River in California,
(4) Rio Grande River in New Mexico,
(5) Rogue River in Oregon,
(6) Saint Croix River in Wisconsin and Minnesota,
(7) Middle Fork of the Salmon River in Idaho, and

(8) Wolf River in Wisconsin.

Why would a state seek to obtain the "Wild and Scenic" designation for its river? The answer is that this is likely a politically popular move that seeks to satisfy the demands for maintaining the natural environment. It satisfies the desires for those who want to interact with our natural river settings, e.g. the kayakers, rafters, fishermen, and hikers who want a lifestyle of natural interactions. The movement for "Wild and Scenic designations" satisfies the progressive body politic for environmental restoration and protection. The demand for the areas where people can experience natural interactions is not just in the West, but it is also currently very strong in the East where the population density is high but natural areas are few. For example, consider the Housatonic River of Connecticut and Western Massachusetts. It is located close to the New York City Metropolitan Area; it provides this densely populated area with scenic hiking, fishing, and whitewater kayaking. The communities along the Connecticut section of the Housatonic are currently seeking to organize the collective action necessary to avoid a local "tragedy of the commons." They seek to have most of the Connecticut portion of the river to be federally designated as "Wild and Scenic" according to the National standards outlined above.

As of April, 2021, the National Park Service (NPS) has evaluated and judged that there is strong support for the Housatonic's protection and the "Wild and Scenic" designation. The local governments, state agencies, elected officials, and environmental organizations all strongly support the granting of this designation. For example, the *Housatonic River Commission* was formed in 1979 for the purpose of providing collective guidance on river management issues for the towns along the upper Housatonic, all of whom participated in the *Commission*: Canaan, Cornwall, Kent, New Milford, North Canaan, Salisbury, and Sharon. The *Commission* produced a management plan in 1981, updated it in 2006, and routinely influences land-use policies along the river. The plan for the "Wild and Scenic" section of the Housatonic includes its northern 40.3 miles from the Massachusetts-Connecticut border to the Boardman Bridge in New Milford. This area is known for its charming rural character, historical heritage, and remarkable natural beauty in the midst of the northeastern densely populated corridor. The management plan consists of controlling three zones: an inner boundary, an outer boundary, and village centers.

(These boundaries are reviewed below.) There are two spectacular waterfalls within this proposed section: the Great Falls in Falls Village, and the Kent Falls in the Town of Kent. For the latter, a series of cascades drop 250 feet in less than a quarter mile of the river. The highest of the falls within this area is 70 feet. Also in Kent, the Bull's Bridge area has a scenic gorge.

As indicated above, a "Wild and Scenic River" must be "free flowing," which means "flowing in a natural condition without impoundment, diversion, straightening, rip-rapping, or other modification of the waterway." A single dam, or some other diversion need not eliminate a river from consideration, but the character of free-flowing must predominate over the section considered, and new modifications must not occur once the designation is granted.[7]

Also as indicated above, a "Wild and Scenic River" must have one or more "river related outstandingly remarkable natural, cultural or recreational resource values" (ORV). To be so classified, the ORV resource value must:

1. be located in the river or within one-quarter mile of either side of the river,
2. contribute substantially to the functioning of the river ecosystem,
3. owe its location or existence to the presence of the river.

In addition to the above criteria, all the river's segments must be classified as either "Wild, Scenic, or Recreational" according to the criteria listed above. The river can, therefore, be segmented where each segment qualifies as one or the other of the three classifications. The overall classification, however, depends on the National Park Service judging the "degree of naturalness" of the overall connected segments given that the river should not be "excessively segmented" in order to qualify.

[7] See "Housatonic River: Wild and Scenic River Evaluation," National Park Service, Boston, Massachusetts, April 2021, p. 10.

Restoring a Rustbelt River: the Housatonic[8]

The Housatonic River originates close to Pittsfield in the Berkshire Mountains of Western Massachusetts (see Fig. 11.1). It flows 149 miles south through Connecticut. Its watershed drains approximately 2,000 square miles before it flows into Long Island Sound between the municipalities of Stratford and Milford in Connecticut. It has numerous tributaries, the largest being the Naugatuck, which contributes 312 square miles of drainage to the Housatonic's watershed. The River and its tributaries are impeded by numerous dams in Massachusetts and five hydroelectric dams in Connecticut, three of which form a chain of lakes: Lake Lillinonah, Lake Zoar, and Lake Housatonic (see Fig. 11.1). Others are on the Naugatuck River tributary. All of these are being prepared for removal and/or modifications to allow fish passage.

The Housatonic River's most significant pollutions stem from three sources: (i) non-point storm runoff, (ii) from legacy PCB contaminations, and (iii) from the legacy mercury contaminations that originated from textile manufacturing. Considering the first source, twenty-four roads cross the Housatonic over its course. Route 7 parallels closely to the river's banks through much of its length. These roads, plus numerous hardened parking lots and adjacent buildings, all channel storm runoff into the river. This runoff causes streambank erosion, and also washes salts and chemicals from snow and ice removal, and vehicular chemicals (motor oils and fuels), plus plastics and other debris into the river.

Between 1932 and 1977, the Housatonic was polluted by PCB discharges from General Electric's (GE) plants in Pittsfield, Massachusetts. In 1997, the US Environmental Protection Agency (EPA) designated the GE facilities and several miles of the river as a "Superfund Site," and it continues to supervise GE's cleanup efforts.[9] As of 2020, remediation of 10 manufacturing facilities within the City of Pittsfield has been completed, but downstream areas have yet to be cleaned. The most polluted of these areas is Woods Pond in Lenox, just south of Pittsfield, which contains most of the toxic sediment associated with this river. The

[8] Much of this section is based on Robinson (2023, Chapter 10, Sect. 4).

[9] These Super Fund Sites are judged by the EPA as having a high priority for cleanup of its poisons. See Robinson (2023, Chapter 2, Sect. 7) for a review of this agency and its "Superfund" list.

birds and fish of the Housatonic area continue to contain high levels of PCBs.

The Connecticut section of the Housatonic contains high levels of mercury. This resulted from the hat manufacturing industry of Danbury, Connecticut, which operated from the mid-nineteenth through the mid-twentieth centuries. This industry used mercury nitrate in the felting process. The mercury continues to pollute the river's sediment, and leaks into the water flowing downstream during storms. The river's discharge into Long Island Sound continues to contain mercury as well as PCBs.

The primary citizen's organization for generating the political impetus for restoring the Housatonic Watershed is the *Housatonic Valley Association* (HVA). Formed in the 1960s, this organization's accomplishments include:

- the dismantling of dams in the watershed,
- the remediation of sewerage flows into the River,
- the restoration and preservation of stream banks throughout the watershed,
- the identification and prevention of illegal toxic waste dumping into the watershed,
- the identification and prevention of illegal construction in the watershed,
- the restoration of fish habitat in the watershed, and
- the prevention of unsustainable land-use planning within the watershed.

The HVA played a pivotal role in GE's PCB cleanup of the Housatonic by facilitating the negotiations between GE and the EPA. The HVA also played the crucial role in monitoring the River for sources of pollution and lobbying the relevant state agencies for remediation. The HVA's efforts also protected more than 19,000 acres of natural wetlands and stream banks within the watershed through its "Housatonic River Greenway" preservation program. It also has effectively promoted a "least-build" land-use planning policy throughout the watershed as the appropriate and sustainable development technique.

GE's settlement with the EPA committed $15 million of restoration funding to be split between Massachusetts and Connecticut. These funds were combined with state, local, and private contributions to finance restoration projects in each state. The plans for these projects include:

- dam deconstructions,
- habitat restorations for fish and wildlife,
- land acquisitions along the Housatonic and its tributaries for trails, wildlife refuges and river access as in the "Housatonic River Greenway" mentioned above,
- floodplain restorations, and
- environmental education initiatives.

Overall, the Housatonic River Watershed is a rustbelt urban-suburban area. Its restoration activities provide examples that illustrate the potentials of well-organized ***environmental advocacy organizations*** (NGOs) for initiating substantial and effective remediation efforts as documented in this Chapter.

The Four Segments of the Housatonic

The four segments of Connecticut's Housatonic have separate identities in its Management Plan:

- The "scenic" segment between Connecticut's northern border and Falls Village flows primarily through farmland, and has a few small-craft access points for kayaks and canoes.
- The Appalachian Trail crosses the river at Amesville Bridge in Falls Village.
- The wetlands of Robbins Swamp, located east of the river in Canaan and North Canaan, are a popular duck hunting area.
- The Blackberry and Hollenbeck Rivers enter the Housatonic in this segment where the Hollenbeck Preserve, owned by *The Nature Conservancy*, encompasses 182 acres along the river. It provides a habitat for significant rare plant and animal species.

The "recreational" section between Falls Village and Kent is heavily used for boating—especially kayaking and fishing. The Meadows State Park is on the west side to provide access and parking. A "Trout Management Area" extends for nine miles through Sharon and Cornwall. It includes a 3.5-mile fly-fishing only area. Macedonia Brook State Park and Kent Falls State Park are also along this segment. Also, the Appalachian Trail extends for five miles along the west side of the river between Kent and Cornwall Bridge.

A "scenic" segment lies between Kent and Bulls Bridge. There the river is wide and flat. The Appalachian Trail runs parallel and west of the river and offers vistas looking down on the river (Fig. 11.1 indicates this Trail). Route 7 is along the east side, but it is sufficiently separated from the river to make access difficult. It also offers scenic views.

The most southern segment of the Housatonic is "recreational," particularly for kayaks. But protecting this sensitive area from overuse is a major concern. The Bulls Bridge Gorge area, however, is environmentally and historically significant and contains rare species as well as historic resources such as the site of the 1826 Bulls Bridge Iron Furnace. Scenic overlooks are located along this section. The Appalachian Trail follows the river in the area of Bulls Bridge (see Fig. 11.1).

The "Wild and Scenic" Management Plan

To be classified as a Wild and Scenic River, the Housatonic must be managed by a state or political subdivision agency. This is accomplished by the *Housatonic River Commission Coordinating Committee* (HRCCC), the purpose of which is the long-term protection and management of the river. The HRCCC includes representatives from the Housatonic River Commission, the Northwest Hills Council of Governments, the National Park Service, the Connecticut Department of Energy and Environmental Protection (DEEP), and the Housatonic Valley Association.

The *Housatonic River Commission* (HRC) was formed in July of 1979 with the goal of managing the river. The objectives of the HRC were to retain the free-flowing and scenic character of the Housatonic. For the purpose of protecting and improving the water quality, it sought to:

1. eliminate toxic waste discharges and sewerage discharges that lacked tertiary treatments,
2. adopt standards for proper performance of septic systems,
3. adopt measures for aquifer protection, ridgeline protection, erosion and sediment controls, and regulation of gravel and topsoil mining,
4. adopt non-point runoff standards for farms within the watershed,
5. adopt measures for control of other non-point source pollution.

To accomplish the above, the Commission sought to facilitate inter-town coordination, particularly for regulating recreational use of the river, and for monitoring development. The HRC also became the agency of contact for the various federal and state agencies that affect the river. It also facilitated land conservation easements and land trusts, and began a public education program that concerned the issues indicated above. The Commission's first management plan, and its 2006 revision, emphasized management of the recreational use of the river.

Since 2006, all towns along the river have adopted zoning updates which establish regulations using the "inner-corridor, outer-corridor, and village center" scheme (see below for details). In addition, the Northwest Hills Council of Governments has created a model guide for low impact development for use in the river corridors. In addition, towns have now instituted the standards for "no build flood zones" for zoning regulations, and also an eradication of invasive vegetation program.

In order for a river to be federally designated as *Wild and Scenic*, it must have a workable long-term protection plan that does not rely on active federal management other than FERC regulation of hydropower projects. The Housatonic's (Connecticut portion) protection mechanisms include:

- local and state regulations,
- non-governmental initiatives, programs, and partners,
- land purchases for conservation protection.

All seven of the river corridor's towns approved the formation of the HRCC and its management plan. The Commission is cooperatively establishing an "overlay zone" system of land-use restrictions. This "Housatonic River Overlay Zone" includes protections for flood plains and environmentally sensitive areas, and also for historic sites. It does so by first defining the "inner corridor" as including the 100-year flood zone and also that includes certain types of soils as defined by the US Department of Agriculture's *Natural Resource Conservation Service*. This is the most environmentally sensitive section of the river's corridor within which zonal regulations are designed to:

- prevent alterations to the natural flow of the river,

- establish standards for environmental conservation of the land such as those designed to prevent erosion or non-point pollution runoff,
- establish regulations that prevent pollution discharges into the river or its tributaries,
- establish shoreline buffer zones,
- preserve floodplain,
- generally protect fish and wildlife, and
- generally conserve the natural conditions and environmental quality of the overlay zone.

The "outer corridor" is the portion lying between the inner corridor and the top of the river valley's ridge as delineated by the lines drawn from peak to peak along the ridge line. Regulations for this corridor are primarily aimed at guarding against pollution, erosion, and sedimentation, and at establishing safeguards on development. For both the inner and outer corridors, projects within the overlay zone are reviewed by the HRC as well as local government agencies according to their ordinances.

The purpose of the Housatonic Valley Association (HVA) is to conserve the natural character, environmental health, and the economies of the region by protecting and restoring its land and waters. It provides expertise in order to assist landowners along the river who might wish to donate land for conservation restrictions. Some of the significant accomplishments of the HVA include protecting 1,500 acres of wilderness, 200 acres of Conservation Fund land, and 100 acres of prime farmland in the heart of the scenic Housatonic corridor from Kent to Cornwall and Sharon. The HVA is also active in leading river cleanups and water quality monitoring efforts.

In 1979, the only permanently protected lands adjacent to the Housatonic were part of the Meadows State Park and the Housatonic State Forest, a small section of the Appalachian Trail and a few conservation easements held by local or regional land trusts. Since then, acquisition of land and easements by the National Park Service, the State of Connecticut, and the HVA, and a robust land trust network has established an impressive thirty-four miles of protected riverfront within the proposed Wild and Scenic area. The HVA also led negotiations that protected more than five miles of Appalachian Trail along the river's west bank in Kent and Sharon, the longest stretch of riverfront trail on the entire Appalachian National Scenic Trail. In addition, the HVA led negotiations to conserve another five miles along the east bank of Kent and Cornwall (see Fig. 11.1).

The Popularity of the Housatonic

As stated above, the Housatonic attracts kayakers from the metropolitan area. Class I waters begin at Falls Village. But most of this river is Class II and III whitewater. Class V is at Bulls Bridge, and there is a Class VI also at Falls Village. Lake Lillinonah and Lake Housatonic are used for crew rowing (Yale University).

The Housatonic is also a popular fly-fishing body, one of the most popular trout streams in the East, especially between Falls Village and Cornwall Bridge. Of course, this river is within a day trip of the metropolitan area. The Appalachian Trial and the covered wooden bridge at Bulls Bridge are nearby. The demands for recreational use are therefore high, and without limitations on access, their high degree of usage would destroy the natural interaction experiences of the users. The HRC is confronting the problem by restricting use through permits.

Naugatuck Dam Removals

The Naugatuck River is the largest tributary of the Housatonic. Prior to the onset of the industrial revolution, the free-flowing Naugatuck supported the southernmost Atlantic salmon runs, striped bass runs, plus shad, and herring runs. But in the late eighteenth century, the Naugatuck became Connecticut's first industrialized "working river" with dams that provided the power, cooling, and waste dissipation needed for brass manufacturing. These industries and localities (especially the Cities of Waterbury and Naugatuck) discharged untreated wastewater into the river. The fisheries were destroyed.

The industrialization of the Naugatuck River began in the late eighteenth century when the Grilley brothers began manufacturing tin and pewter buttons in Waterbury (see Fig. 11.1). This was a convenient location for button manufacturing because the City of Waterbury was close to the clothing industry in the New York City area. But the early 1790s saw a shift in fashion toward brass buttons—brass being created from the melting of copper and zinc. In the early nineteenth century, the

Naugatuck River Valley became the Nation's center for brass production, and Waterbury was called "Brass City." The dams that facilitated this industry included the Union City Dam, and Platts Mill Dam (both in the Town of Naugatuck) plus the Anaconda Dam, Freight Street Dam, and the Chase Brass Dam (all in Waterbury).

With the twentieth century's ending of the old industrial revolution's production methods, the Naugatuck's dams became obsolete and were abandoned to become derelict structures. They stood without flood control, power generation, or other economic purpose. One-by-one the dams have been taken down in recent decades, and the Naugatuck River is being restored to its free-flowing state. The Chase Brass Dam was removed in 2004; the Union City Dam, the Platts Mill Dam, the Anaconda Dam, and the Freight Street Dam were all removed in 1999. At the time of their removal, all five of these dams had no flood control or other economic purpose, but all five disrupted the ecology of the Naugatuck River. Today, only the Kinneytown Dam in Seymour (the farthest downstream dam), the Plume and Atwood Dam in the Town of Thomaston (north of the City of Waterbury), and the Tingue Dam (south of the Town of Naugatuck) remain. The plan is to remove the Plume and Atwood Dam before 2026, and to expand the fish passage on the Tingue Dam so that fish migration is possible along the entire length of the Naugatuck River except for one lone blocking structure, the downstream Kinneytown Dam which has a modern fish passage facility. But plans for totally removing even this final obstruction are also being formed.

Dam deconstructions along the Naugatuck River are not sufficient to restore the area to being a desirable location for clean dwellings and business. Until very recently, if one were to float downstream through the towns of Waterbury or Naugatuck, one would observe only derelict factories, crumbled facilities, and waste dumps. Restorations of the "brownfields" along the river are required and have begun. In October of 2023, on Freight Street in Waterbury the final building of the Anaconda American Brass site was taken down. There are more than a dozen brownfield sites in Waterbury, and $30 million in federal and state funds are financing their rehabilitation. Six rusting buildings spanning forty-three acres in Naugatuck are also being remediated. But the processes of restoration of these old industrial relicts are long-term undertakings that only begin with the derelict dam removals.

The Dams of the Housatonic

The West Branch of the Housatonic flows through the City of Pittsfield in Western Massachusetts. The City aims to establish a free-flowing West Branch which it will call the "Westside Riverway." This section of river will connect two parks (the Wahconah and Clapp Parks) with the purpose of increasing residents' access to the river for recreation. In particular, kayaking and canoeing will likely be popular from the Wahconah Park to Woods Pond in Lenox.

The removal of the Tel Electric Dam in Pittsfield in 2020 was a key to this restoration of the West Branch. This dam once provided power to the Tel Electric Mechanical Player Piano Manufacturing Company for over a century, but it was a derelict dam at the time of its removal. The removal's cost of $4.3 million involved $1.7 million from federal sources, and $2.5 million from State of Massachusetts sources, but also local sources. The Tel Electric Dam was a derelict "high hazard" structure that posed a drowning hazard, and also posed a risk of complete failure during heavy rains. This would cause downstream flooding. It had accumulated polluted sediment which was removed to an off-site safe disposal facility.

The Bel Air Dam is also on the West Branch of the Housatonic River. Its removal is financed by a $20 million federal grant. It is also an abandoned "high hazard" dam. Removal will facilitate the Westside Riverway, a recreational section of river that is appropriate for canoeing and kayaking. The Bel Air Dam will be removed in 2024. Its removal is motivated by the heavy flooding and damage that occurred during the Summer of 2023.

The Rogue River and Its Whitewater

The Rogue River is one of America's most iconic whitewater bodies. It flows two-hundred and fifteen miles through southwest Oregon on a course through deep western canyons lined with tall firs and pines. It is one of the initial group of eight rivers granted "Wild and Scenic" status in 1968 by the passage of the *Wild and Scenic Rivers Act*. Its headwater is high in the Cascade Mountain Range at Boundary Spring near Crater Lake. From Crater Lake, it flows southerly past Medford where it then turns to flow westerly past Grants Pass and then through federally classified "wilderness" areas. It finally empties into the Pacific Ocean at Gold Beach.

In 1926, author Zane Grey bought a miner's cabin at Winkle Bar near the River. He fished the Rogue for steelhead, and he wrote some of his western books at this location including his 1929 novel *Rogue River Feud*. In 1928, he wrote *Tales of Water Fishing* which included a chapter based on his 1925 drift boat trip down the Rogue. The area then became a popular fishing destination. The Trust for Public Land purchased the Winkle Bar property in 2008 and transferred it to public access. In the 1930s and 1940s, movie stars such as Clark Gable, Tyrone Power, Myrna Loy, and Bing Crosby visited the lower Rogue for its scenery, fishing, and boat trips. Authors Jack London and William Faulkner, journalist Ernie Pyle, and Football Hall of Fame great Norm Van Brocklin were among the many celebrity visitors who fished the Rogue. Baseball Hall of Fame's Bobby Doerr made his home along the Rogue. Actress Ginger Rogers owned a one-thousand-acre Rogue River Ranch along the Rogue, and actress Kim Novak also had a ranch on the River.

The Three Segments of the Rogue

The Rogue River should be envisioned as composed of three distinct segments. The seventy-mile upper section flows southerly before turning west between Shady Cove and Medford (see Fig. 11.2). Over this segment, the North, Middle, and South Branches all flow through steeply walled canyons and thick forests of conifers. A large flood control dam impounds a ten-mile segment on this upstream section to create Lost Creek Lake. From there, the middle section of the Rogue flows westerly for sixty-four miles from Shady Cove to Hellgate Canyon. The surrounding area is commonly called "Rogue River Valley." There the terrain is gentle agricultural lands with few rapids. The tributaries include Little Butte River, Applegate River, and Bear Creek which join the Rogue along this segment where it passes the small City of Grants Pass. The downstream section then flows eighty-two miles from Hellgate Canyon to Gold Beach. This segment generally consists of deep canyons carved through the rugged rainforests of the Siskiyou Mountains. The Illinois River joins the Rogue along this segment. This downstream section has challenging whitewater that is popular for rafting and kayaking.

In 1902, two dams were constructed on the Rogue: the hydroelectric Ament Dam on the upper segment of the Rogue, and the Gold Ray Dam above Bear Creek on the middle segment. But the former was repeatedly attacked with dynamite by fishermen until it was removed in 1921. The Gold Ray Dam was reconstructed in 1941, but is was removed in 2010. Between 1928 and 1931, however, hydroelectric dams were constructed on each of the forks of the upper Rogue.

The middle section of the Rogue flows through the agricultural Rogue River Valley; its tributaries of Bear Creek and Little Butte Creek were dammed by the US Bureau of Reclamation for agricultural irrigation during the 1950s and 1960s.[10] The Savage Rapids Dam was also constructed for irrigation in 1921, but it was removed in 2009 and replaced by direct pumping from the river.

To control periodic floods, the US Army Corps of Engineers (ACE) constructed two large flood control dams in the 1970s and 1980s—the Lost Creek Dam above Shady Cove on the upper section, and the Applegate Dam on the Applegate River—a tributary of the middle segment. The Corps also partly constructed a dam on Elk Creek but it was not completed, and the remaining structure was eventually removed for environmental reasons.

It is the Rogue's downstream segment that was classified as "Wild and Scenic" in 1968. It is heavily used for white water rafting, kayaking, and drift boat fishing with thousands using this section each year. The forty-mile "Lower Rogue River Trail," and the "Wild Rogue Wilderness" (established in 1978) also attract thousands of hikers who explore the lower river's canyons. Also, the "Upper Rogue Wild and Scenic River" was designated in 1988. The forty-mile "Upper Rogue River Trail" (readily accessed from Crater Lake highway), also brings thousands of hikers along the River every year.

[10] Chapter 4 gives a review of the US Bureau of Reclamation.

The Rogue's Dams

A popular "Paddler's Guide to Oregon's Rivers" reviews whitewater runs along all three segments of the Rogue and its tributaries.[11] The thirty-five-mile section on the main branch downstream of Grants Pass is one of the best known and popular whitewater runs in the US. It consists of "Class III+" rapids separated by gentler stretches of deep pools. All of this stretch is within the "Wild and Scenic" section. A maximum of 120 users per day are allowed through the required permit process. Jet boat excursions are also allowed on the lowest section of the Rogue but are also limited in numbers.

Since the removal of the Gold Ray Dam in 2010, only two dams remain on the Rogue River: the Lost Creek Dam and the Applegate Dam. Both are large hydroelectric and flood control dams. They block salmon, steelhead, and other fish migrations at 157 miles upriver from the Pacific Ocean. This leaves the lower Rogue for whitewater excursions, plus fishing and hiking through its canyons and wilderness areas. But there are numerous dams still blocking the Rogue's many smaller tributaries which still block salmon and steelhead from accessing spawning beds. In general, the rivers of the Pacific Northwest have become a battleground for dam deconstruction in part because of the endangered Chinook and Coho salmon.

With respect to Coho, the US National Marine Fisheries Service (NMFS) has identified seven specific populations, called "Evolutionary Significant Units" (ESUs) of Coho in Washington, Oregon, and California. These are wild stocks of salmon that officials are trying to save in order to strengthen the genetic future stock. Four of these ESUs are listed under the Endangered Species Act (ESA). These four include the stocks of the Lower Columbia River (listed as threatened), the Southern Oregon and Northern California Coasts including the coho of the Rogue River (listed as threatened), and Central California Coast (listed as endangered). The long-term trend for all four of these populations is downward. All of the Chinook Salmon are listed under the US Endangered Species Act as either threatened or endangered. The Rogue River Basin's Coho and Chinook Salmon are weak, but they are still thriving, and the Rogue's stocks of salmon are essential for recovery of these species. For example,

[11] See *Soggy Sneakers: A Paddler's Guide to Oregon's Rivers* published by the "Willamette Kayak and Canoe Club" at www.wkcc.org.

the NMFS estimated that in the 1940s, there were 150,000 to 400,000 naturally spawning Coho in the Southern Oregon and Northern California Coastal Rivers, but now there are only 10,000 naturally spawning fish.[12]

Between 2005 and 2020, twenty-two small dams were removed in the Rogue Basin, and fourteen culverts were replaced or repaired, all with the aim of improving fish passage. But in 2020, the State of Oregon listed seventy-four dams and other fish barriers that still should be removed in the Rogue Basin.[13] These were generally small structures on tributaries.

Significant Tributaries of the Rogue

There are three large tributaries to the Rogue River, all flowing northward: the Applegate River, the Illinois River, and Bear Creek. The Applegate River is named after Oregon Pioneer Lindsay Applegate who started the gold mining industry that thrived along this River in the 1850s through the 1890s. The Applegate River is a fifty-one-mile tributary of the Rogue that drains the Siskiyou Mountains, including the Siskiyou National Forest. The Applegate's watershed is approximately seven-hundred square miles. This River originates in the Mountains close to the Oregon-California border. It flows north to join the Rogue close to the small City of Medford. The Applegate River flows through the farms, ranches, and more recently established wine vineyards located around the small towns of the Applegate Valley (Ruch, Applegate, Williams, Murphy, and Wilderville). This scenic Valley has grown over recent decades with new residents building large homes on wooded ten to twenty-acre lots.

The Applegate Valley saw significantly large floods in 1964 and 1974. Consequently, the Corps of Engineers (ACE) constructed the Applegate Dam with its reservoir of Applegate Lake located close to the Oregon-California border. But the Applegate is an important spawning river for salmon, steelhead, and lamprey, especially along its tributaries: the Little Applegate, Williams Creek, Cheney Creek, and Jamison Creek. Given the farming and ranching nature of the Valley, conflicts arose between the

[12] See "Species of Concern," at www.nmfs.noaa.gov/pr/species/concern/ Nmfs.noaa.gov, July 9, 2021.

[13] See Jeff Duewel (August 17, 2020), "To improve habitat, Williams Dam taken out," *Northwest Sportfishing Association*, at https://nsiafishing.org/general/to-improve-habitat-williams-dam-taken-out/.

environmentalists who sought preservation and the farmers, ranchers, and loggers. These conflicts were particularly strong in the 1990s. Fish habitat was a major point of contention given that the farmers wanted irrigation dams and environmentalists wanted free-flowing water for salmon and steelhead spawning beds. In response to these conflicts, the *Applegate Partnership & Watershed Council* (APWC) was formed in 1992. The founders of this environmental advocacy organization sought to resolve these conflicts through their expertise in establishing technical solutions to irrigation problems, and their ability to organize the necessary funds to finance these solutions. Several of these dam removals, stream restorations, and improved irrigation solutions are reviewed here.

Forest Creek joins the Applegate River in the small town of Ruch. The Forest Creek Dam was a small concrete obstruction that substantially prevented the migration of Coho salmon, steelhead, and cutthroat trout. The dam once powered a sawmill, but the sawmill was long abandoned before the APWC arranged the finance from public and private sources for removing the dam and restoring the Creek. The negotiations with reticent landowners began in 2015 and the dam's deconstruction was achieved in 2018.[14]

In the Rogue River watershed, an earthen "push-up-dam" is one constructed each April for summer irrigation, and then removed with October rains. A "push-up-dam" was constructed each year on the Williams Creek, a tributary of the Applegate River at Provolt. The purpose was to divert water into the "Lower Bridgepoint Irrigation Ditch." This method of irrigation diversion began in the nineteenth century in Oregon, but it blocks salmon and steelhead summer migrations to and from spawning beds. By lowering water levels in streams, the water temperatures also rise (a significant problem during this era of global warming) which destroys young salmon and steelhead trout since both are cold-water fish. But APWC has demonstrated that these diversion dams are inefficient for the irrigation farmer, but they can be efficiently replaced by buried piping with screens at the gate-head that block fish from entering the pipe. This piping can divert water during high flows into cisterns or ponds for later irrigation when needed. The piping also eliminates seepage and therefore saves water for irrigation. Also, since the push-up dams need not be constructed each year, the erosion of the

[14] See Kevin Sweeney, "An unlikely partnership removes a dam and restores Forest Creek," *Applegater*, Winter of 2019.

stream bed caused by the construction is eliminated. Two organic farms have benefitted from these efforts of the APWC: Whistling Duck Farm and Blue Fox Farm. The dam changes have been financed by APWC arranged federal, state, and private funds.[15]

The Illinois River is a fifty-six-mile long tributary of the Rogue. Its watershed is a little less than a thousand square miles of southwest Oregon and northern California, an area that constitutes about one-fifth of the Rogue River watershed. In 1984, 50.4 miles of the Illinois River was classified as a "National Wild and Scenic River." The twenty-nine miles of the Illinois that flows through the Kalmiopsis Wilderness is a stretch that offers one of America's premier whitewater streams for rafting or kayaking. It has one-hundred and fifty whitewater rapids, but it is also the most inaccessible by either trail or road of any river section in America's lower forty-eight states with the possible exception of the Olympic Peninsula. This stretch has eight Class IV or IV+ rapids, and a Class V section (Greenwall) below which lies three miles of difficult whitewater rapids. There are no dams on the Illinois River, just whitewater and rapidly flowing water that is so clean that one can see salmon and trout fourteen feet down in some chasms along the Illinois.

Four factors make the Illinois River one of America's most important refuges for salmon and steelhead:

1. The Illinois is free-flowing along its entire length. There are no dams along the Illinois, nor along the Rogue from its confluence with the Illinois to the Rogues's mouth at Gold Beach.
2. The Illinois River's salmon, steelhead, and cutthroat trout are genetically pure native stock. There has never been a hatchery program for the Illinois, and the few hatchery fish that accidently find their way into the Illinois are stopped from upriver migration by the long canyons and Illinois River Falls. They do not reproduce in the upper basin. The same is also true for the Illinois tributaries given that all have steep gorges and falls. Consequently, the native fish of the Illinois Basin are genetically pure so that this Basin is the anchor for wild fish recovery within the greater Rogue Basin.

[15] See Kevin Sweeney, "Williams Creek restoration project set for summer," *Applegater*, Summer 2020.

3. Eighty percent of the Illinois River Watershed is National Forest, and one percent is Bureau of Land Management area. Most of this is in the Kalmiopsis Wilderness. Therefore, almost all of the Illinois River Basin is federally protected.
4. The Illinois River Basin is relatively unpopulated and very unpolluted.

For these reasons, the Illinois River Basin presents a high priority for preservation of healthy and strong native salmon and trout.[16]

The threats to the ecology of the Illinois River Watershed come primarily from logging in the upper basin and its associated road building. This area is steep terrain, and erosion is substantial in logging areas and along logging roads. In addition, there are proposals to establish nickel strip mines in the areas outside the designated "Wilderness" borders. (Applications for the mines are currently in judicial dispute and review.) The US Forest Service has, however, a twenty-year-old proposal for five new "Wild and Scenic Rivers" in the Kalmiopsis Wildlands. It also has a proposal for an expansion of the Kalmiopsis Wilderness.

Bear Creek is the third significantly large tributary of the Rogue River. Bear Creek drains four-hundred square miles of Rogue Valley. It originates at Emigrant Lake and flows through approximately twenty-nine miles through the Valley's population centers of Medford and Ashland. The Bear Creek watershed is a curved valley of thirteen miles by twenty-eight miles. The Creek joins the Rogue River in the northwest corner of the Valley.

The geography of the watershed that surrounds the Valley consists of steep canyons prone to flooding from Spring melts. These high Spring flows cause problems for Bear Creek. To resolve these floods, the federal Bureau of Reclamation constructed Agate Dam on Dry Creek, a tributary of Bear Creek about ten miles northeast of Medford. Agate Dam is a nearly ninety-foot high earthen dam with a reservoir (Agate Lake) that provides irrigation to 30,000 acres of Rogue Valley farm land.

Currently, the small cities of Medford and Ashland—both along Bear Creek—face increasing development pressure, particularly from real estate developers. This sort of development fragments the riparian habitat of

[16] This argument is presented in "The Illinois River Basin, a Wild Salmon Stronghold," at https://kalmiopsisrild.org/explore-kalmiopsis-wildlands/the-river/illinois-river-basin/.

the watershed, and interrupts the ecology of the Creek. Environmental conservation requires more restricted development for the Rogue Valley.

Little Butte Creek is the third significant tributary of the Rogue that joins it in the Rogue River Valley. It is a seventeen-mile long tributary that drains slightly more than three-hundred and seventy square miles of the eastern portion of the greater Rogue River Watershed. This Creek has two major branches: the North Fork and the South Fork. Both have headwaters in the Cascade Mountains. They both flow west and meet near Lake Creek to form Little Butte's main stem, which then continues to flow west until its confluence with the Rogue three miles southwest of Eagle Point (see Fig. 11.2).

Large amounts of irrigation water is diverted from the Little Butte to the Rogue Valley's Howard Prairie Lake and Agate Lake, but also directly into Rogue Valley farms. This Creek, however, is moderately polluted. Despite several dams, the Little Butte is still one of the better Coho and Chinook Salmon producing streams in the Rogue River Basin. A newly designed fish ladder was installed at the Eagle Point Dam (rebuilt in 2008) that helps to facilitate salmon runs.

The North and South Branches merge about fifteen miles northwest of Medford. Water is diverted from the main stem into the "Joint System Canal" for storage into Agate Lake and for irrigating the farms around Medford.

Of the three-hundred and seventy square miles of Little Butte's watershed, approximately two-thirds is forested, and one-third is irrigated farmland. The area's forests have giant Douglas firs, ponderosa pines, and red firs. Almost half of this land is federal, and a little more than half is privately owned. Steep and rugged western canyons are common here. The human population in the Little Butte Basin is approximately ten thousand.

There were almost six-hundred water rights recorded in the Little Butte Watershed. Of these, almost four-hundred involved federal irrigation. In typical years, water is over allocated, and consequently frequent shortages occur in the Little Butte.

Some of the irrigation water for the central Rogue Valley comes from Little Butte Creek by way of its diversion into Agate Dam and also into Fish Creek Dam (see Fig. 11.2). The latter was first constructed in 1908, and rebuilt in 1922 and 1996. It is a fifty-foot tall earthen structure that is nine-hundred and sixty-feet long. A series of canals and tunnels constructed largely under supervision of the Bureau of Land Management facilitated the irrigation use of the Creek's water.

The Oregon Department of Environmental Quality (DEQ) has monitored the water of the Little Butte Creek for its temperature, oxygen saturation, pH, nutrients, bacteria, and chemicals such as pesticides, metals, and turbidity. The entire main stem and its branches were listed as deficient in temperature, oxygen saturation, bacteria, and turbidity. The temperature problem is due to excessive water withdrawal and deteriorating stream bank. All of these deficiencies threaten the salmon and steelhead spawning runs, and also threaten the survival of the juvenile fish. The bacteria problem is from agricultural runoff. Since 2022, the Rogue River Watershed Council has been fencing critical areas of the Little Butte to keep livestock out of the Creek, and has also been restoring the Creek's beds and stream banks.

One of the key institutions working to remedy the water quality problem of the Little Butte is the environmental advocacy organization *Southern Oregon Land Conservancy* (SOLC). Founded in 1978, the conservancy uses "conservation easements" purchased from land owners to protect 12,840 acres of the Rogue's Basin, of which 2,208 acres are along the Little Butte Creek's North and South Forks and its main stem. These easements protect the riparian forests that provide shade to the Creek thereby having a cooling effect. This is one contributing factor that attempts to preserve the anadromous fish populations. The SOLC also has provided fencing along the Creek to protect it from roaming livestock.

Conclusion: Whitewater, Fishing, and River Restoration

The Housatonic River was an "industrial working river" in America's Northeast. It is being restored to a much cleaner body that is in demand for the natural interactions of visitors from the nearby metropolitan areas. These interactions include whitewater recreation through kayaking, canoeing, rafting, fishing (especially trout fishing), and hiking along its banks. It is a river that meets considerable demands for local recreation.

The Rogue River is in America's Pacific Northwest. Its restoration and preservation are essential for the longer-term survival of genetically healthy Pacific Salmon and steelhead. The Rogue is also a paramount Western American whitewater river for rafting, kayaking, and fishing, all in significant demand. Its pollution is very limited as compared to rivers in the old northeast rustbelt area, but it does have some agricultural-related pollution that is being addressed by local advocacy organizations. The Rogue watershed does, however, have several lengthy sections of amazingly clean water flowing fast through remote canyons of classic Western settings. These attributes have attracted recreators to travel considerable distances to experience unique whitewater rafting and kayaking, and also unique native-stock steelhead and salmon fishing.

Dam removals and river restorations have occurred (and are occurring) along both the Housatonic and Rogue River Basins. These two classic rivers illustrate the significant changes in public demands for river management. Industrial and agricultural demands are rapidly diminishing, and recreational demands are greatly expanding. In both cases, it is clean flowing water in natural settings that society demands and uses. These demands bring considerable visitations to these rivers, and these visitors are foundational to the political movement for river restoration and the dam deconstruction movement. The two rivers reviewed in this Chapter illustrate this current American phenomenon.

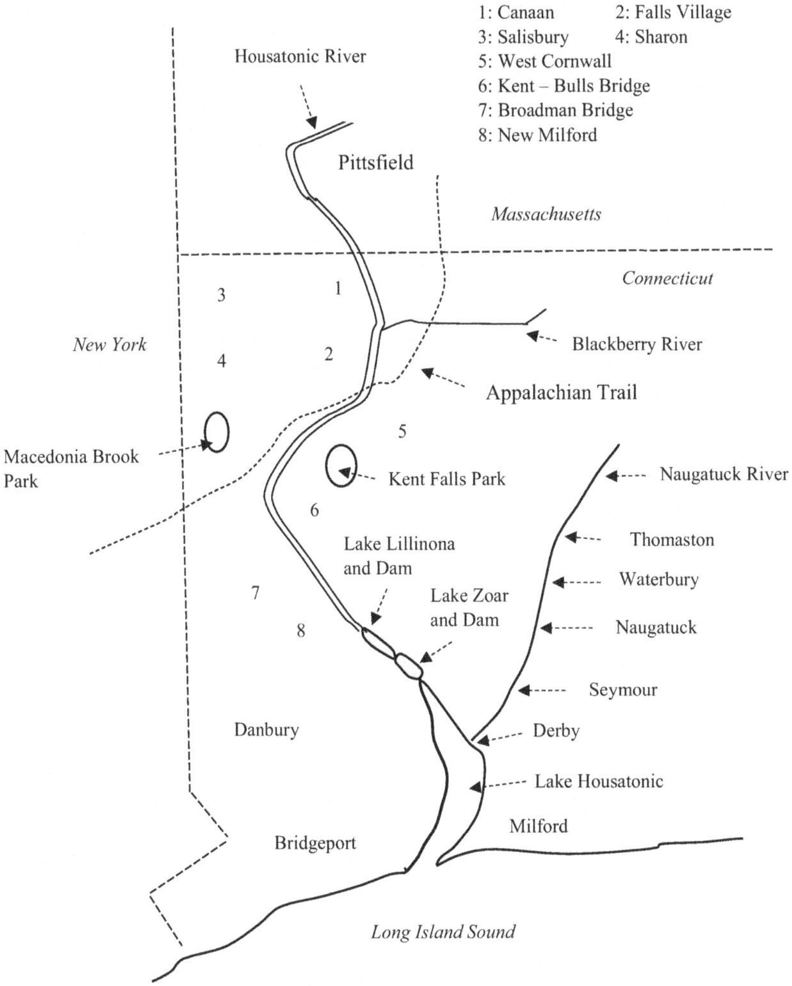

Fig. 11.1 Housatonic River

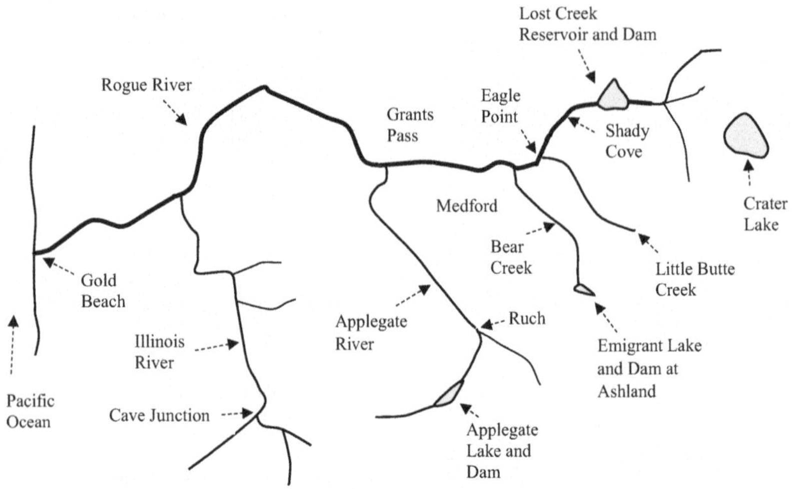

Fig. 11.2 Rogue River

References

Brown, D. (1970). *Bury My Heart at Wounded Knee*. Holt Rinehart and Winston.

Kesey, K. (1964). *Sometimes a Great Notion*. Viking Press.

London, J. (1913). *Valley of the Moon*, Project Gutenberg, ebook can be found at https://gutenberg.org/ebppks/1449

Muir, J. (1911). *My First Summer in the Sierra*. Houghton Mifflin.

Muir, J. (1918). *Steep Trails*. Houghton Mifflin.

Robinson, R. (2021). *Environmental Organizations and Reasoned Discourse*. Palgrave Macmillan.

Robinson, R. (2023). *Environmental Advocacy and Local Restorations*. Palgrave-Macmillan.

Roosevelt, T. (1888). *Ranch Life and the Hunting Trail*. The Century Company.

Roosevelt, T. (1893). *Hunting Grisly and Other Sketches*. The Century Company.

CHAPTER 12

Over Engineering the "Big Muddy"

Introduction: The Great Plains and Its River

Mark Twain had it all wrong! It is not the Mississippi River but the Missouri that is "too thick to drink, but too thin to plow!"[1] Along with the Colorado and the Mississippi, the Missouri is yet another of America's great rivers converted to the twentieth-century's notion of being a tamed, useful and "working" asset. In these cases, however, the human adulteration now primarily serves agriculture, not manufacturing industry as occurred in the nineteenth- and twentieth-century developments.

The Missouri River drains America's *Upper Great Plains*; to be specific, the Missouri's watershed drains Eastern Montana (most of Montana), most of North Dakota, South Dakota, Nebraska, and Wyoming, and large sections of Colorado, Kansas, Iowa, and the State of Missouri. This is one-sixth of the total area of the lower forty-eight US states. It even drains the southern sections of the Canadian Provinces of Alberta and Saskatchewan. Twelve million people live in the Missouri River's watershed.

The key characteristic of this *Upper Great Plains* is its prairie grasslands, which provide massive amounts of silt to America's "Big Muddy" River. Prior to the twentieth-century's damming, and constricting, and

[1] Twain reportedly said, "Mississippi water: Too thick to drink, too thin to plow." But with advances in hydro-agriculture, and even more evaporation due to global warming, some crazy farmers will certainly consider trying to plow the Missouri, so who knows?

R. M. Robinson, *Restoring America's Rivers*,
https://doi.org/10.1007/978-3-031-81758-8_12

channeling of the Big Muddy, it transported 175 to 320 million tons of silty sediment per year to the lower Mississippi River. This was more than half of the sediment the Mississippi once deposited annually in the Gulf of Mexico.[2] The Mississippi Delta was largely formed from this Missouri River silt. But the twentieth- century's large dam constructions, flood control dikes, and channeling structures all reduced this transported silt to only 20 to 25 million tons per year. This past century's restructuring of America's longest river now prevents it from depositing almost all of its natural sediment to its formerly natural floodplains. The reservoirs formed by the Missouri's large earthen dams are now estimated to trap approximately 37 million tons of silt on average per year.[3]

The Missouri River is 2,341 miles long, the longest river in the US (Fig. 12.1 presents a schematic of its course). Its headwaters are in the Bitterroot Mountains in southwestern Montana. From there, it flows north to the eastern edge of the City of Helena, the capitol of Montana. Helena is in the Rocky Mountains very close to the Continental Divide to the east of which water generally flows inevitably toward the Mississippi. The Canyon Ferry Dam blocks the Missouri just to the east of Helena. From there, the River flows northerly and northeasterly through Montana's Great Falls—a series of five large waterfalls. It continues flowing easterly through the scenic "Missouri Breaks" area, a rugged series of canyons where the River forms the large reservoir of the Fort Peck Dam. From there, the Missouri continues to flow easterly into North Dakota and past the small City of Williston, which is close to the Montana/North Dakota border. It then reaches the reservoir of the Garrison Dam more than 150 miles to the southeast of Williston. The Garrison Dam forms Lake Sakakawea, a large two-hundred-mile long reservoir created by the Garrison Dam's blockage of the Missouri. After this dam, the River then flows southerly for another 80 miles to pass the City of Bismarck in North Dakota.

The Missouri River has ninety-five large tributaries and hundreds more smaller feeder streams. Along its northern section, the more significant tributaries include the Musselshell River which meets the Missouri at Sakakawea Lake, and the Yellowstone River which meets the Missouri

[2] See US Army Corps of Engineers (2006, January 1), "Missouri River Mainstem Reservoir Master Water Control Manual," at http://digitalcommons.unl.edu/cgi/viewcontent.cgi?article=1070&context=usarmyceomaha.

[3] Ibid.

close to the North Dakota/Montana border. This latter river is the Missouri's largest tributary measured by discharge and the second longest by length.

Past Bismarck, the Missouri flows southerly, and crosses the South Dakota border sixty miles south of Bismarck. Approximately another 100 miles south is the Oahe Dam at the small city of Pierre, South Dakota. It forms Oahe Lake, a reservoir that stretches back the entire hundred and sixty miles to Bismarck.

Past Pierre, the Missouri flows southeasterly through the Big Bend area and its Big Bend Dam. Then approximately 150 miles southeast of Pierre, the Missouri River forms the border between the States of Nebraska and South Dakota. The Fort Randall Dam is in South Dakota close to the Nebraska border. After flowing another 150 miles, the River turns southerly and forms the Nebraska/Iowa border. The Gavins Point Dam is half-way along this stretch. About one-hundred miles farther south, the Missouri passes through the City of Omaha.

The Platte River, a most significant river for irrigation in the Great Plains, joins the Missouri south of the City of Omaha. It is the longest of the Big Muddy's tributaries, but water extractions drastically reduce its flow to the Missouri. The eastern border of Nebraska is entirely formed by the Missouri River, and this border is approximately 300 miles long before meeting the State of Kansas. At this point, the Missouri River continues to flow southerly to form the border of Kansas with the State of Missouri. Kansas City is a border town along the Kansas/Missouri border. At Kansas City, the River turns to directly flow easterly through the entire State of Missouri to the City of Saint Louis on the eastern edge of the State. There the River meets the Mississippi just north of Saint Louis. The Osage River is another significant tributary of the Missouri River. It meets the Missouri River about half-way along its course through the State of Missouri.

The Missouri River Watershed's four largest cities include *Omaha* in Nebraska which is just north of the Missouri's confluence with the Platte, *Kansas City* at the confluence of the Missouri River and the Kansas River, *Saint Louis* at the confluence of the Missouri River and the Mississippi, and *Denver* in the State of Colorado's "Front Range" of mountains. The latter is in the Platte River drainage area. These cities form the transportation centers of the Missouri River Watershed's vast agricultural industry which encompasses one-fourth of the US agricultural land, and that produces one-third of its key grain production (wheat, flax, barley, and oats). But these are primarily non-irrigated crops. Only 11,000 square miles is irrigated of the 170,000 total square miles that produce crops in the Watershed, while 281,000 square miles is used for livestock grazing.

The Dams of the Missouri River

The Missouri River can be envisioned as having three sections: (1) the mountainous upper section, (2) the canyonlands section in the middle, and (3) the lower prairie-meandering section. Each of these three sections has subsegments and each are examined in turn below. Dam construction on the Missouri River's "main stem" occurred during the late nineteenth through the mid-twentieth centuries. Dams have transformed approximately a third of the longest river in America's "Great Plains" into a chain of reservoirs, the motivation for this primarily being flood control, irrigation, and hydropower generation. Hydropower projects began in the 1890s for powering the City of Helena, and to also serve other small towns in the Upper Missouri River area, and to also serve Montana's mining interests. In addition, large flood control and storage dams were constructed on the middle section of the Missouri in the 1940s and 1950s. Table 12.1 presents relevant statistics for each of the dams reviewed.

The Mountainous Dams Above the Great Falls

The Town of Helena was established in 1864 during southwest Montana's gold rush. It is the site of "Last Chance Gulch" where Montana's "gold rush" began. In the 1880s, it was a wealthy town with more than fifty millionaires, well-established western bars, and Victorian architecture. It is close to the Continental Divide and sits at 4,000 feet in elevation. It became Montana's Capital in 1875. Its population is approximately 32,000 according to the 2020 US Census; its metropolitan area is approximately 83,000. The topology around Helena is steep and narrow Rocky Mountain canyons. In 1889, Canyon Ferry Dam was constructed on the eastern slope of the Continental Divide, close to Helena to provide hydroelectric power to the small city.

There are four dams along the Upper Missouri that are close to Helena in Montana. All four are in steep mountainous canyons in the 3,300 to 4,000-foot altitude levels. These are all fairly narrow hard-rock canyons upriver of Great Falls. They cover a gradient of falling 700 feet in 90 miles (7.8 feet per mile). The highest elevation is the farthest upriver Toston

Dam at slightly over 4,000 feet. All four dams (Toston, Canyon Ferry, Houser, and Holter) are concrete structures anchored in the hard-rock narrow canyons. They range in length from 1,364 feet to 200 feet, and height from 50 to 225 feet. In power capacity, they vary from only 10 megawatts to 50 megawatts. Table 12.1 presents the other specifics of these dams.

In 1940, Montana's "Department of Natural Resources and Conservation" (DNRC) completed a small "run-of-the-river" dam close to Toston in Western Montana (see Fig. 12.2). The State of Montana used $820,000 of Works Progress Administration (WPA) funds to construct this Toston Dam along with the much larger downriver Fort Peck Dam, which was also constructed in 1940 for the purpose of diverting flow into irrigation canals. The Toston Dam is a concrete construction built on ledge. In 1989, a ten megawatt hydroelectric generator was added to help pay for other DNRC projects. The irrigation provided is 42,000-acre feet distributed across 300 contracts.

Table 12.1 Missouri River Main Stem Dams (upstream listed first then downstream)

Dam	*State*	*Construction*	*Dimensions (length by height)*[a]	*Power capacity in Megawatts (MW)*
Toston	Montana	Concrete	705 ft. by 56 ft	10 MW
Canyon Ferry	Montana	Concrete	1,000 ft. by 225 ft	50 MW
Hauser	Montana	Concrete	700 ft. by 80 ft	19 MW
Holter	Montana	Concrete	1,364 ft. by 124 ft	48 MW
Black Eagle	Montana	Concrete	782 ft. by 34.5 ft	21 MW
Rainbow	Montana	Concrete	1,055 ft. by 29 ft	36 MW
Cochrane	Montana	Concrete	753 ft. by 59 ft	64 MW
Ryan	Montana	Concrete	1,336 ft. by 61 ft	60 MW
Morony	Montana	Concrete	883 ft. by 94 ft	48 MW
Fort Peck	Montana	Earthen	3.98 miles by 250 ft	185 MW
Garrison	N. Dakota	Earthen	2 miles by 210 ft	515 MW
Oahe	S. Dakota	Earthen	1.77 miles by 245 ft	786 MW
Big Bend	S. Dakota	Earthen	2 miles by 95 ft	493 MW
Fort Randall	S. Dakota	Earthen	2 miles by 165 ft	320 MW
Gavins Point	S. Dakota and Nebraska	Earthen	1.9 miles by 74 ft	132 MW

[a]This uses 5,280 feet per mile

Above the Great Falls' dams, and below Helena, the next downstream dam is Canyon Ferry, first constructed in 1898 (see Fig. 12.2). This was a "run-of-the-river" dam that produced only seven and a half megawatts for Helena.[4] Note that in 1949, the US Bureau of Reclamation built a replacement Canyon Ferry Dam a half-mile downstream, and this inundated the original structure.

Whereas the Toston Dam is the most upstream structure, the Canyon Ferry Dam is the second most upstream structure. Both were constructed in the narrow canyons close to Helena's east slope. The newer Canyon Ferry Dam is a 1,000-foot long and 225-foot high concrete dam that creates Montana's third-largest reservoir. This dam provides flood control for the downstream Great Falls area. It is only twenty miles east of Helena. It was constructed as part of the World War II era "Pick-Sloan Missouri Basin Program," as reviewed below. It replaced the much smaller rock-crib structure that is now inundated 1.5 miles upstream of the current Dam. It also provides irrigation and almost fifty megawatts of hydroelectric power for the Helena area. The Canyon Ferry Reservoir has a length of thirty miles, and a width of five miles. Its surface covers 33,500 acres with a volume of more than two million acre-feet. But this is small as compared to the much larger dams and reservoirs of the Great Falls area.

The original Houser Dam is the next construction downstream from Canyon Ferry but above Great Falls. This was a flawed steel dam built in 1907, flawed because of its faulty ceramic base, not because of the steel design or material. During flood conditions, this faulty base failed in 1908, and the downstream Black Eagle Dam at the Great Falls was consequently dynamited to save nearby factories from the flood. The Houser Dam was rebuilt as a concrete structure in 1910, and that rebuilt structure remains in place today. It is fourteen miles northeast of Helena. The Houser Lake reservoir is twenty-five miles long with a surface of 3,800 acres and a storage capacity of 98,000 acre-feet when full. It has a hydro-power capacity of seventeen megawatts.

Construction of the Holter Dam began in 1908 to eventually provide power to the copper company that was to become Anaconda Copper. This company started as a silver mine near Butte in Montana, but it was purchased in 1881 by an Irish immigrant and converted into a copper

[4] A "run of the river dam" refers to one that simultaneous discharges any flow coming into its reservoir. It does not store this flow for later discharge beyond the initial buildup of its reservoir.

mine. The original Anaconda Copper Mining Company included other mines in the Butte area, a local smelter, and other processing plants begun in Great Falls.

The Holter Dam was forty-five miles downstream of the Houser Dam (see Fig. 12.2). The Montana Power Company and United Missouri River Power Company began construction of the Holter Dam at the "Gates of the Mountains." This was formerly a spectacular canyon that Meriwether Lewis described as "the most remarkable cliffs that we have yet seen."[5] Construction was abandoned, however, in 1910 after only part of the concrete foundation was poured. In 1912, Montana Power purchased Canyon Ferry Dam, and the Houser Dam, and the partly completed Holter Dam.[6] Montana Power then completed Holter in 1918. The Holter Reservoir is twenty-five miles long with a storage capacity of 243,000-acre feet. This completed structure was actually slightly larger than the originally planned dam. Its power capacity is 48 megawatts of power.

Powering Great Falls

The Missouri River's "Great Falls" are on its upper section in Montana but below the more mountainous section of the Helena area. The "Great Falls" is a series of five waterfalls over which the River drops approximately a hundred feet in a ten-mile stretch (ten feet per mile). At the dawn of the electric age, this drop together with the hard-rock canyons of the area provided natural settings for hydroelectric dams that could power copper smelters to serve the output of the Anaconda Copper Mine located to the west of Helena (see Fig. 12.1).

The five dams along the Great Falls section (Black Eagle, Rainbow, Cochrane, Ryan, and Morony) have lengths that range from 782 feet to 1,336 feet, and heights that range from 29 to 94 feet. (Fig. 12.2 indicates the position of the five dams, and Table 12.1 presents their specifics.) Their power capacity is 21 to 64 megawatts. The hard-rock canyons of this section are also narrow and similar to the upstream canyons. But over

[5] Montana Bureau of Mines and Geology (2009), "Gates of the Mountains," Lewis and Clark—A Geological Perspective, at https://web.archive.org/web/20090205162342/http://www.mbmg.mtech.edu/gmr/lewis_clark/lewis_clark-gates.asp.

[6] See Malone (2006, p. 204).

the entire canyonlands of Eastern Montana, the gradient has a lower slope (1,200 feet over 271 miles) as compared to the upstream section.

Black Eagle Dam, completed in 1891, was constructed at the town of Great Falls. It was the first dam on the middle section of the Missouri. This was a rather small weir structure that channeled only a small portion of the River's flow into an electric power generating plant. This weir was eventually replaced by another structure in 1926. But before this replacement, the original "Boston and Montana Smelter" was built nearby to utilize the power generated by the Black Eagle Dam. The then small town of Great Falls grew to provide domiciles for the smelter's employees, and Great Falls promoted itself as the "Electric City." Between 1890 and 1915, this expansion was considerable. Following this rapid development, the Boston and Montana Smelter facility planned for an expansion with new copper and zinc smelters, and consequently new hydropower facilities were constructed to provide the necessary electric power.

In 1915, the Ryan Dam was erected ten miles downstream of the Black Eagle Dam. It is the largest of the dams along the "five falls," and was constructed directly upstream of the eighty-foot Big Falls. This was built by the Montana Power Company. Its reservoir can store 5,000-acre feet, a rather small surface. It was erected around Ryan Island but is primarily concrete on each side. Following the Ryan Dam's construction, the Cochrane Dam and then the Rainbow Dam were constructed upstream; the next dam downstream of Ryan is the Morony Dam.

The Rainbow Dam is six miles downstream from Black Eagle Dam. This structure was erected at Rainbow Falls. Its capacity is thirty-six megawatts. The Cochrane Dam was constructed in 1958 eight miles downstream of the Black Eagle Dam. It is a concrete "run of the river" structure.[7] It provides a capacity of sixty-four megawatts of power. The Morony Dam was completed in 1930, and it is located downstream of the Ryan Dam. It was also constructed to provide power to the copper and zinc smelters of Great Falls, forty-nine megawatts of power.

The five dams constructed in the Great Falls section of the Missouri River together deliver a capacity of 229 megawatts of electric power. Most of this is utilized to process mining ores at smelters in the Great Falls area. As the Missouri flows downstream from Great Falls, its gradient becomes a bit flatter, but it still flows through canyonlands around Fort Peck. After

[7] A "run-of-the-river" dam does not store water. The amount of flow of the river equals the flow past the structure.

these canyons, the River flows through the "Great Plains Prairie Lands" where the problems concern flood control and irrigation.

The Canyonlands Below the Great Falls

During the late nineteenth century and through the 1920s, the most significant barrier to the economic development of the Missouri River Basin was its frequent large-scale flooding. In particular, the 1926–1927 massive flooding in the Mississippi-Missouri Valleys was a major factor in restructuring America's heartland.[8] The US Army Corps of Engineers provided much of America's response to this significant national-level natural disaster. In 1933, as part of the "New Deal" of the "Great Depression" of the 1930s, the Corps began the Fort Peck Dam, a major effort at controlling the flood prone river. This large Dam was completed in 1940 as the first constructed on the Upper Missouri below Great Falls. It is the highest of the six large dams constructed along the Missouri's main stem (see Table 12.1). Its location is at the old Fort Peck trading post, a site previously used for docking paddle-wheeler steamboats in the mid-nineteenth century. This site is close to the town of Glasgow located north of the River. The Fort Peck Dam is 21,026 feet in length and 250 feet in height which is much larger than the Missouri's upstream dams. This is an earthen dam with concrete spillways; it creates Fort Peck Lake, the sixth-largest dam-created reservoir in the US. It stretches one-hundred and thirty-four miles through the "big canyon country" of the Missouri River, and Fort Peck Lake backs up the Missouri into these winding deep canyons. The Dam has a hydropower capacity of more than 185 megawatts, more than three-fourths of the total combined capacity of all five of the hydropower dams of the Great Falls area.

Fort Peck Dam was another WPA project. Construction started in 1933 at the height of the Great Depression, but it was not completed until 1940 because of emergency delays in completion. Construction problems involved the Dam's shale foundation and its liquification (Eight men were killed during a partial collapse during construction). The hydropower generation was not completed until 1951. This facility was constructed by the Corps of Engineers which operates it today.

[8] See Barry (1977).

Fort Peck Lake is a major Montana reservoir located just west of Glasgow in Montana. It extends 134 miles with canyon inlets that create a shoreline of 1,520 miles. This reservoir was initially created to assure adequate flow storage for navigating the downstream river during the normal lower flows of summer months. But it was also created to store the high flows during the spring flooding periods in order to limit downriver flooding. Erecting this Dam was not, however, adequate for controlling the floods of the early 1940s and their consequent destruction of manufacturing plants in downriver Omaha and Kansas City. These destructive floods disrupted the manufacturing of World War II military supplies. In response, Congress passed the Flood Control Act of 1944, which gave the Corps of Engineers authorization for a large expansion of flood control structures in the Missouri River's Basin, i.e. the "Pick-Sloan Missouri Basin Program." The Sloan Plan emphasized irrigation expansion through building eighty-five smaller dams on the Missouri's tributaries. But the Pick Plan emphasized flood control and hydropower development through large storage dams along the Missouri River's main stem below Great Falls. In the 1950s, this plan led to the construction of the five significantly large dams along the Missouri River below Great Falls: the Garrison, Oahe, Big Bend, Fort Randall, and Gavins Point Dams.

After five years of construction, the Garrison Dam was completed in 1953. At over two miles in length, it is the fifth-largest earthen dam in the world.[9] Its height is 210 feet. It has a hydropower capacity of 515 megawatts, the second largest of the Missouri River's dams. Its reservoir is Lake Sakakawea, which extends westward beyond Williston and the River's confluence with the Yellowstone River, one of the Missouri's more significant tributaries. The Yellowstone flows into the Missouri in North Dakota, close to the North Dakota border with Montana. The Dam is named after the town of Garrison, which is located midway between the Cities of Bismarck and Minot in North Dakota.

In the early eighteenth century, the "Three Native American Tribes" of the Hidatsa, Mandan, and Arikara left what is now the central plains of Iowa and Nebraska, to settle as farmers in permanent villages along the Upper Missouri River in the Dakotas. When the first French-Canadian fur traders arrived in the northern Great Plains in the mid-eighteenth

[9] See US Army Corps of Engineers (2004), "Garrison Dam and Powerplant," www.nwo,usace.army.mil/html/Lake_Proj/garrison/dam.html.

century, they encountered the "Three Tribes" with domiciles close to present day Garrison in North Dakota. These three tribes were farmers and traders with earthen lodge villages that provided centers for trading with the nomadic tribes of the Upper Plains. In 1804, the Hidatsa hosted the Meriwether Lewis and William Clark expedition that was commissioned by President Jefferson to explore the Louisiana Purchase from Saint Louis to the Pacific Ocean. While at the Hidatsa Village, Lewis and Clark enlisted the trapper Toussaint Charboneau as an interpreter and guide. His wife Sacajawea (Sakakawea in Hidatsa which means "Bird Woman") was a Shoshoni whose knowledge of the to-be-explored land was invaluable to Lewis and Clark. Lake Sakakawea is named after her.[10]

Following the Lewis and Clark expedition, during the mid-nineteenth century, Fort Clark, Fort Abraham Lincoln, Fort Stevenson, Fort Randall, and Fort Buford were all established along the Missouri to protect the settlers and to accommodate steamboat traffic. During the post-Civil War era, farmers from Europe settled the Missouri River area of North Dakota primarily raising cattle in the Little Missouri River region (see Fig. 12.1). Theodore Roosevelt established his Elkhorn Ranch in the Little Missouri region. But today, rather than cattle ranching, recreation has become the major industry of the area.

Lake Sakakawea divides the Berthold Indian Reservation where it inundated one-sixth of the Reservation, and ninety-four percent of the Three Tribes' agricultural lands (those of the Hidatsa, Mandan, and Arikara). They were forced to accept only $7.5 million in compensation. Some of the tribal towns were also inundated by Lake Sakakawea.

Oahe Dam is the next downriver structure. It was also authorized by the 1944 "Flood Control Act" and the Pick-Sloan Plan. Oahe Dam is just six miles north of Pierre South Dakota. Like the Garrison Dam, it is a very large earthen structure: 9,360 feet in length and 245 feet in height. Lake Oahe is the fourth-largest dam-created reservoir in the US. It stretches north for 231 miles to Bismarck in North Dakota. This lake covers 370,000 acres with 2,250 miles of shoreline. The Dam generates 786 megawatts of power, the most of any dam on the Missouri. This power serves much of the north-central plains. In this section of the Dakotas, the lands to the west of the Missouri River are primarily cattle grazing prairie, and to the east it is dry-land winter wheat.

[10] See Meyer (1977) and also Lawson (1982).

From Bismarck in North Dakota to Pierre in South Dakota, the Missouri River flows almost directly south. But at Pierre, it bends to flow toward the southeast. Sixty miles downstream of Pierre is the Big Bend Dam, a "run of the river" earthen structure by Fort Thompson on the Crow, Creek, and Sioux Tribal lands. There is actually a large "Big Bend meander" on the Missouri that now has its water level partly raised by the Big Bend Dam which lies seven miles to the south. This giant meander is approximately twenty-two miles across. The land to the west is in the Lower Brule Reservation, and to the east is the Crow Reservation. The "Big Bend meander" was actually explored by William Clark on his expedition as noted in his journal.

At 95 feet in height, the Big Bend Dam is small as compared to the Oahe or Garrison. It has a hydroelectric capacity of 494 megawatts, which is substantial for dams on the Missouri. Chronologically, it was the last of the six dams authorized by the 1944 "Flood Control Act." Its construction began in 1959 and was completed in 1963.

Past Big Bend, the Missouri continues to flow toward the southeast where it forms a portion of the eastern border between South Dakota and Nebraska. Six miles above this border is the Fort Randall Dam, named after one of the old forts constructed along the Missouri River. Fort Randall Dam is an earthen structure of slightly more than two miles in length. Its hydroelectric capacity is 320 megawatts, the fourth-largest capacity on the Missouri. Its reservoir is the eleventh-largest dam-created lake in the US. This Lake is approximately 107 miles long with 540 miles of shoreline. It has a 102,000-acre surface area. The Dam was completed in 1953.

Gavins Point Dam is the most downstream dam on the Missouri's main stem. Similar to the other four upstream dams authorized by the 1944 "Flood Control Act," it is an earthen structure. It is 1.9 miles in length, and was completed in 1957. It is located 811 miles upstream from Saint Louis, close to the Gavins Point bluff on the north bank of the River. Its reservoir is the 25-mile long Lewis and Clark Lake. Its hydroelectric capacity is 132 megawatts.

These five large dams, along with the Fort Peck Dam, form an enormous storage capacity of 74.1 billion acre-feet, the largest reservoir storage system in the US. This is equivalent to more than three years of Missouri River flow as measured at Gavins Point, the most downstream dam on the River. The six dams below the Great Falls area have a combined capacity of 2,431 megawatts of power. The five dams in the

Great Falls area, and the four other upstream dams produce only 356 megawatts combined.

The nine most upriver dams on Missouri's main stem are either concrete or "almost all concrete" structures. They were built between steep hard-rock canyons of the sort that were appropriate for concrete erections and anchoring. But the six downriver dams (Fort Peck to Gavins Point) are constructed in severely eroded canyons of far less depth, i.e. more gentle canyons of the Great Plain prairie. These settings are appropriate for the much broader rolled-earth and chalk structures of the downriver dams. Table 12.1 presents some of the descriptive data for all fifteen main stem dams.

Channeling the Missouri River for Transportation

Most of the Missouri River's natural floodplain has now either been converted into agricultural lands or developed for human home or business domiciles. As a result, there is now a high risk of inundation of infrastructure by flooding. Levees have been constructed along a third of the River's banks with the result being a faster flow velocity and an increase in peak flow amounts downstream. More than 300,000 acres of formerly natural flood plain have been converted into agricultural lands through older federally funded programs.[11] But fertilizer runoff from agriculture and *concentrated animal feeding operations* (CAFOs) in Iowa, Kansas, Nebraska, and Missouri drain animal and crop fertilizer wastes into the Missouri River and then the Mississippi River. All of this severe pollution adds to the "dead zone" problem in the Gulf of Mexico.[12] Such is the cost of this rapid expansion of negative agricultural externalities in the Upper Great Plains. One of these negative externalities stems from the over engineering of the Missouri River for agricultural expansion, an expansion that was not economically justified as reviewed in this section.

Slower flowing rivers, especially those that carry large amounts of sediment, "meander" in that they curve and twist back and sling their sediments. These sediments then build up along the river's meanders. "Channeling" involves straightening these natural river meanders by

[11] See Kendle (1970).

[12] See Robinson (2021, pp. 219–221) for review of the CAFO problem.

building levees and wing dams that increase the flows velocity. Channeling usually also involves deepening the river's course so that larger barge traffic can navigate. *Channeling* of "river meanders" reduces the volume of sediment built up in natural marshy areas along the Missouri River, and this has eliminated natural fish and aviary habitat.

Seeing the decline of native species because of the channeling of the Missouri River, the US Fish and Wildlife Service has recommended riparian habitat restoration of the River.[13] In the early years of the twenty-first century, the Corps of Engineers began restoring riparian habitats destroyed by *channeling*. This sort of river engineering (the straightening of meanders and deepening the channel's depth) occurred along much of the Missouri River because it was considered necessary for facilitating the shipping of agricultural goods and also mining ores via barge traffic. But forecasts that were in excess of the actual transportation needs consequently resulted in the Missouri being over channeled. The heightened demands for shipping did not materialize. (The text presented below reviews this forecasting problem.) Because of the low use of the Missouri for barge shipping, the Corps of Engineers is now removing many of the levees, dikes, and wing dams that were designed and constructed to restrict the River's flow and increase its velocity for channeling purposes. In 2001, there were 87,000 acres of natural riverside floodplains restored by channel removals. But at that time, there was still almost a third of the Missouri flowing through artificially straightened channels.[14] An ecological restoration of the Missouri requires that the previous slowly meandering and wide river be restored to a considerable extent, and this requires having its channel constructions removed. They no longer serve the area's transportation needs but they are destroying the riparian ecology.

Prior to the damming of the Missouri, powered river transportation was by steam-driven paddle-wheel boats, especially the shallow drafted vehicles of the sort that then traveled the Mississippi. By the 1890s, however, the Northern Pacific Railroad availability eliminated even the small amount of river traffic on the Upper Great Plains. In 1912,

[13] See US Army Corps of Engineers and US Fish and Wildlife Service (2012, March 21), Executive Summary," at www.fws.gov/mountain-prairie/pressrel/03-87.htm.

[14] US Geological Survey: Columbia Environmental Research Center (2010, May 27), "The Missouri River Story," at https://web.archive.org/web/20100527124725/http://cr.usgs.gov.the_River//MORstory.htm.

for Missouri River navigation purposes, Congress ordered the Corps of Engineers to maintain a six-foot river depth between Kansas City and Saint Louis, a length of 368 miles. The Corps accomplished this by constructing levees and wing dams that directed the River into a straight narrow channel that scoured the river bed. In 1925, the Corps widened this section of the River to 200 feet. Then in 1927, the Corps extended this navigation channel northward to Sioux City in Iowa. This straightening of the River reduced its overall length by approximately 200 miles.

In 1945, the Corps began the "Missouri River Bank Stabilization and Navigation Project," which aimed to increase the width of the Missouri's navigation channel to 300 feet, and to extend its depth to nine feet. The 735 miles between Sioux City and Saint Louis was also channeled with rock dikes that straightened the River's flow, and that also increased the velocity of the flow so that it would scour any sediment buildup. This also entailed cutting off meanders and side channels. But since the Missouri carries large amounts of sediment, where would all this silt go? The answer is that the Corps consequently must annually dredge a considerable amount of sediment to maintain the Missouri's navigation channel, but the amount necessary appears too much for even the Corps. In 2006, barge tows frequently "ran aground" because this Sioux City to Saint Louis section was so silted.[15]

In the 1960s, the Corps was predicting substantial increases in barge traffic. The prediction was that 12 million tons would be shipped per year by 2000. This was the basis of the Corps' subsequent channeling and dredging efforts. Actual shipments, however, were only 3.3 million tons in 1977, and then only 1.2 million tons in 2000. For example, winter wheat, the Great Plains primary agricultural crop, is now largely shipped by railroad. Also, large destructive floods in 1993 and 2011 substantially disrupted river traffic. Given the potential that floods have for causing shipment damage, and also their disruption of shipping facilities, then as a result the "Big Muddy's" future as a transportation route is limited.

[15] See "Corps failing to keep Missouri River navigation open as promised," at http://ago.mo.gov/newsreleases/2006/072606.htm. Also see "Letter from Missouri Attorney General Jeremiah W. Nixon to Brigadier General Gregg F Martin," at http://ago.mo.gov/pdf/2006/MartinLetter072606.pdf.

The Platte River Tributary

The Platte River is formed by the confluence of its North and South Branches that meet near the small city of North Platte in southwestern Nebraska, close to the Wyoming and Colorado borders. Both the North and South Platte originate in the Rocky Mountains of Colorado on the eastern side of the Continental Divide. The headwaters of the North Platte are in the North Park Valley (northwest of Denver) which drains twelve mountain peaks of over 11,000 feet near the town of Walden. It flows north into Wyoming where it meets the Seminoe Dam and Reservoir on that State's southwest corner. The Bureau of Reclamation constructed the Seminoe Dam as part of the "Kendrick Project," which was a 1930s Depression Era public works effort authorized by Congress in 1933. The purpose of this project was to expand irrigation and hydroelectric power in southern Wyoming. The Seminoe is a concrete hydropower dam of 295 feet in height, and 530 feet in length completed in 1938. (See Figs. 12.1, 12.3, and 12.4 for schematic maps of the Platte River.) Its reservoir contains almost one million acre-feet of water and the dam has a hydropower capacity of 45 megawatts (The Medicine Bow River also flows into the Seminoe at its southeast corner). From this Reservoir, the North Platte continues to flow north for two miles to meet the Kortes Dam. This concrete dam is 244 feet in height and holds 4,700-acre feet of water, and has a hydropower capacity of 36 megawatts. These upriver structures are each in deep hard-rock canyons appropriate for anchoring large concrete dams.

After the Kortes Dam, the North Platte flows into the Pathfinder Reservoir and Dam (see Fig. 12.3). The "Pathfinder Dam and Reservoir" was named after the great western explorer John Fremont, the "Pathfinder," who explored this Wyoming area in 1842.[16] Farmers and ranchers had previously attempted to irrigate the semi-arid sagebrush desert of Wyoming and Nebraska during the 1880s by using diversion dams on small rivers and streams. In Wyoming, the initial idea was to use the Sweetwater River to provide the needed irrigation by using diversion dams. But it was apparent that since the Sweetwater flowed into the North Platte forty-five miles south of Casper (the location of the Pathfinder), a dam located there would facilitate a large reservoir. The Pathfinder Dam has a capacity to store more than one million acre-feet, and the Dam has a

[16] See Rolle (1991).

hydroelectric capacity of 67 megawatts. This Dam was one of the Bureau of Reclamation's first constructions, and it was part of the "North Platte Project" of the first decades of the twentieth century. The "Pathfinder" was located 45 miles upriver from Casper in Wyoming, and completed in 1909. It is made from granite blocks cut from a near-by quarry. It stands 214 feet high and 432 feet in length.

After flowing through the Pathfinder Dam, the North Platte begins a very large curve through the Alcova Reservoir and Dam, and then through the Small City of Casper in Wyoming. This is not the deep hard-rock canyon terrain of the upstream North Platte. It consists of the more gentle rolling sagebrush canyons of central Wyoming. Consequently, the Alcova Dam, and other downstream dams, are earthen with concrete spillways. The Alcova's hydroelectric capacity is 36 megawatts. This Dam is earthen with a height of 265 feet, and length of 763 feet. It was completed in 1938. The North Platte then begins a southeasterly course to the Glendo Reservoir and Dam, and then the Guernsey Reservoir and Dam. The Glendo Dam is earthen with a height of 190 feet, and a length of 2006 feet. It was completed in 1958 by the Bureau of Reclamation. It has a hydroelectric capacity of 38 megawatts. The Guernsey Dam is also an earthen structure. It was completed in 1928 by the Bureau of Reclamation. It is 135 feet high and 560 feet long, but it only produces 6.4 megawatts of power. After the Guernsey Dam, the North Platte then crosses into Nebraska near Scottsbluff. Flowing farther in the southeasterly direction, the North Platte reaches the Kingsley Dam near Ogallala in Nebraska. The Kingsley Dam is a 162-foot high and 3.1 miles long earthen structure that has a capacity of 50 megawatts. It was completed in1941. Sixty miles to the east of the Kingsley Dam, the North Platte joins the South Platte.

The South Platte originates in the Mosquito Mountains of Colorado's "Front Range" southwest of Denver. It drains a series of 12,000-foot peaks and then flows southeasterly through the Antero Dam and Reservoir, the Spinney Reservoir and Dam, and the Eleven Mile Reservoir and Dam. There it turns north to flow through the Cheesman Reservoir, Strontia Springs Reservoir, and Chatfield Reservoir. All of these reservoirs provide municipal water to the Denver metropolitan area. After the Chatfield Reservoir, the South Platte flows north through Denver, and it continues northerly to the small City of Greeley. There it turns to flow easterly and then northeasterly to cross into Nebraska near Colorado's northeast corner where it joins the North Platte in Nebraska to form the Platte River.

The mountainous area south and west of Denver has deep rocky canyons appropriate for concrete dams and deep reservoirs. The reservoirs along the South Platte above Denver primarily provide municipal water to the many smaller cities and towns in the metropolitan Denver area. But this water has been substantially augmented by other reservoirs constructed on the West Side of the Continental Divide in the Mountains above Denver. These include the "Grand River Ditch" which pipes an additional 20,000-acre feet annually into the South Platte, and the Bureau of Reclamation's "Big Thompson Project" that pipes 200,000 acre-feet annually from the West Slope of the Divide into the South Platte, plus Denver's Dillon Reservoir which also diverts water from the West Slope into the South Platte. Previously, much of this water fed agricultural irrigation between Denver and Forth Collins. But this irrigation has decreased by a third as municipal demand has outbid agricultural demand, and farmland has been replaced by suburban development.

The section of the South Platte downriver from Greely is primarily an irrigation canal. It serves irrigation diversions for the northeast agricultural section of the State, i.e. the section that provides approximately seventy percent of the value of Colorado's agricultural output. Along this section, irrigation water is diverted into canals and smaller storage reservoirs for growing crops and raising animals, primarily cattle (*Concentrated animal feeding operations* (CAFOs) predominate here and are fully reviewed below). The "South Platte River Compact" between Nebraska and Colorado apportions the contentious water rights between these two agriculturally oriented States. Except for flood periods, during irrigation season of April to October, the South Platte is a braided shallow stream with low flows due to the voluminous irrigation withdrawals.

This "South Platte Compact" is a 1923 agreement between the States of Nebraska and Colorado. This "Compact" established the right to withdraw water from Lodgepole Creek (a close by river in southern Wyoming and Nebraska) and the South Platte River.[17] The agreement specifies that both States must maintain a stream gage on the South Platte at the Julesburg Bridge, which is close to the border between the two. It also specifies that at this gage, Nebraska is entitled to a flow of 120 cubic feet per

[17] See https://coloradoencyclopedia.org/article/south-platte-river.

second during the irrigation season of April to October, and 500 cubic feet per second during the rest of the year. The Compact further specifies that if the flow is not sufficient, then Nebraska has the right to establish its own canal in Colorado that would enable storage of irrigation water during the offseason so as to enable the otherwise minimum required flow during the irrigation months. In 2022, Nebraska announced its intention to establish this diversion at the "Perkins County Canal Project." The diversion is accomplished via Nebraska's construction of the "Prewitt Diversion Dam and Canal" in Colorado.

Between Greely and the South Platte's confluence with North Platte, numerous diversion canals distribute the River's flow to irrigation districts, their agricultural fields, and CAFOs. The same is true for Wyoming's numerous diversion canals along the North Platte between Casper and Scottsbluff. This is also true for the entire length of the Platte's main stem as it flows through Nebraska. The water distributions between Nebraska, Colorado, and Wyoming are contentious because the total water amounts are ultimately limited; therefore the distributions appear to be a zero-sum game. But there are, however, other sources of irrigation water, in particular from the gigantic Ogallala Aquifer. The issue of this alternative source is examined next.

The Ogallala Aquifer as a Savior

The Ogallala Aquifer is America's largest. It is a shallow groundwater aquifer surrounded by sand, silt, clay, and gravel located below the Great Plains. It underlies 174,000 square miles of surface beneath portions of eight states: southeast Wyoming, south-central South Dakota, almost all of Nebraska (except its far eastern edge), southeastern Colorado, western and central Kansas, the panhandle of Texas, western Oklahoma, and eastern New Mexico. This aquifer provides irrigation to the agriculture of much of the Great Plains, especially in northwest Texas, western Kansas, and western Oklahoma where it is the only source of irrigation, and/or water to *concentrated animal feeding operations* (CAFOs). Note that CAFOs are reviewed in detail below.

The water of the Ogallala Aquifer lies between only a few feet down to 1,200 feet down. Before its depletion, it was generally down about 400 feet in most of the northern portions, but only 100 to 200 feet in most of the southern portion. This aquifer's water was geologically formed and if depleted would require 6,000 years to be completely replenished, i.e. a very slow process. In 2005, it was estimated that almost 3 million acre-feet of water was stored in the Ogallala. Since its advent in 1940, center-pivot irrigation has been the new technology that quickly drew down the aquifer by as much as five feet per year in many areas that are without surface water sources. Almost ten percent of the Aquifer has been depleted since World War II, and sections of western Kansas, Texas, and west Oklahoma have already been effectively depleted. In 1950 Kansas, irrigated agriculture only covered 250,000 acres, but by using center-pivot irrigation, 3 million acres of cropland became irrigated.[18]

Center-pivot irrigation was invented in 1940 by a farmer (Frank Zybach) in the eastern Colorado prairie. It uses a center pump to extract water from the aquifer and send it into a large pivoting system with sprinkler heads. The system circles the pump to create large circular irrigated crop fields. This provides a type of more efficient usage of irrigation from the aquifer. The resulting rapid expansion provides another of many examples of Jevons Paradox, i.e. an increase in the efficiency of use of a resource leads to an increase in total usage.[19]

From 2001 through 2008, the Ogallala's depletion is about one-third of the entire previous twentieth-century's depletion, and since then, the withdrawal problem has been accelerating. According to the Kansas Geological Survey's "Annual Report" of 2022, in Kansas the Ogallala dropped by more than a foot in 2021. The problem was more acute in southwest Kansas where the decrease was 2.17 feet for 2021.[20] The Ogallala provides seventy to eighty percent of all water used in Kansas, and eighty to ninety percent of this usage is due to irrigated agriculture and CAFO usage.[21] If this extraction continues, more than two-thirds of the Aquifer under Kansas will be extracted within forty years.[22]

Warnings concerning the establishment of irrigated agriculture existed since the 1890s, especially from the distinguished Director of the prestigious US Geological Survey, John Wesley Powell. He was the great

[18] See Wines (2013).

[19] This paradox was named for William Stanley Jevons in 1865 with respect to coal usage. Some modern examples include (i) increasing road capacity can make road congestion worse, and (ii) increases in computing power leads to new software consuming the increase in computing power.

[20] See Condos (2022).

[21] Ibid.

[22] Ibid.

explorer of America's Grand Canyon. He was a geologist, ethnologist, and Professor of Geology at Ohio Wesleyan University. By rowboat, he traveled the length of the Mississippi River (1856), the Ohio River from Pittsburgh (1857), the Illinois River (1858), and the Des Moines River (1859). He was a cartographer, topographer, and military engineer during the American Civil War (1861–1865). In 1869, he organized and led the first expedition down the Green River in Wyoming to the Colorado River and then through the Grand Canyon. In 1871–1872, he organized the second trip down the Colorado to photograph and map the Grand Canyon. Powell was appointed by President Garfield as the Director of the US Geological Survey (1881–1894).

For all of his accomplishments, Powell gained prestigious notoriety. One would expect that his warnings about agricultural development of the "arid American West" would not be ignored. But the "would be exploiters" of that development made sure that Powell's warnings were ignored, at least for a time period sufficiently long to enable the agricultural development and exploitation. Powell warned that for land west of the one-hundredth meridian, only agricultural lands close to significant rivers—about two percent of lands excluding those west of the Sierras or Cascade Mountain rages—should be considered for irrigated crops. The rest—about ninety-eight percent of arable land in the arid west—should only be utilized for low density grazing.[23] The hundredth meridian extends roughly through Bismarck in North Dakota, Pierre in South Dakota, North Platte in Nebraska, and Dodge City in Kansas. All of these cities are within this arid region, and all are also within the Missouri River Watershed.

The opposition to Powell's recommendations to limit western agricultural development came from the railroads. They owned vast tracts of western land obtained as compensation for their rail lines extending into the unpopulated nineteenth-century West. The railroads wanted the business that would result from western development. They lobbied Congress to reject Powell's proposals in favor of more federal support for widespread irrigated agriculture. In the late nineteenth and early twentieth centuries of American westward expansion, the myth was popular

[23] See Powell's recommendations in his *Report on the Lands of the Arid Regions of the United States* (1894), reprinted in 1962 by Harvard University Press, Cambridge, MA.

that "rain would follow the plow," that once the grasslands of the Great Plains were tilled, precipitation would follow.[24]

The "Dust Bowl" period of the 1930s resulted from one of the frequent repetitive droughts that affect the arid West. Periodic and frequent drought is the norm throughout the arid portion of the West. The Kansas Geological Survey estimates these is a seventy percent probability that a severe drought will occur in any twenty-year span.[25] The "Dust Bowl" years were an environmental and economic crisis of enormous magnitude. During the 1930s, two and a half-million people migrated out from the Great Plains. Powell's recommendations were proven prescient. But then center-pivot irrigation swept much of the Great Plains and the depletion of the Ogallala Aquifer began. In much of the central and west Kansas, the Aquifer is approaching depletion, and water conservation measures need to be adopted, i.e. the adoption of low-level sprinkler systems that conserve water and crops that demand less irrigation. But even these small and simple conservation measures meet with resistance from many of the area's farmers.

In the 1970s, Kansas established five groundwater management districts. By 2026, all five of these districts will be required to establish plans approved by the State's Chief Engineer that are aimed at reducing water consumption by at least fifty percent over the subsequent years. But farmers in the southwest of Kansas—District 3—do not like government control, and consider their legacy water rights as sacred.[26] In response, an old idea from the early 1980s has been reconsidered, that is the construction of the "Kansas Aqueduct."

The Great Kansas Aqueduct?

In 2015, The Kansas Water Office and the Army Corps of Engineers re-examined a 1982 study by the Corps concerning the feasibility of constructing an aqueduct that would divert Missouri River water from a point ninety miles upriver from Kansas City to west-central Kansas, i.e.

[24] See Stegner (1954).

[25] A "severe drought" was defined as a lack of any significant precipitation over at least twenty months.

[26] See McCoy (2024, June 30).

the *Kansas Aqueduct*. It is supposed to only divert water during high-flow periods along the Missouri. This Aqueduct would divert water into a large reservoir in northeast Kansas, then pump the water 375 miles to a second large reservoir. This would utilize a series of sixteen pumps since the elevation that must be overcome would be 1,750 feet. The canal would be 137 feet wide and 23 feet deep. The plan is to divert five percent of the total of Missouri's annual flow. The construction cost would be $18 billion and have $1 billion in annual operating expense. This would deliver to western Kansas 1.4 million acre-feet annually. This massive project is aimed at saving ten percent of the State's economy including 240,000 jobs. It would supposedly resolve the problems of the depleting Ogallala Aquifer.

The Kansas Secretary of Agriculture indicates the "Great Aqueduct" is unlikely to be built (see Warady, 2015). It is too expensive and the cost would likely be about $470 per acre foot of water delivered, way too expensive to support the area's irrigated crops. But an additional reason for not building this aqueduct is that some of the water districts in western Kansas have yet to enforce various water conservation measures such as shifting to crops that demand less irrigation. But in Kansas, the water drawn from the Ogallala is used in support of concentrated animal feeding lots for fattening cattle. The major industry of West Kansas consists of the cattle feeding lots and slaughter houses that need the production of the area's irrigated agriculture.

Opposition to this massive water diversion comes from the State of Missouri. The Missouri River flows through Kansas City and then easterly through the entire length of the State of Missouri to Saint Louis. It provides a major source of irrigation water to the State of Missouri's agriculture and its own CAFOs. This neighboring State is, of course, suspicious of any significant diversion from "its River." But in addition, Kansas' eminent domain problems of buying the necessary property, and obtaining the federal subsidy required for the $18 billion construction project's costs, all pose considerable political problems. But another large portion of the political problems of obtaining the subsidies necessary for the survival of West Kansas agriculture is the resistance some farmers pose for overcoming the *tragedy of the commons* through organizing the Aquifer's conservation.

The Tragedy of the Ogallala Commons

The *tragedy of the commons* is an economic phenomenon that concerns the depletion of *common property resources* (CPR) such as the Ogallala Aquifer, or various other bodies of water that have free access, or other commonly used resources that do not exclude access.[27] Elinor Ostrom (1933–2012) won the 2009 *Nobel Prize in Economics* for her work on this "tragedy" phenomena, specifically on the evolution of rules for effectively governing CPRs, i.e. *the commons*. She argued that centralized rings of authority that evolve for managing a CPR were the most effective, with the inner ring being that closest to the resource users, and the outer ring being the central authority. Her work showed that the *tragedy of the commons* can be avoided, and often is avoided by the resource users naturally developing the institutions for governance without coercion from the higher-level authorities. The users can be from the lowest level of actual authority, and their arranged institutions for governance might be more effective than rules dictated from higher authorities. These institutions, Ostrom argued and illustrated with some examples, are best organized from the ground up by face-to-face communications among users and are ultimately built on trust. Reasoned discourse therefore should be a foundation of this governance.

The *tragedy of the commons* is the principle problem of environmental analyses and discourse. This is a social-economic problem in that any CPR will be overused and underinvested provided there are no governing arrangement established by its users.[28] For example, a common grazing pasture (a "commons") will be overgrazed without the users reaching an appropriate institutionalized arrangement. If they are acting only from their own self-interest, then those users faced with the problem that any further grazing will deplete the resource beyond its capacity to regenerate, will reason that if they do not allow their livestock to graze, others will. They therefore may as well be the one to cause the overgrazing. The commons will consequently be depleted. But Olson (1965) argued that those with common interests will act to further these interests; they will seek to form a community for that purpose. The institutions for governance of a CPR should, and might, emerge from a collective community

[27] Robinson (2021, Chapter 5) examines this "tragedy of the commons" phenomenon in depth.

[28] In modern literature, Garrett Hardin (1968) first pointed out the problem.

of users that will seek to place requirements on its members restricting its use. To be effective, however, coercion in the case of misuse must be possible. This coercion generally takes the form of exclusion in case of abuse. This introduces the problem of *free riders*, those who avoid the obligations of community membership, but who use the resource without meeting those obligations. Exclusion of usage must therefore be possible in order to avoid overuse of the CPR.

What pre-conditions would lead users to form a CPR community and reach an agreement on its use? This "community agreement" is likely to occur when:

i. information is available and recognized as indicating that the CPR is depleting;
ii. users realize that acting independently yields an expectation of lower net benefits than acting in coordination;
iii. the difficulties and costs of communicating with fellow users are sufficiently low as to be potentially overcome with practical efforts.

These three conditions are all information problems associated with initiating *reasoned social discourse*. The first of these pre-conditions was satisfied concerning the Ogallala; the third should have been resolved by the establishment of water districts in Kansas, and other states, that disseminate information concerning the condition of the Ogallala. (Five of these water districts exist in West Kansas.) Nonetheless, the use of the Ogallala provides an illustration of the difficulties of reaching such an agreement as explored here. To explore these "difficulties" we note that the agreements to be reached must specify:

i. the access and appropriation rights of the users;
ii. the conditions and range of the adaptive management strategies for the resource, including possible contingent strategies.

In the case of aquifer use, the "adaptive strategies" should allow for variations in allowable withdrawal depending on the measurable state of the aquifer. "Contingent strategies" consider broader changes in the resource, and should allow for substantive changes in governance of the aquifer. For example, the monitoring system itself might need change if it is perceived as ineffective for applying exclusion rules.

Any CPR agreement must create "working rules" (practical and effective rules) with credible commitments from users. It also requires realistic and effective monitoring arrangements.[29] A possible problem with these agreements is that users may appear to commit in order to induce others into the agreement, all the while intending to violate the agreement. Each user, knowing that this is possible, and wanting to avoid being the "sucker" who is the last to actually keep the agreement, must be suspicious of the effectiveness of the monitoring provision.[30] Effective practical monitoring, therefore, is the key to these CPR governance commitments. How would the water districts in Kansas monitor withdrawals from the Ogallala? Any system of monitoring might be subverted. The system of monitoring, therefore, must ultimately rely on trust which might be difficult to obtain among widely spaced farmers who are suspicious about the aquifer withdrawals of others.

To understand the organizational problems, consider the grazing problem reviewed above. The common sense solution is that the users reach a governance agreement for rationing the grazing rights. Ostrom pointed out that this grazing problem was solved by farmers in a Swiss village in part by establishing an agreement to limit grazing to the number of cattle that each farmer could care for during the winter months. Spontaneous, practical, and evolved rules for monitoring the rationing with possible exclusion as a threat were reached and administered by the users. These rules evolved over considerable time, and as a result they asserted the rational solution. Such arrangements spawn Ostrom's law, "A resource use arrangement that works in practice can work in theory." She explored case examples to illustrate that over time, rules could be established for CPR use that promote sustainability.[31] These rules pose self-governance systems, and evolved from face-to-face communications, and were based on reciprocity, i.e. "I'll limit my use if you'll limit yours." But with respect to the Ogallala, this giant aquifer is not a Swiss village with a relatively small commons where monitoring is a simple problem. The Ogallala, however, is a giant common aquifer with users who are dispersed by considerable distance.

[29] See Bates (1988).

[30] See Elster (1989).

[31] See Ostrom (1990).

In examining the institutions of governance for CPRs, Elinor Ostrom studied common pasture lands in Switzerland and Japan, agricultural irrigation systems drawing from commonly shared rivers in Spain, and commonly shared irrigation systems in the Philippines. Each of these manifested institutional agreements among users that evolved over lengthy time periods. From these background observations, Ostrom derived and posed eight rules for managing a CPR[32]:

i. The user group boundaries must be clearly defined.
ii. The rules governing CPRs should match local needs and conditions.
iii. Those affected by the CPR rules should be included in any modifications of those rules.
iv. The rules for governing CPR use should be respected by outside authorities.
v. The community of users should develop a monitoring system for CPR use.
vi. Graduated sanctions should be used for rule violations.
vii. Dispute resolutions should be of low cost and accessible.
viii. The responsibilities for governing CPRs should be nested into tiers from lowest to highest, with the lowest being closest to the users.

All eight of these rules depend on meaningful reasoned discourse (whether face-to-face or otherwise) among the community of users. They also depend upon inclusivity among users and those affected in forming those rules. The rules for CPR use, however, should also meet some other standards of inclusivity concerning the information to be gathered and the otherwise fairness of the decision criteria (rules) to be used. But how can "local needs and conditions" be applied to the very large regional aquifer, or any other large regional irrigation system? In such a large case, the effective enforcement authority cannot be local; it must be regional where the "region" must be large enough to cover the aquifer. It must include the states involved, and it must develop a regional compact. The CPRs that Ostrom cites, however, have the following characteristics:

[32] See Ostrom (1990, p. 90).

i. The using groups are not significantly disparate in their uses. The users also tend to be relatively small in number.
ii. The agreements evolved over considerable time while adjusting as new problems and demands needed to be resolved. They did not emerge from significant and immediate crises.

Note that these are not characteristics manifested by the Ogallala and its users, who are large in number, and where regional agreements have not begun to be established.

Cooperation Versus Autonomy, and the Information Problem

Irrigated farming exists in a world of limited resources. There are costs and possible benefits resulting from its associated decisions, and these costs and benefits include externalities such as pollution. But one typical externality results from the inhibitions against cooperative action, and in favor of atomistic competition.[33] Consider, for example, the *tragedy of the commons* as illustrated by the Ogallala Aquifer. This resource has become close to being completely eliminated through over depletion. At some point, the aquifer could reach levels below the minimum level needed for a reasonable cost extraction, and the irrigated farming would then completely collapse. If the individual farmers know this is occurring, and they believe that if they do not pump the water from the aquifer then others will, then they reason that they may as well pump. Under this circumstance, depletion will continue beyond the critical level. For this destruction to occur, any one of three conditions might be causal assuming farmers believe that the continuance of the aquifer is desirable:

1. The farmers are either not knowledgeable, or perhaps are in willful denial, about the effects of over pumping. (This knowledge might not exist among the users who are widely dispersed by distance because of the difficulty in communication.)
2. The possibility of reaching a cooperative agreement to limit extractions is considered remote, when perhaps it is not. (Again, the

[33] This bias for atomistic competition, i.e. competition among many small entities, and against cooperative action can be an expression against giving up control associated with the latter, and in favor of the individual entrepreneurial independence of the former. See Immanuel Kant (1784, pp. 8, 21) for an exploration of the tendency toward isolation.

problem might be the dispersed distance and its effects on communication.)
3. The discovery of the potential aquifer collapse comes too late to save it. (This might be true for the Ogallala given its relatively accelerated depletion.)

In the third case, there is a timely-knowledge problem, one that can be potentially solved through collective scientific action based upon information about the water levels. Knowledge about optimal extraction rate is a business problem, and in these *tragedy of the commons* cases, a solution generally requires collective action and cooperation. That is the rational business solution, but knowledge about the willingness of other farmers to join a cooperative agreement may be lacking. As explained in the section above, if it is known that others are willing to join, then it is more likely that each individual farmer will also be willing to join. In the case of West Kansas, that should be a function of the water districts, i.e. the promotion of the long-term benefits to all by agreeing to the necessary withdrawal restrictions.[34] To be knowledgeable means to become familiar with the relevant facts concerning the likelihood of depletion in the absence of effective restrictions. This farming example illustrates one of many cases of the moral obligation to be rational and informed with respect to environmental resource concerns. This case is associated with cooperative action, but a lack of familiarity with the relevant facts stimulates a failure of this cooperation under the first two conditions listed above.

Besides the possible knowledge problem illustrated above, there are four other inhibiting and destructive biases associated with this aquifer's problems. Each is briefly reviewed next.

The Bias Caused by Abundance

If the aquifer provides abundant water at little if any current cost, then high levels of extraction might occur until the water becomes scarce. If, however, the extractors discover that current high levels of extraction mean much lower levels for the future, then it is rational that they will lower their current levels of extraction. The narrow economic solution

[34] See Kuebrich (2019, January 23).

for a farmer, or collective of farmers, is to maximize the present value of the net income stream to the farmers. Restrictions consistent with this objective would be collectively placed on the current extraction rates. But this requires cooperative behavior from the farmers, i.e. no "free riders." There might be, however, a bias against giving up the individual freedom for submitting to the collective effort and rules. Nevertheless, an easy abundance to extract might divert attention from the finite nature of the resource and lead users to not properly ration.

The Bias Due to an Overly Narrow Vision

Farmers could learn to use and adopt new water conserving technology, but bias against the "new" can exist, and this might include a bias against changing from high irrigation-demanding crops to alternative low water-using crops. The answer to the "envisioning" problem could be that farmers, and society in general, should be searching to develop substitute crops and cost saving technology, but this search might be truncated by a bias against exploring the "new."[35]

The Bias Due to "It's Gone!"

"Out of sight, out of mind!" "When its gone, its gone!" These adages may apply to those environmental resources that are either completely eliminated or so degraded as to not be recognizable as an environmental asset; yet this asset might still be valuable if properly used and perhaps restored. Changes in an aquifer's management, for example, can lead it to become a more valuable farming asset. This "its gone" bias is merely another version of the vision problem.

These environmental degradations would likely not have occurred if (i) participants had the necessary information to anticipate the ultimate effects of their isolated actions, (ii) all those who would be negatively affected were knowledgeable about these facts, and were given the opportunity to participate in the public discourse and decision-making, and (iii) actions that degraded the environment, thereby imposing external costs on others, were prohibited. Collective management of an aquifer would best avoid all three of these adverse deficiencies.

[35] Ibid.

The Anti-Intellectual, Anti-Elite Bias

Lives and commerce are often affected by external forces whether exerted by nature, or market forces, or epidemics, or political movements, etc. These forces can be indicated by and analyzed through use of scientific and/or political socio-economic information. In these cases, those Americans affected have demonstrated a tendency to criticize the "elites," i.e. those with more intellectual backgrounds, or who are perhaps political authorities, or who merely deliver the message. Since America is not a class-stratified society, such as in Britain and some other nations, perhaps education has replaced class as the stratifying factor, i.e. there are those with higher education degrees, and those without. Those of the scientific and political sectors become classified by blue-collar citizens as "elite" who might make biased judgments that act as external forces of unfairness against the interests of the "non-elites."[36] Perhaps the anti-elite claims are merely manifestations of a *fight or flee syndrome* when confronted by the objective facts. Perhaps they are purposely mendacious attempts at self-serving personal resistance and avoidance of actions that if taken would be in society's broader interests.

Consider blue-collar jobs lost through what some argue are recently established environmental regulations: (i) coal miner positions, (ii) steel worker positions and associated coke manufacturing positions, and (iii) fishing boat positions. The first two are actually due to newer technology associated only indirectly and weakly with environmental regulations, and the third is due to drastic species reductions and consequent restrictions that enable the species to rebound. The blame asserted by the blue-collar workers affected has been placed on (a) the quality of the science supporting the need for change, and (b) the elite nature of the decision-makers with authority. The evidence in each case, however, indicates the need for a regulatory change. The problem is not personal (as in claims of "elite") but in technical circumstances and changes. Using the example of the Ogallala, the aquifer is almost depleted, but technology can extend its life. It is not the "elitest" messenger that is to blame (be it a hydrologist, engineer, or political representative delivering the message), it is the objective circumstances.

The depletion of the Ogallala Aquifer indicates that some biases exist that prevent a rational solution. This aquifer might not be preserved,

[36] See Nichols (2019, pp. 18–19).

and therefore the current state of West Kansas agriculture may not be sustainable.

The Battles for the Yellowstone River

As measured by the flow that it contributes to the Missouri River, the Yellowstone River is the largest of the Missouri's tributaries (see Figs. 12.1 and 12.5). At 692 miles, the Yellowstone is the second longest of the Missouri's tributaries next to the Platte. It drains more than 37,000 square miles. It is one of America's most classic Western rivers with respect to its fish, wildlife, recreation, and lore. It is also America's longest free-flowing river in the contiguous forty-eight states, i.e. it does not have storage reservoirs.

The Yellowstone rises in the Absaroka Range on Wyoming's Continental Divide in the Shoshone National Forest, close to Younts Peak within the Yellowstone National Park's southeastern corner. It flows north into Yellowstone Lake, and then exits the Lake to flow over the spectacular Upper and then Lower Falls. It then flows through the equally spectacular vista of the "Grand Canyon of the Yellowstone." It exits the National Park flowing through Montana's "Paradise Valley" between the Gallatin and Absaroka Ranges. (Much is written about this area below.) At Livingston in Montana, it turns east to flow by Billings in Montana where it is joined by the Bighorn River, and then the Tongue River farther to the east, and then the Powder River in far Eastern Montana where there are significant coal reserves and strip mining. The Yellowstone then joins the Missouri River near the North Dakota border and Lake Sakakawea.

The Yellowstone River drops approximately 11,000 feet over its almost 700-mile length (from 12,850 feet in altitude to 1,850 feet). It has no storage dams on its main stem, only run-of-the-river diversion structures, but it does have dammed reservoirs on its tributaries, the largest being the Yellowtail Dam of the Bighorn River which forms the 72 miles long Bighorn Lake. By comparison, the reservoir on the Tongue River, another tributary of the Yellowstone, is much smaller at 12 miles long. Since the 1860s, the Yellowstone has been used extensively for irrigation especially through its tributaries.

As indicated above, the Yellowstone is unique as the only major river in the lower forty-eight that flows unimpeded by dammed reservoirs. It has been, however, continually threatened with large dam constructions, but every threat has inspired successful defenders to rise as though they had a patriotic duty to defend its free-flowing nature. It is the history and grandeur of the region the River drains that requires us to believe it should freely flow over its 700-mile course from the spectacular Yellowstone National Park, through the grandeur of "Paradise Valley," and then through the rest of the canyonlands of Montana.

America's first and most popular preserve is the Yellowstone National Park. This American treasure is named after the Yellowstone River and its unique Yellowstone Lake, which is actually the caldera of a gigantic dormant volcano. More than half of the world's geysers and hydrothermal features are in Yellowstone Park. Grizzley bears, cougars, elk, wolves, and sizeable herds of grazing bison are all typically encountered in this preserve, which has been classified as a UNESCO World Heritage Site since 1978.

This Park was established in 1872, but it was not well protected from poachers and various exploiters until 1900 when the law changed to allow prosecutions of violators. In 1883, the Northern Pacific Railroad constructed a terminal at close by Livingston in Montana, and brought visitors into the Park, 5,000 visitors in 1883. The National Park Service was created in 1916 to better manage the Park (versus the US Army that managed it in previous years). After World War II, the US Interstate Highway System together with the advent of modern automobiles, brought a radical increase in visitors. Now, more than 4.25 million visit annually.

The Allenspur Gap is a narrow canyon along the Yellowstone River in the "Paradise Valley," a few miles upstream from Livingston. There the River cuts through hard-rock limestone cliffs which provide a paramount location for a dam on the Upper Yellowstone. Such a dam was first proposed in 1902. It failed to obtain approval because it was uneconomic as either a hydroelectric facility, or a flood control dam, or an irrigation dam. It also failed because of conflicting government bureaucratic evaluations. In 1920, however, another dam was proposed at "Fishing Bridge" that would block the River at Yellowstone Lake to serve irrigation interests outside the Park. The environmentalist George Grinnell successfully rallied opposition to block such an exploitation of the National Park.

With the OPEC "oil embargo" in the 1970s, and President Nixon's consequent drive for an "energy independent US," the exploitation of Wyoming's coal deposits became a popular political suggestion. In 1971, the Bureau of Reclamation issued its "North Central Power Study" which proposed the construction of forty-two steam-generated coal fueled electricity power plants on the northern plains, all drawing water from the dammed reservoirs along the Yellowstone and its tributaries. One of these dams would be a proposed 380-foot structure at the Allenspur Gap in Paradise Valley. This proposed power generation would consume enormous amounts of water (thirty percent of the Yellowstone's water would be diverted) that could only come from the Yellowstone River and/or its nearby tributaries such as the Powder River and the Bighorn River. In opposition to this exploitation, a coalition of ranchers, farmers, recreationists, and environmentalists formed "The Northern Plains Resource Council." Its purpose was to oppose the coal mining interests and energy companies from raiding the Yellowstone's water. The political debate in the media was fierce concerning the strip mining of coal and destruction of Montana's natural streams. Montana's Fish and Wildlife department composed a 32-minute film—*The Yellowstone Concerto*—and *Life* magazine issued a ten-page color photo and essay dedicated to saving the Yellowstone.[37]

Montana's Legislature altered this debate when in 1974 it passed its "Water Use Act," together with its "Yellowstone Moratorium." The former Act required that major users of the State's waters must obtain a State Permit for any withdrawals. The latter act halted any water withdrawals for three years, and asked the various State agencies how much water should be preserved for future uses. Montana's "Fish and Wildlife Department" responded that ninety percent of the Yellowstone's average annual flow should be maintained to preserve its riparian habitat. This established a new water right for the natural environment of the Yellowstone. Montana's "Board of Natural Resources" agreed. Water for the State's natural environment and its irrigation (a much smaller amount) took priority according to the Board's opinion in December of 1978. The Board voted for the free-flowing Yellowstone and not for dams and reservoirs for feeding steam-generated power plants under the so-called energy independence initiative. (Note that Jevons Paradox—reviewed above—would indicate that feeding energy demands would not satisfy those demanding. It would only solicit further increases in demands.)

[37] See www.arlis.org/docs/vol1/susitna/41/APA4147.html.

Although the Yellowstone River does not have reservoir storage dams, it does have irrigation diversion dams, the largest being the "Intake Diversion Dam" seventy miles upstream from the Yellowstone's confluence with the Missouri River, and eighteen miles downstream, from Glendive in Montana. This is a weir type diversion dam that facilitates the irrigation of 50,000 acres in eastern Montana and western North Dakota. This diversion dam was constructed by the Bureau of Reclamation as part of the "Lower Yellowstone Project." This diversion system is operated by the "Lower Yellowstone Irrigation District Board of Control." It consists of 71 miles of main canal, 225 miles of laterals, three pumping stations on the main canal, and four pumping stations on the Yellowstone River and one on the Missouri River.

A low profile $59 million dam is now planned for the Yellowstone close to the North Dakota Border, and that would replace the Intake Diversion Dam. This is being contested in court by the "Defenders of Wildlife," and the "Natural Resources Defense Council." The issue concerns the Pallid Sturgeon, which has been listed as "endangered" since 1990. The original design of the dam was modified to include a bypass channel for the sturgeon, but the environmental organizations claim that the bypass will not be effective to save the sturgeon. The issue is currently being litigated (as of July of 2024).

The Missouri River Watershed and CAFOs

Concentrated animal feeding operations (CAFOs) are defined by the US Department of Agriculture (USDA) as "an intensive animal feeding operation" with more than 1,000 cattle, or 2,500 pigs weighing more than 55 pounds, or 10,000 pigs of less than 55 pounds, or 125,000 chickens.[38] To fit the USDA's definition, these high concentrations of animals must also be sustained for more than forty-five days. There are more than 21,000 CAFOs in the US as of 2020.[39] These agricultural operations have been rapidly increasing since the 1950s. They use high amounts of water for both animal ingestion and also for cleaning of animal wastes into "anerobic lagoons." These are man-made earthen basins filled by

[38] See "Animal Feeding Operations," *Livestock*, Washington D.C., US Natural Resources Conservation Service, at www.nrcs.usda.gov/wps/portal/ncrs/main/national/plantsanimals/livestock/afo/.

[39] See www.epa.gov/system/files/documents/2024-06/cafo-status-report-2023.pdf.

a manure slurry washed out from beneath animal pens and piped into the lagoons. This organic material naturally generates carbon dioxide, methane, ammonia, and hydrogen sulfide. The material is also often spread on fields that grow corn or other foods for the animals. But these lagoons often leak the slurry-waste material into streams and then rivers. CAFOs have been accurately accused of designing their lagoons to be smaller than necessary for storing the amounts of manure generated so that when significant rain storms occur, they wash the wastes into nearby streams and rivers. In addition, the manured fields are often over-fertilized by excessive amounts so that rain washes the excess into the streams and rivers, a so-called non-point source of contamination.[40] These pose the more significant regulatory problems for state and federal authorities (the US EPA). States with high numbers of CAFOs have an average of 20 to 30 serious water quality emergencies per year as a result of manure management failures.[41] For example, in 1995 a 120,000 square foot "lagoon" ruptured in North Carolina and spilled 25.8 million gallons of waste into the New River. This spill killed ten million fish and led to disease outbreaks.[42]

This method of meat production produces large amounts of greenhouse gases, especially methane and ammonia, and causes a variety of respiratory diseases and other negative medical effects. The associated smell produced by ammonia, hydrogen sulfide, and other gases also substantially and negatively affect nearby property values.

Table 12.2 presents the numbers of CAFOs in the Missouri River Watershed as listed by state. As of 2020, the Missouri Watershed had thirty-five percent of the CAFOs in the US.[43] Southern states had twenty-seven percent of the CAFOs in the US.[44] Michigan, Wisconsin, and

[40] See "Why are CAFOs bad?" the *Sierra Club,* at www.sierraclub.org/michigan/why-are-cafos-bad.

[41] Hribar, Carrie (2010), "Understanding Concentrated Animal Feeding Operations and Their Impact on Communities," *National Associations of Local Boards of Health,* Bowling Green, OH. At www.cdc.gov/nceh/ehs/Docs/Understanding_CAFOs+NALBOH.pdf.

[42] See "Facts about Pollution from Livestock Farms," *Environmental Issues: Water.* New York, NY: Natural Resources Defense Council, as www.nrdc.org/water/pollution/ffarms.asp, 2011-12-07.

[43] See footnote 39.

[44] These Southern States included Alabama, Arkansas, Florida, Georgia, Kentucky, Louisiana, Mississippi, North Carolina, South Carolina, Tennessee, Texas, and Virginia.

Table 12.2 CAFOs in the Missouri RiverWatershed (2020)

State	*Number of CAFOs*
Colorado	210
Iowa	3,896
Kansas	432
Missouri	513
Montana	124
Nebraska	1,743
North Dakota	90
South Dakota	431
Wyoming	49
Total in Watershed	**7,588**
Total in US	**21,465**

Minnesota had ten percent combined, and New York and Pennsylvania combined had four percent. California had five percent alone.

Much of the agricultural water usage in the Missouri Watershed is for CAFO operations. To a considerable extent, the engineering of the Missouri River and its tributaries has led to this support for CAFOs.

The Dams and Water Distributions of the Missouri River

The Missouri River is a classic American institution of a working river that primarily serves agriculture with political forces that will ultimately resist restoration strongly. It flows through the spine of America, originating along the Continental Divide of Colorado, Wyoming and Montana. It flows through the hard-rock canyons of Montana where it powered smelters for that area's mines. It then flows through the prairielands of the Upper Great Plains. There it becomes a heavily silt-carrying river that has been heavily channeled, i.e. straightened and deepened by the Corps of Engineers. Its six downstream dams are among the largest earthen hydroelectric dams (with concrete spillways) in America. They include:

- Fort Peck Dam in Montana—four miles long and 250 feet high that produces 185 megawatts,
- Garrison Dam in North Dakota—two miles in length and 210 feet high that produces 515 megawatts,

- Oahe Dam in South Dakota—one and three-quarters miles in length and 245 feet high that produces 786 megawatts,
- Big Bend Dam in South Dakota—two miles in length and 95 feet high that produces 493 megawatts,
- Fort Randall Dam in South Dakota—2 miles in length and 165 feet high that produces 320 megawatts,
- Gavins Point Dam in South Dakota and Nebraska—2 miles in length and 74 feet high that produces 132 megawatts.

These structures provide flood control and electric power to the Northern Great Plains. But they also essentially trap the heavy silt and consequently create considerable problems for the Corps of Engineers who must remove the massive silt buildup. But one of the Missouri River's two largest tributaries, the Platte River, also involves contentious water withdrawal disputes between Nebraska and Colorado. Coincidentally, the Ogallala Aquifer, which has provided substantial past water withdrawals for these same states, is now close to depletion. Both of these versions of *tragedies of the commons* (involving the Ogallala and the Platte) stem from the rapid growth of CAFOs in the sparsely populated areas of Kansas, Nebraska, South Dakota, Iowa, and Missouri. For the CAFO industrial development to exist, it must find sparsely populated areas with ample degradable water supplies to exploit since people are repulsed by the air borne gas odors and the particulate pollutants that CAFOs produce. As with the degradations of our industrial rivers during the previous two centuries, the negative externalities of this CAFO industry force the real costs of production onto society in the form of degradation of our rivers.[45] As a result, big sections of the Missouri River Watershed have been converted to nothing other than a plumbing system that provides the key "factor of production" (besides the animals themselves) to the American meat industry of beef, pork, and chicken production. Given these agricultural developments of recent decades and current demands for meat, restorations of the Missouri River, and its Platte River tributary are unlikely to occur soon. Indeed, there has been a recent resurrection of the idea for a "Great Kansas Aqueduct" that would transport a significant portion of the Missouri River (five percent diversion is the plan)

[45] Robinson (2021) extensively reviews society's problems with negative externalities in production.

into Western Kansas to keep its agricultural industry alive given that it has already largely depleted the area's Ogallala Aquifer it has depended on. The $18 billion (or more) necessary federal subsidy is, however, not likely to be granted.

But not all of the Missouri Watershed has been so despoiled. The Yellowstone River, the Missouri's second significant tributary, is still without large reservoirs despite over a century of attempts to construct these by various special interests such as the coal and power combination of the 1970s. The Yellowstone does have irrigation diversion dams, but not reservoir creation dams. It remains the only significant US river that still flows freely with only run-of-the-river obstructions. There is, however, a new proposal to change this with a dam seventy-miles upstream from Yellowstone's confluence with the Missouri. As with all previous attempts, this project is being strongly resisted by environmental coalitions. Even within the highly engineered Missouri River Watershed, there is still a free-flowing Yellowstone River that draws the ire of environmental groups with every suggestion that it be dammed. Within this sort of support history, it is likely to remain a free-flowing entity.

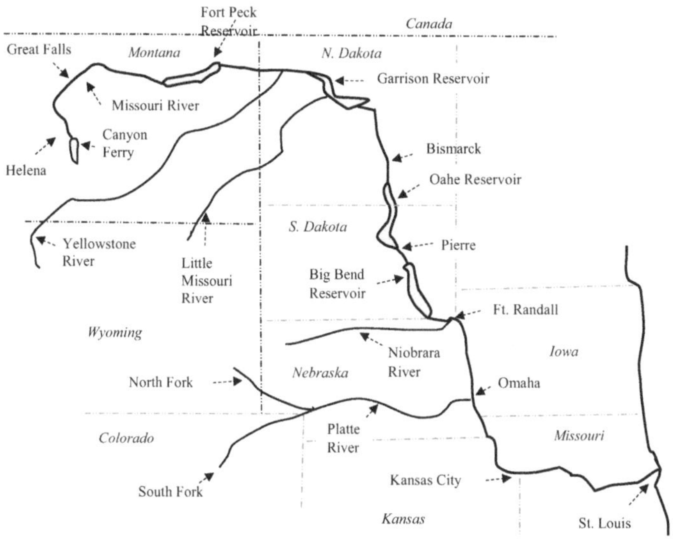

Fig. 12.1 Missouri River

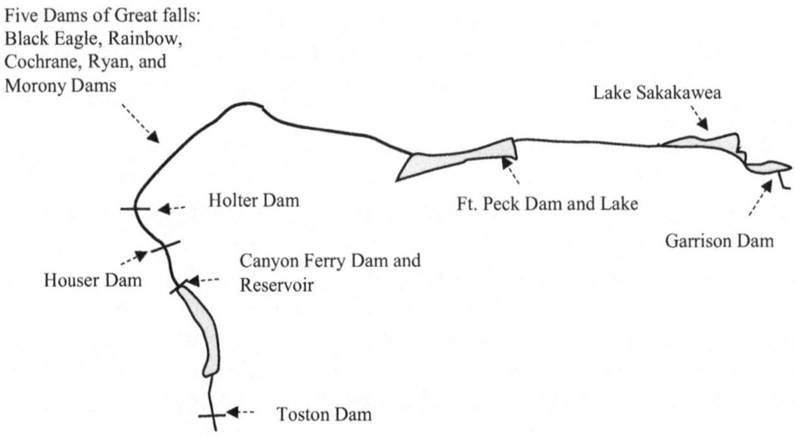

Fig. 12.2 Upper Missouri Dams

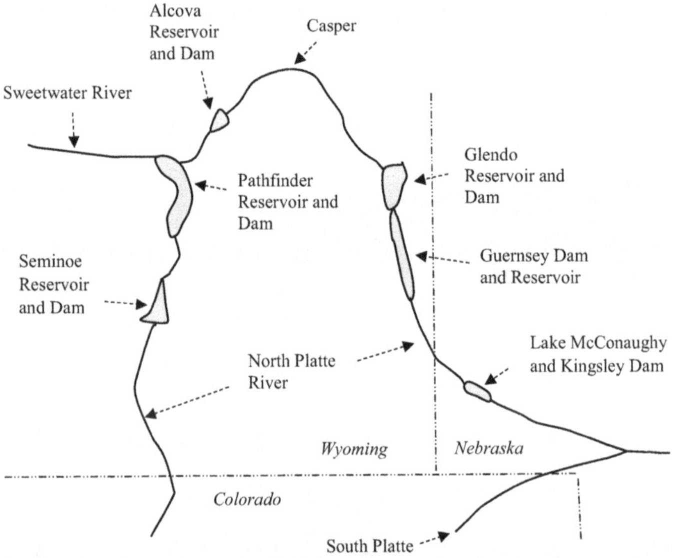

Fig. 12.3 North Platte River and Dams

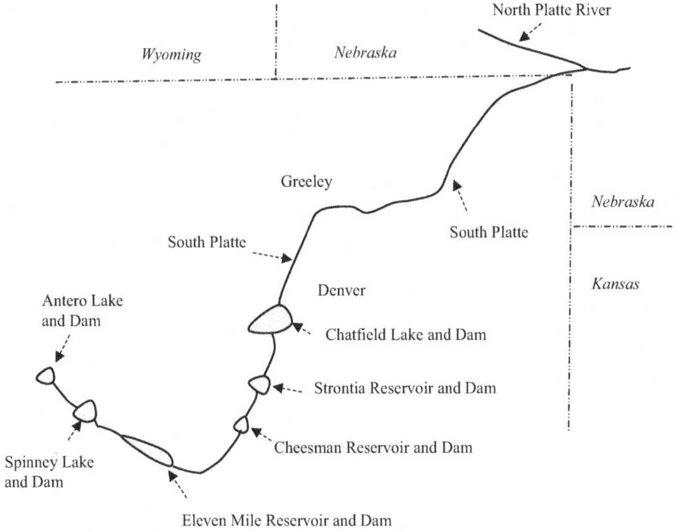

Fig. 12.4 South Platte River and Dams

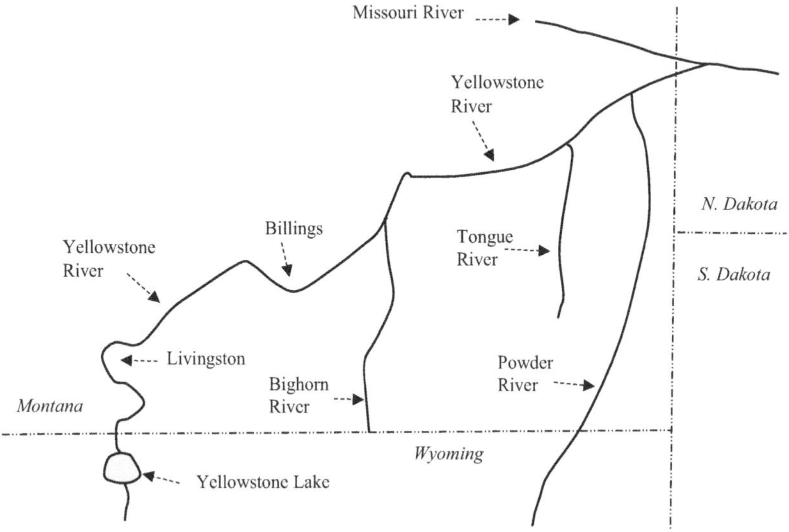

Fig. 12.5 Yellowstone River

References

Barry, J. (1997). *Rising Tide: The Great Mississippi Flood of 1927 and How It Changed America*. Touchstone.

Bates, R. H. (1988). Contra Contractarianism: Some Reflections of the New Institutionalism. *Politics and Society, 16*, 387–401.

Condos, D. (2022, April 9). Ogallala Aquifer Dropped 12+ Inches in 2021. Land Value Could Lose Billions as Water Source Runs Dry. *The Topeka Capital Journal*. www.yahoo.com?news/ogallala-aquifer-dropped-12-inches-120031757.html

Elster, J. (1989). *The Cement of Society: A Study of Social Order*. Cambridge University Press.

Hardin, G. (1968). The Tragedy of the Commons. *Science, 162*, 1968-12-13, 1243–1248.

Kant, I. (1784). The Idea for a Universal History with Cosmopolitan Intent. In A. W. Wood (Ed.), *Basic Writings of Kant*, Modern Library Classics (2001). Random House.

Kendle, E. (1970). *The Effects of Channelization in the Missouri River on Fish and Fish Food Organisms*. Nebraska Game and Parks Commission, Lincoln Nebraska, University of Nebraska. http://digitalcommons.uni.edu/cgi/viewcontent.cgi?article=1022&contex=nedgamewhitepap

Kuebrich, B. (2019, January 23). Are Billions Needed For a Cross-State Canal to Save Western Kansas. *KCUR*. www.KCUR.org/2018-01-23/are-billions-needed-for-a-cross-state-canal-to-save-western-kansas

Lawson, M. (1982). *Dammed Indians: The Pick-Sloan Plan and the Missouri River Sioux, 1944–1980*. University of Oklahoma Press.

Malone, M. (2006). *The Battle for Butte: Mining and Politics on the Northern Frontier, 1864–1906*. University of Washington Press.

McCoy, M. (2024, June 30). *Hoping for a miracle to save the Ogallala Aquifer? Prepare for the next Dust Bowl*. https://kansasreflector,com/2024/06/30/hoping-for-a-miracle-to-save-the-ogallala-aquifer-prepare-for-the-new-dust-bowl/

Meyer, R. (1977). *The Village Indians of the Upper Missouri: The Mandan, Hidatsa, and Arikara*. University of Nebraska Press.

Nichols, T. (2019). *The Death of Expertise: The Campaign Against Established Knowledge and Why It Maters*. Oxford University Press.

Olson, M. (1965). *The Logic of Collective Action: Public Goods and the Theory of Groups*. Harvard University Press.

Ostrom, E. (1990). *Governing the Commons: The Evolution of Institutions for Collective Action*. Cambridge University Press.

Robinson, R. (2021). *Environmental Organizations and Reasoned Discourse*. Palgrave-Macmillan.

Rolle, A. (1991). *John Charles Fremont: Character as Destiny.* Oklahoma University Press.

Stegner, W. (1954). *Beyond the Hundredth Meridian: John Wesley Powell and the Second Opening of the West*. Houghton Mifflin.

Warady, M. (2015). The Great Kansas Aqueduct: Solution or Folly from a Bygone Era?. *Associated Press*. https://waterfm.com/great-kansas-aqueduct/

Wines, M. (2013, May 19). Wells Dry, Fertile Plains Turn to Dust. *New York Times*. www.nytimes.com/2013/05/20/us/high=plains-aquifer-dwindles-hurting-farmers.html?smid=pl-share

CHAPTER 13

Our Age of Restoration

Where Is This Movement Going?

The previous twelve chapters review the state of US dam removals, rehabilitations, and associated ecosystem recovery management. These chapters indicate that the US is developing a culture of dam removals, but accomplishing a removal is shown to usually be a lengthy process. Environmental restoration and safety concerns demand many of these removals, but *ex post* ecosystem management is also warranted. Note that global climate change exacerbates significant weather events, and these events increase the stress applied to America's systems of dams. This chapter attempts to summarize some aspects concerning the state of our dam removal movement, and perhaps to also explore where it is heading.

To begin this final exploration, we expand upon some recent developments concerning America's first significant dam removal, i.e. the 1999 events in Maine as explored in Chapter 1.

After Initiating the Movement, What Happened in Maine?

Our story of America's dam removal movement began almost at the very end of the twentieth century with the deconstruction of Maine's Edwards Dam on the Kennebec River (see Chapter 1). The people involved in this deconstruction of July, 1999, meant to create a movement for dam

R. M. Robinson, *Restoring America's Rivers*,
https://doi.org/10.1007/978-3-031-81758-8_13

removals elsewhere on the Kennebec and throughout its watershed. That energy did spread and also brought about needed additional dam deconstructions on the Kennebec River. This energy then moved to Maine's Penobscot River and its much larger watershed. From there, it spread to the Olympic Peninsula of Washington, and then elsewhere in the West. The movement then circled back to the Northeast to bring down derelict low-head dams, and finally other larger dams throughout the US. Very recently (Fall of 2024), the Klamath River Dams were deconstructed in southern Oregon. This dam removal segment of the environmental movement is documented in the previous twelve chapters.

This final chapter begins by exploring the events in Maine after its initial dam deconstructions. How did the removal movement evolve after those first stages of river restoration occurred in Maine, i.e. during the last five years of the quarter century since the movement began?

Maine's typology is appropriate for hydroelectric generation since it has 30,000 miles of rivers. Sixty of its rivers empty into the Atlantic Ocean, which means there are sixty distinct watersheds in this State. Given this fact, one expects the attitude of the State's citizenry to be that some of these water bodies can remain working rivers; they need not all be restored to their natural ecology of being suitable for Atlantic salmon and herring runs, and the like. But Maine's rivers are substantially unique. In most of America's rivers, climate change and periodic drought have produced significant variation in flows. But Maine's rivers have drained a relatively consistent forty to fifty inches in annual precipitation. Its mountain snowpack keeps its summer river levels flowing rather evenly. This would normally be beneficial for its Atlantic Salmon runs and its overall ecology except that these same even flows also facilitate hydropower generation.

In 2023, Maine's fifty-one hydroelectric plants generated 726 megawatts of renewable electricity. This was thirty-one percent of Maine's overall electricity generation, and forty seven percent of its renewable generation. These proportions have been fairly constant over the previous quarter century. From 2002 through 2023, electricity generation in Maine actually decreased by approximately one-half. But this was mostly due to the decrease in generation from natural gas, and its corresponding increase in imported electrical power from Quebec. (This is reviewed below.) The amount of hydroelectricity generated has been fairly constant. Various renewable sources (hydro, wind, solar, wood, and biomass) generated sixty-five percent of Maine's electricity in 2023 according to the US

Energy Information Administration. Maine plans to have eighty percent of its electric generation from renewables by 2030, and one-hundred percent by 2050.[1] Hydropower could be essential for meeting these goals.

Hydropower is particularly desirable if the dam has a substantial reservoir for storage. A reservoir creates a battery-type storage, whereas a run-of-the-river dam where the "flow through" equals the river's natural flow, does not. Reservoir storage creates a complementary energy source for supplementing wind power in that the storage can be used when the wind is not generating power. There are, however, opportunity costs associated with hydropower in that it degrades the river's ecology, and thereby interrupts the Atlantic salmon and other fish spawning runs.

Maine imports a significantly large portion of its electric power from Hydro-Quebec, the provincial government-owned corporation that operates Quebec's sixty-one hydropower plants. In part, this imported power allows Maine to avoid investing in new hydropower facilities, and even to avoid refurbishing its current facilities. (See below for the importance of these imports of Quebec's hydro power.) Also, as with other northeast states, Maine's program of assuring the safety of its dams is underfunded and understaffed with only two engineers to inspect over 500 dams. (See Chapter 10 for a review of safety requirements.) But in September of 2024, the "Bipartisan Infrastructure Law" allocated $33.8 million in US Department of Energy funding for twenty-one of Maine's privately owned hydro facilities. This funding provides needed maintenance and upgrades, and the construction of new fish passage technology. The safety problem was therefore resolved at least temporarily.

According to American Rivers, between 1912 and 2024, Maine lost fifty hydroelectric dams. At an average age of 104 years, Maine's active hydro facilities are at the longer end of America's age distribution of its dams since they were mostly constructed between 1900 and 1960. Consequently, Maine is currently planning on a broad expansion of wind and solar power generation. It plans on 3,000 megawatts of floating offshore wind capacity by 2040, and 750 megawatts of small-scale solar arrays. But these renewable sources are sometimes-on and sometimes-off, and consequently they need large storage capacity. If Maine wants to continue to pursue its river restorations, then electric battery storage

[1] See "An Assessment of Maine's Renewable Portfolio Standard," by Maine Governor's Energy Office at www.maine.gov/energy/sites/maine.gov.energy/files/maine-RPS-impacts-and-procurement-Policy-Options-Report-Master-FINAL.pdf.

appears to be a needed answer. To this end, last August the US Department of Energy awarded a $147 million grant to develop an 85-megawatt project in Lincoln, Maine, for multi-day storage.[2] In Southern Maine, a site has already been cleared for a 175-megawatt electric battery storage facility in Gorham that is expected to "come online" in mid-2025. This is the "Cross Town Energy Storage Facility," a $100 million project with 156 tractor-trailer type containers of lithium-ion phosphate batteries. The trailers will be lined up across five acres of Gorham Industrial Park, a sort of trailer park of batteries. It is being built by Houston-based "Plus Power," which currently has sixty energy storage projects online or in development in the US and Canada.[3]

Large lithium-ion storage projects are now the trend across Canada and the US. In 2018, FERC rule changes allowed battery projects to compete with conventional power generation in the wholesale electricity markets.[4] More than 1,000 megawatts of new battery storage has now been approved to supply electricity in the next few years in New England, some as stand-alone projects while others are linked to wind or solar facilities. In Maine and Massachusetts, the charging of the batteries can also be from the Quebec-Hydro source during off-peak lower cost periods. (This Quebec-Hydro source is explained below.) Maine already has six small battery projects functioning, the largest being 20 megawatts, and the smallest being 0.5 megawatts. These already help Maine's electric grid to absorb excess capacity as more wind and solar come online. When drops in the renewable sources occur, the battery storage comes online in milliseconds as needed. When demand ramps up because people come home on cold or hot days, the battery storage fills the need. The State of Maine will have 175 megawatts of battery storage by mid-2025, and plans for 400 megawatts of storage capacity by 2030.

Industry developers are working on creating longer duration batteries that can discharge at full capacity for several hours. ISO-New England

[2] "Multi-day storage" means it could supply multiple days of energy demands.

[3] See Tux Turkel (2024, January 7), "Maine's first giant battery project set for construction in Gorham," *The Maine Monitor,* at https://themainemonitor.org/giant-battery-project-gorham/.

[4] FERC Order No. 841. See William Kissinger and Arjun Ramadevanahalli (2024, March 4), "How Recent FERC Orders are Regulating Electric Storage, QFS, and Inverter-Based Resources," at www.Morganlewis.org/pubs/2024/03/how-recent-ferc-orders-are-regulating-electric-storage-qfs-and-inverter-based-resources.

has proposed a Massachusetts facility of 700 megawatts. Plus-Power's "Cross Town" storage park is using battery systems supplied by a Chinese energy company. There are 3,072 battery cells in each 30-foot long trailer with 156 trailers over the five-acre facility. The containers have safety features that include heating and cooling equipment, thermal sensors, fire detection and suppression equipment, and 24/7 monitoring.

Unlike other power generation and/or transmission proposals, the Cross Town proposal did not attract opposition or even any of the public's attention. Being in an old industrial park, it was away from homes and close to a substation. Also, battery storage has attracted little public discussion. At Gorham's Planning Board meeting about this project, only one citizen rose to speak about it, and that was to ask about safety concerns.

These battery storage facilities were aimed at complementing wind and solar facilities. There has been, however, a pushback against solar development in Maine—a "moratorium" on solar construction. Maine's environmental agencies have set aside forests and farmlands from any solar array development. Also, Maine's plan for offshore wind development was discouraged when it failed to obtain a $456 million federal grant that would have enabled a planned port to be developed on Sears Island for assembling wind turbines.[5] Two companies did purchase leases in the Gulf of Maine that could be developed into 6.8 gigawatts of offshore wind power, but four other leases were unsold. There is, therefore, a hesitancy to further develop wind power in Maine. President Trump has also voiced his opposition to wind power in favor of his "drill baby drill" petroleum philosophy.

Meanwhile, dams such as the Shawmut on the Kennebec are still being considered for either removal or installation of newly improved and expensive fish passage facilities.[6] But Maine's dams are increasingly criticized for blocking Atlantic salmon and other fish passing to their spawning grounds, and also for their general degradation of river ecologies. Both recreationists and environmentalists, however, are energizing the American dam removal movement. Maine's waterways are perceived as America's last viable habitat for the endangered Atlantic salmon, and

[5] See Emmett Gartner (2024, November 10), "Maine dams face an uncertain future," *The Maine Monitor*, https://themainemonitor.org/uncertain-dam-future/.

[6] See Chapter 3 for a review of this new fish passage technology.

Wabanaki officials are advocating for the deconstruction of Maine's dams (see Chapter 1).

Landes Hudson of the *environmental advocacy organization* (EAO) Maine Rivers indicates that this EAO is focusing on removing "dinky dams" that no longer produce, or perhaps never did produce electricity, in order to restore access for native sea-going fish to upriver spawning grounds. Because of the dams' effects on these essential fish stocks, fish passages must be established at a minimum on Maine's larger dams, and "dinky dams" must be removed.[7]

Four years ago, Maine's Governor Mills recommended removing the Shawmut Dam and other dams on the Kennebec River. Their hydropower generation was slight, but they did block fish passage up most of the River. The federal officials recommendation for relicensing generated considerable dissatisfaction among environmentalists and recreationists. The case is still under adjudication.

The "New England Clean Energy Connect Project" and Hydro-Quebec Power

The New England Clean Energy Connect Project (NECEC) is a transmission line from Quebec in Canada. It provides up to 1,200 megawatts of renewable hydropower from northern Quebec. The 150 miles of high voltage power line crosses from Quebec into Maine at Beattie Township, which is on the southern border of Quebec with Maine. This transmission line then runs to Lewiston, which is in southern Maine close to the New Hampshire-Massachusetts border. In Lewiston, the transmission line connects to the electric grids of Maine and Massachusetts. Along with its targeted renewable generations, Maine will use the NECEC to meet its goal of avoiding fossil fuel energy reliance by 2050. The Hydro-Quebec power originates in the "James Bay Project" as explained next.

The Land and Waters of the Cree

One of the largest hydroelectric generating developments of the last half-century is along the rivers flowing into James Bay in Quebec. It generates power partly sold to New England, thereby affecting the dam removals

[7] Ibid.

and modifications in Maine and Massachusetts. Chapter 1 explains America's US dam removal movement as originating in Maine in July of 1999. Other chapters review the dam protest movement such as those aimed at development along the Tennessee River, the Colorado River, and the Columbia River, but the Canadian protests of the 1970s concern Hydro-Quebec's developments. These had substantial impacts on America's generation of hydropower. One of those protests involved a very consequential canoe trip that is reviewed here.

This story begins with an examination of these Canadian events and impacts, especially on the development and usage of battery storage in Main and Massachusetts as explained above. This story completes, or perhaps "complements," the narrative of Maine's river restorations.

Hudson Bay is an inland sea on the margin between the Arctic Ocean and the vast land mass of Canada. It extends more than a thousand miles south from Canada's Nunavut Province to well into the Provinces of Quebec and Ontario, an extension that reaches approximately six-hundred miles north of Montreal. Hudson Bay is about seven hundred miles in width, approximately the distance between Detroit and Quebec City. This Bay is the drainage basin for almost half of Canada, approximately 1.5 million square miles, but it also drains a northern portion of the Great Plains of the US. Hudson Bay's surface is almost a half-million square miles, but of particular importance is that it has a sizeable appendage that points southeasterly off its southern tip—James Bay.

In August of 1610, Sir Henry Hudson explored the Artic Sea into Hudson Bay on his ship *Discovery*, his fourth exploration of North America. Not realizing the effects of winter, Hudson allowed the *Discovery* to become trapped in ice at the southern tip of James Bay. In the Spring of 1611, when the ice finally freed the *Discovery*, Hudson sought to continue his exploration, but his crew mutinied and abandoned him adrift in a small boat with a few other crew members. There is no record of Hudson's survival even though in a 1612 expedition, the British did search the area for him. In 1668–1669, the British ketch the *Nonesuch* explored Hudson Bay for trading possibilities and actually did trade for beaver pelts with the indigenous of the area.[8] This ketch was a smaller vessel designed to sail upriver into the Bay's tributaries, and since it was only fifty-four feet long with a draught of six feet, it was also designed

[8] A "ketch" has one main mast and one smaller aft mast.

to be dragged onto a dry dock of tree trunks to prevent the vessel from being crushed by winter ice.

The *Nonesuch's* trade in beaver pelts led to the Hudson Bay Company (HBC) being established as a monopoly charter granted by the English crown in 1670 for trade throughout the watershed of Hudson Bay. This trading basin was then called "Rupert's Land." The monopoly granted "the right of sole trade and commerce." Rupert was the nephew of King Charles I and the first "governor" of the HBC.

France contested the monopoly granted to the HBC by sending several military expeditions to the region. The British–French conflict ended with the Treaty of Utrecht signed on April 11 of 1713. French concessions to the British included the Hudson Bay region ("Ruperts Land"), Nova Scotia, and Newfoundland. The HBC then constructed several forts and trading posts at the mouth of Hudson Bay's significant tributaries (Severn, York, Churchill, and the Prince of Wales Fort) (see Fig. 13.1). Being on the larger tributaries, these strategic positions facilitated inland explorations and trading posts for commerce with the indigenous, particularly for the fur trade.

The HBC's monopoly was abolished in 1870. Starting in 1913, the Bay was extensively charted by Canada for the navigation development that continued up to recent years.

The indigenous of this area who traded with the HBC were the Cree. Their name is from the Ojibwa term for all those who live around the Hudson Bay. They were a hunter-gatherer group who were also active traders. (See Chapter 12 for a brief review of the Cree who settled around Fort Peck.) The Cree are one of the larger groups of Canada's "First Nations" indigenous tribes. They number more than 220,000 who mostly reside in the vast area north of Lake Superior and west of James Bay. The Bay was named after an early Welsh captain who explored the area in 1630–1631. It is split politically at the southern tip with the west half being in Ontario and the east half being in Quebec. This north–south division also divides Hudson Bay with the east half being in Quebec, and the west half being in Ontario, Manitoba, and Nunavut in the north. The world's largest herds of caribou migrate along the Bay. The land along the Bay is rocky and wet with hundreds of rivers flowing into James Bay, the largest being the La Grande. On the east side of the Bay there are four coastal communities belonging to the Cree: Waskaganish, Eastman, Wemindji, and Chisasibi. On the west side of the Bay, there are five coastal communities: Moose Factory, Moosonee, Fort Albany, Kashechewan, and Attawapiskat (see Fig. 13.2).

Table 13.1 gives the annual climate data for James Bay as recorded at La Grande Airport. The mean temperatures plus the average rainfall and snowfall all indicate the close-to-artic climate. The James Bay area is cold with lots of snow and rain as indicated. The area's population density is also low. Prior to the 1970s, the Hudson Bay–James Bay area was desolate, largely barren, and appeared to be appropriate for large-scale hydroelectric projects.

Approximately 15,000 Cree, however, lived in the Quebec area east of James Bay and Hudson Bay. The Cree originally organized into "lodges" who usually were of two families interrelated by marriage who lived in either domed tents (wigwams) or conical tents (tipis). A group of lodges composed a "band" who hunted or fished together. Differing bands were also interrelated by marriage and would hunt and socialize together. According to "First Nations" registrations, there are 135 registered Cree Bands. Those Cree who resided around James Bay were also called "Innu." The poverty of the Innu was apparent in the early 1970s. They might have willingly accepted the resources generated by hydroelectric development had it been initially offered.

Table 13.1 Climate data for La Grande Riviere Airport, 1981–2010[a]

Statistic	*Annual means*
F^0	$-3\ F^0$
Mean Minimum	$8\ F^0$
Record Low	$-48\ F^0$
Average Precipitation in Inches	27 inches
Average Rainfall in Inches	18 inches
Average Snowfall in Inches	103 inches
Average Annual Days of Precipitation	189 days

[a] Temperatures are in F^0. Precipitation is given in inches

Source Environment Canada

James Bay Project

In Chapter 2, the problems associated with private electric utilities in the US during the 1920s and early 1930s, especially in providing power to rural America, led to the establishment of the *Tennessee Valley Administration* (TVA) in 1933. The financial problems revealed by the early Great Depression also led to the *Public Utilities Holding Company Act* of 1935, and also the *Rural Electrification Administration* of Franklin Roosevelt's New Deal. Similar problems occurred in Canada, and in response to the so-called electricity trust, with its high rates and excessive profits, a successful public utility nationalization movement occurred in Ontario in 1906. But this had not extended into Quebec.

Before Adelard Godbout was elected Premier of Quebec in 1939, he was outraged by what he perceived as the corrupt and inefficient power system dominated by the privately owned companies Montreal Light, Heat & Power (MLH&P) and Shawinigan Water & Power which he accused of being a duopoly of "economic dictatorship, crooked and vicious."[9] In April of 1944, the Quebec Legislative Assembly created the *Quebec Hydroelectric Commission* (Hydro-Quebec), a Crown corporation with an electric and gas monopoly in the Montreal area mandated to serve its customers "at the lowest rates consistent with a sound financial management."[10] The purpose was to speed up rural electrification in those areas with either no or limited electric service.

In 1944, Hydro-Quebec only had 600 megawatts of generating capacity, but it somewhat expanded its generation capacity during the Post-World War II period.[11] When Robert Bourassa, an economist, was a back-bench member of Canada's *National Assembly*, he met over lunch with Roland Geroux, the head of Hydro-Quebec. They discussed the James Bay Project as an alternative to the Canadian Government's nuclear vision, *Atomic Energy of Canada*. At that time, nuclear energy was being promoted as the energy foundation for Quebec's future. Bourassa, however, envisioned the development of Quebec's hydropower resources

[9] See Gilles Gallichan (1995), De la Montreal Light, Heat and Power Hydro-Quebec," at https://archive.org/details/hydroqyuebecautre0000unse/page/63.

[10] Ibid.

[11] See Hydro-Quebec (2009, April), "Powering Our Future: Annual report 2008," at www.hydroquebec.com/publications/en/annual_report/pdf/annual-report-2008.pdf.

as the energy source for its future. Of course, Hydro-Quebec supported Bourassa's alternative vision.

Bourassa was elected Premier of Quebec in April of 1970. He had campaigned on a job creation and natural resource development platform, especially on the development of Quebec's hydropower capacity. A year after his election, Bourassa announced the "James Bay Project," the construction of a 10,000-megawatt hydroelectric complex somewhere on the eastern tributaries of James Bay. At the time of the project's announcement, the sites and details had not been selected, but by May of 1972, the diversion of the Eastman and Caniapiscau Rivers into the La Grande was selected as the main element of the project. There would be a series of three dams and four power generating stations along the rivers. The options selected would be more cost effective for generating the 10,000 megawatts, have a smaller forest impact, require less flooding, and thereby have a smaller impact on Cree hunting and fishing.

The river areas on the east coast of James Bay, and the nearby Hudson Bay coast, were the traditional hunting and fishing areas for 5,000 Cree and 3,500 Inuit residing north of the Cree. In 1971, there were no access roads in the area that would facilitate construction of the dams and electric facilities. Also, the Cree and Inuit had not been informed about the hydro project until construction of the access roads had begun. The indigenous believed that the government of Quebec was violating treaties by unlawfully expropriating and destroying their traditional hunting and fishing areas. The Federal Indian Affairs Minister Jean Chretien sided with the Cree and Inuit and hired lawyers to litigate their claims. Also, the conflict between Federal Prime Minister Pierre Trudeau's nationalist political philosophy versus Bourassa's provincial-centered philosophy added to the dispute. Chretien charged that Quebec did not have the right to construct the James Bay Project on Cree land. In 1973, the federal lawyers won a court injunction stopping the Project until a treaty could be signed with the Cree and Inuit. The injunction was overturned by an appeals court. But when this case was to be heard by Canada's Supreme Court, Bourassa agreed to negotiate with the affected First Nations' tribes.

Prior to the negotiations concerning the James Bay project, the Cree had no centralized structure for handling the defense of their Treaty rights. For this purpose, in 1974, the Grand Council of Cree was formed with Billy Diamond as their first Grand Chief. The Cree then began negotiations with the provincial and federal Governments which eventually resulted in their first land claim settlement, the "James Bay Northern Quebec Agreement."

After fourteen years of construction, the first phase of the James Bay Project was completed in 1986. This included four generating stations with 10,800 megawatts of power. Two-hundred and fifteen dams and dikes were constructed for this project plus the extensive transmission lines. The second phase began in the late 1980s with five additional power plants that added 5,200 megawatts of generating capacity. A new high voltage transmission line was being constructed to export power to the New York Power Authority, plus a 750-mile long transmission line with a capacity of 2,000 megawatts was constructed to cross into New England. But during this second phase, Hydro-Quebec proposed an additional project on the Great Whale River just north of the La Grande River (see Fig. 13.2). The Cree opposition was strongly vociferous.

In 1990, the Grand Chief of the Cree organized a canoe trip from Hudson Bay to New York's Hudson River. This canoe trip was to rally opposition to this extension of the James Bay project. It was aimed at pressuring Hydro-Quebec. The Cree had already experienced significant culture shock due to the road constructions from the south, and few Cree were employed on the project's construction. Poverty and social problems remained among the indigenous of the area.

Given these protests, in 1990 the New York Power Authority canceled its US $5 billion power contract with Hydro-Quebec. In 1994, the Great Whale Project was also canceled as not being necessary for Quebec's power needs.

The Cree and other Inuit tribes were joined by *environmental advocacy organizations* (EAOs) to protest the project. Its environmental impact was severe. Ten thousand caribou drowned in 1984 while crossing one of the newly dammed rivers on their migration route. The graves of the Cree who had lived for centuries along the rivers were inundated. The ecology of the James and Hudson Bays were degraded by the disruptions in the spawning cycles of migrating fish. The dead trees and other decaying vegetation produced by the inundation of large river valleys also produced mercury which was absorbed by the fish. Now a majority of the indigenous in the villages where fish was the primary diet showed symptoms of mercury poisoning. In 1993, the Cree, together with the EAOs, filed suit against the New York Power Authority in the US District Court of New York. The suit challenged the legality of the agreement between

Quebec-Hydro and New York Power as based on Constitutional grounds and the 1916 Migratory Bird Treaty between the US and Canada.

In February of 2002, nearly seventy percent of Quebec's James Bay Cree endorsed a US $2.25 billion settlement with the Province of Quebec to allow hydropower development on their land. The agreement was a vindication of the long campaign (since 1975) to have Cree rights respected. The James Bay Cree received $16 million in 2002, $30.7 million in 2003, and a promise of $46.5 million annually for forty-eight years thereafter. The Cree dropped their lawsuit and also agreed to hydroelectric plants along the Eastman and Rupert Rivers.

Although the New York Power contract was canceled, the NECEC transmission to New England enabled Maine's deconstruction of its hydropower dams to continue. The passages of Atlantic salmon and other fish runs continued to improve because of the combination of Quebec's hydropower and the advances in lithium-ion battery storage.

The Elements Motivating America's Dam Deconstruction Movement

In reviewing the previous twelve chapters, it is apparent that in addition to the general ecological restoration of our rivers, there are four specific other and perhaps surprising issues motivating our dam deconstruction movement (These motivations do overlap). The first is the extensiveness, popularity, and strength of the "justice for the indigenous" movement. The second is the extent of our problems with low-head dams in the northeast of the US. The third concerns our woeful lack of safety inspections and remediations. The fourth concerns the demand for nature-interacting recreation that brings large numbers of paddlers, fishers, other boaters, and other recreators to our rivers. They desire clean flowing rivers in unpolluted natural settings.

The Indigenous Justice Movement

Except for our nomadic northern and central plains' tribes who hunted buffalo herds, our native tribes were largely riverside and/or lakeside peoples. Their sustenance mainly came from our rivers, streamsides, and bays. Damming our rivers destroyed their sustenance and their cultures. Removing the dams on the Elwha River, for example, has begun to allow the Klallam to again fish for Pacific salmon, and to practice their festival ceremonies. The two dams on the Elwha River were in clear violation of the Klallam's treaty rights (the *1855 Treaties* entered into by the indigenous tribes of the Pacific Northwest) which were finally upheld by the *1974 Boldt Decision*. (All of this was reviewed in Chapter 3.) Enforcing this decision was, however, contentious. Without the threat of salmon extinction, enforcement might not have occurred. The two offending dams on the Elwha were removed in 2014, and the restoration has been significantly successful.

The Klamath Basin Tribes (Karuk, Yurok, Hupa, and Shasta) have had a similar struggle. They also have been successful in removing dams along the Klamath River in recent months. They are currently using the lessons from the Elwha's successful restoration for restoring the Klamath.

Dams have also been removed in the Columbia River Watershed with the aim of restoring the Pacific salmon to health. In this basin, numerous tribes had their *1855 Treaties* rights violated (Colville, Spokane, Coer d' Alene, Yakima, Nez Perce, Cayuse, Plus, Umatilla, Cowlitz, Kootenay, and Warm Springs Confederated Tribes), but some efforts for restoration have occurred such as on the White Salmon River. Dam removal conflicts remain on four dams along the lower Snake River, and on constructing effective fish passage technology along the main stem of the Columbia River.

We have also reviewed the success of the indigenous Penobscot Nation in leading the dam removal movement along the Penobscot River of Maine. There, as with other dam removals, the FERC hydropower licensing process has substantially assisted in enabling hydropower dam delicensing and subsequent removals. This same FERC process also enabled delicensing of the Klamath's and Elwha River's dams.

The Problem of Low-Head Dams

The second surprising issue is the extent of America's low-head dam problem. Throughout our Northeastern states, there are thousands of these usually abandoned low-head structures on our rivers, particularly our smaller tributary rivers. In some states, removal is difficult because the required permit process is cumbersome, and some people consider the waterfalls that flow over many of these small structures as scenic. These dams are, however, ecologically destructive and dangerous to recreators who mistakenly view the turbulence before and behind these small structures as innocuous until they are trapped. These dams also create dangerous flood problems associated with failure risk because they are seldom maintained.

The number of low-head dams (those less than fifteen feet in height) is surprising, e.g. eight thousand in Pennsylvania, three-thousand in Massachusetts, and many thousands more in other Northeast US states. These structures tend to be remnants of colonial and nineteenth- century attempts to provide public and industrial water supplies to relatively small establishments (towns and industrial businesses) (see Chapters 7 and 8).

The many low-head dams of New England provide examples of the quaint waterfall structures that generate resistance to removal because they are often picturesque center pieces of New England scenes. The Montgomery Dam in Camden Maine is along the Megunticook River. It is an example of a low-head that creates one of these New England scenes. The South Middleton Dam and the Ipswich Mills Dam, both on the Ipswich River, are also examples of the quaint waterfall dams of New England Scenery (see Chapter 8).

To overcome resistance to removals of these New England structures we need (i) a clear presentation of an attractive alternative of a clean flowing stream of equal appeal to the proposed to be removed waterfall, and (ii) an adequate explanation of the ecological improvement benefits to be derived from the restoration. These are often just public relations problems, but effectiveness in these communications have become a necessary component of the environmental movement.

One significant difficulty with the number of old low-heads in the Eastern US is that many are on private property and are unknown to officials. All low-heads pose flood dangers, and are also dangerous to paddlers, fishermen, and swimmers (even waders) because they look so innocuous while causing such turbulence. Recreators who are caught in this turbulence, both on the upstream side and the downstream side of the dam, have difficulty escaping. Children who seek to swim in what appears to be a perfect spot are particularly susceptible to the danger. Knowing how dangerous these dams are, state authorities are removing significant numbers of these structures.

Safety Standards

The third surprising problem concerns our safety inspections and soundness assurance programs. According to the *National Inventory of Dams* database kept by our Army Corps of Engineers, we have over 92,000 large dams in the US, and many thousands more of smaller dams. These can be dangerous structures. As reviewed in Chapter 10, a significant series of dam failures occurred in the 1970s, most notably the Teton Dan on the Teton River of Idaho, the three Buffalo Creek Hollow Dams in West Virginia, and the Kelly Barnes Dam on Toccoa Creek in Georgia. These failures led to the organization of the *Association of State Dam Safety Officials* (ASDSO) in 1983. Since then, the ASDSO has been organizing our non-federal dam safety standards for inspections. They have developed a two-category classification system: (a) what is the condition of the dam, and (b) what is the extent of the consequences in case of failure. Our problem is that the state agencies are woefully understaffed and under funded so that there are few inspectors available for coverage of large numbers of dams often at great distances across our states. We are generally far behind in keeping to our schedules for inspections, and for necessary re-inspections when deficiencies are supposedly fixed after operators have been warned that remedial actions must be taken. Because of this underfunding, dam failures are likely. When dams lose their usefulness and are not maintained, they should be removed.

Despite the state inspection process, two significant dams recently failed (or came very close to catastrophic failure in the case of one of these dams): the Oroville Dam in California in 2017, and the 2019 Spencer Dam in Nebraska. The first failed because of faulty design and especially poor construction of its spillway. The second failed because the responsible state engineers who knew the history of the dam's difficulties in dealing with ice jams in its gates had abandoned their supervision. Replacement operators did not know of this history and so the dam's gates were closed and unable to be opened when ice jammed them. Safety inspectors were unable to overcome these difficulties. Perhaps a more rigorous inspection program that (i) examined the design of the dam, (ii) supervised the operation of the dam, and (iii) provided more frequent inspections of the dams under its domain, could be more effective.

Many states have large numbers of older dams, but few inspectors to assure their safety. For example, New York State has more than 7,000 state dams that should be inspected. It only claims to inspect 1,920 of these, i.e. only the "high hazard" dams are inspected with inspections being every other year.[12] But 93 of these are also classified as being in either

[12] See Karlin and Munson (2021, December 17).

"poor" or "unsatisfactory" condition. Because of having too few trained inspectors engaged by the State, and because the State is large in size with considerable distances between sites, there is difficulty in maintaining New York's schedule of inspection.

In the State of Vermont, there are more than 1,000 dams to be inspected by the State. One third are in "poor" condition[13] (see Chapter 10). During the flood conditions of July, 2023, three of these dams failed. Part of the problem is that Vermont has difficulty identifying abandoned dams and consequently arranging the legalities necessary for dam removals.

There is a federal "Dam Safety Program" under which for the 2024 fiscal year, there was $211 million of funding spent for dam rehabilitation. Under the *National Dam Inspection Act* of 1972, the Corps of Engineers is directed to maintain its "National Inventory of Dams." It has 92,275 dams identified with seventy fields of data entry for each dam. This includes the dam's safety status, and "inundation maps" in case of dam failure. In 2023, the ASDSO estimated that repairing the nation's non-federal dams would cost $157.5 billion, and that $34.1 billion is needed just for the "high hazard" classified dams. A high priority does not appear to exist for assuring the safety of the dams of the US.

Recreation Demands

We now have a national movement of recreators who paddle, boat, and fish on our rivers, or hike and bike along our riverbanks. They form a considerable component of our political movement for river restoration. Chapters 1, 8, and 11 review strong examples of the demands for this recreation and river preservation and remediation. For example, the demand for restoring Maine's rivers for fishing and paddling has greatly helped this state's impetus toward restoring the Kennebec and Penobscot Rivers. But this is also the case for the Rogue and Illinois Rivers in Oregon as reviewed in Chapter 11.

[13] See Weinstein (2022, February 14) and Duria (2023, July 30).

The First Gatherings

This volume largely concerns the restorations of the main stems and/or larger tributaries of our rivers. This is certainly a major current concern, but we must also worry about the conditions of our watersheds, whether they are filled with poisons from lawn runoffs, or agricultural contaminations, or poorly functioning septic systems, or whatever modern civilization could seek to just wash away downstream to become the problem of others. It is as though we give up on our collective civilization, so in some way we isolate ourselves from it, and dump our wastes on the collective and away from ourselves. In opposition to this attitude, I present a few closing remarks.

In Central Oregon, the *Three Sisters Wilderness* surrounds the three mountain peaks of the same name. On the east side of the 11,000-foot South Sister, and west of a smaller chain of peaks called Broken Top, there is a series of small moraine ponds—Green Lakes. They are formed from rivulets that flow from the glaciers down over the volcanic red-scree soil of South Sister. These rivulets are the *first gatherings* of the water molecules from the glaciers. Flowing down, they form small brooks and then streams that create the green moraine ponds and small lakes popular for summer wilderness exploration. From these ponds, streams flow south to Soda Creek, and eventually into the Deschutes River, the quintessential western desert river that flows unobstructed northerly along the edge of the Cascade Mountains. All along its route, the Deschutes gathers water from the streams that flow easterly from other peaks in the Cascades. These streams are all formed from other *first gatherings* high on the steep sided glaciers. The Deschutes eventually flows past the east side of Mount Hood, another 11,000-foot peak, and then into the Columbia River just upstream of the Dalles Dam. This is a structure of extraordinary length on a wide section of the Columbia River. It was created by, and is operated by the US Army Corps of Engineers. The water of the Deschutes originates with the rivulets from the Cascade Mountains which are largely composed of federally designated wilderness and national forest areas. This water is clean of human generated poisons.

While in my twenties and early thirties, this volume's author was dedicated to hiking, climbing, and exploring the spectacular Cascade Mountains of Central Oregon. But I was born and raised in Eastern Massachusetts where there are no mountains, and no volcanic red scree, but only a very different soil composition. There the rivulets, brooks, and streams are not protected from pollution.

New England's ground "has more granite in it than soil," or so one comes to exclaim when taking pick and shovel to it. All those quaint granite stone walls we see when viewing New England's rural landscapes usually have a purpose besides establishing the border of a farmer's field; they are the accumulation of the hard effort of tossing and dragging granite rocks and boulders to the sides of a field so there remains some rock-free soil for growing the farmer's produce. There are three combining elements to any of those rural New England landscapes, elements that combine to give us picturesque scenes: the granite rock walls, the maple and other hardwood trees, and the brooks of clean water. Those brooks are natural constructions from the rivulets of *first gatherings* of water flowing over that ground of soil and granite stone. The granite stones come to form the brooks' beds as the water washes the soil down into ponds which become their own small ecological systems, essential entities of nature that are not very different from the moraine ponds of Green Lakes.

Those granite bedded New England brooks are special places worthy of examination in any season, but I particularly loved what they provided in winter. There was an old and small farm across the street from my boyhood residence of the 1950s where New England's snow-storms kept everyone else indoors. Snow storms created perfect conditions for heavy dress and exploration because in those pristine snowy conditions all the tracks of the different animals were clearly visible. Those tracks inevitably led either back to the animal's lair or to one of those rock bedded brooks. Ice would be attached to the rocks on the sides of the brook, but the animals would still be about. The water would be flowing past those boulders and stones, and then down into the ponds and ultimately the reservoir that provided the town's water supply (Crystal Lake in Wakefield, Massachusetts). The land between the farm's border and the reservoir was acreage the town owned. We called it "the cove." It supposedly had restricted access. To protect our water supply, nobody was to be there. But that restriction just made it the province of those few kids who were natural sneaks, explorers of nature who had the special rights of natural exploration that kids should have. Those brooks had to be explored even in winter, and as a young boy, I explored them.

In the 1970s and 1980s, some good friends and I would ski into wilderness areas such as the Green Lakes Basin of Oregon's "Three Sisters

Wilderness." We would do this in the early Spring when the snowpack was still high. The motivation for these explorations was similar to my boyhood's snowy ventures into "the cove" area of Crystal Lake. Other people were not there, but the wilderness area's wildlife were always there to be seen. Just as with my boyhood ventures, these explorations always were worth the effort.

Today, when I visit back in Massachusetts, the farm is long gone, developed into housing lots. The "nobody can be there" acreage is no longer restricted since the lake is no longer the town's water supply. It is just public parkland. The brooks now drain suburban lawns, and the ponds are filled with algae during the summer. Fertilizer and pesticides transform the ponds into something less than healthy nature. Our brooks and creeks form our ponds, streams, and rivers. When the brooks and creeks become polluted, then what remains? Restorations should always begin with the brooks and creeks because our watersheds should be protected.

In the 1950s, there was a pond (Doyle's Pond) adjacent to "the cove" area of Crystal Lake. During the winters, it was used by generations of neighborhood kids for skating. (The pond was filled in during the early 1970s to create an elementary school, a parking lot, and play area for the school). Doyle's Pond was on public land that bordered the other side of the above-mentioned restricted acreage and reservoir, but it did not have restricted access. It had a couple of feeder streams coming in on the north side, and it had a small concrete dam with a heavy wooden sluice gate on the south side. This was the first dam I remember noticing. I was perhaps six years old, and with my much older sister when I first stood on its concrete. The dam was maybe eight feet wide and two feet thick. When the small gate was opened, the flow created a creek that traveled about three-hundred yards to reach the town's reservoir. I once saw the closing of that wooden sluice gate. I was only seven or eight at the time when standing by Doyle's Pond, I first heard the term "Corps of Engineers."

Each year, when the weather was really cold, on school days we would walk as fast as we could from Saint Joseph's School to get home to grab our skates and hockey equipment and head for Doyle's Pond. In December, when the temperature began to turn really cold, a sizeable grouping of us formed on the way home from school at the edge of the pond to judge whether the ice was thick enough to skate on. (Doyle's Pond was only a short diversion from our route home, and "Yes!" we walked the distance.) One year, when I was seven or eight years old, we were performing our annual duty of judging the thickness of the ice on the Pond. A truck pulled up at the nearby dirt road and a man got out and used a lever to close the sluice gate. One older friend exclaimed, "He must be from the Corps of Engineers!" For me this was a mysterious exclamation from a boy a few years older than I. He said this with such assurance that he must have actually known who controlled our pond. I'm sure that the mysterious sluice operator was just someone from the town's public works department, but the closing of the sluice gate seemed like such an important and official act that the man with the lever had to also be important. That term, "Corps of Engineers," sounded extremely important. The older boy seemed to indicate that our world was greatly influenced by this "Corps of Engineers," and he then told us that with a few more years of maturity, we would be pondering the very existence and importance of the "Corps." This was an obvious youthful exclamation of such significance that we would surely be greatly concerned with it for the rest of our days. (That boy did grow up to become an engineer for the Corps.) Personally, at the time I was just sure that what was important in "our world" consisted of the rolling open fields with a few clumps of evergreens, oaks, and maples, plus granite boulder walls and brooks, and especially the ponds. Who would want to affect that? What was this external force called the "Corps of Engineers?".[14]

If we're going to try to restore our environment, it seems to me that we should not ignore the forests, fields, brooks, and ponds of the watersheds that feed our rivers. That is where the health of our inland ecological systems begins, i.e. our watersheds. Our forests, fields, brooks, and ponds are not under the official domain of the Corps or any other federal government institutions, but that means we need those smaller institutions that can and will protect them, e.g. the local land-use laws and enforcement mechanisms such as those developed for the Housatonic River reviewed in Chapter 11, and the Columbia River Gorge reviewed in Chapter 6.[15] Those common property resources must be collectively controlled or they will be degraded, i.e. suffer the "tragedy of the commons" phenomena. This collective control means protection for the brooks, streams, ponds, and riparian areas where the "first gatherings" occur. In order to not be abused, these resources must be managed

[14] In the late 1970s, I was the principal researcher for the Corps' "Columbia River Basin Water Withdrawal Environmental Review."

[15] Robinson (2021, Chapter 10) reviews these institutions of local land-use and enforcement.

intelligently. As explained in previous chapters, intelligent management has not always been present for our rivers.

Some Observations on the State of Dam Removals and Rehabilitations

We can date the beginning of the US dam removal movement to 1999 with the deconstruction of the Edwards Dam on the Kennebec River in Maine. Twenty-five years of removals have progressed to include the 2024 removal of dams along the Klamath River. These dam removals are essential and important parts of our river restoration efforts. After two-hundred years of industrial and agricultural degradation, our rivers are currently being ecologically renewed. One motivating aspect has been the pursuit of justice for our indigenous tribes who were decimated by two centuries of river degradation. As documented in previous chapters, this degradation was built on treaty violations and willful defiance of the just law. As our dam removals and restoration efforts develop further, however, the politics of reaction are likely to become stronger. Concerning this reaction, we need to emphasize two points: (i) we must continually emphasize that our legal agreements and court decisions are the basis for our environmental justice pursuits, and (ii) these decisions to provide justice also lead to the ecological restorations that benefit us all. Indigenous justice is not a separate movement from our environmental movement, it is a necessary component. That is why "environmental justice" has such broad appeal to environmentalists.

This final chapter is written during the early days of the second Trump Administration when the wildfires of Pacific Palisades and other Los Angeles locations are burning. President Trump has been exclaiming that we should "turn on the valve and send the waters of Northern California" down south to solve the wildfire problem. He even referred to sending water down from Canada. This extreme ignorance encapsulates the illusion of Western water abundance, that our Western rivers could be replumbed to create a "paradise in the desert." This was the illusion that created great resource reallocation and severe environmental degradation. This illusion periodically emerges from the body politic to repeatedly require us to argue the facts; that we must point out the direct expenses required in pursuit of this illusion; that the opportunity costs of what we are giving up are enormous; and that the degradations that will follow if we pursue this "illusion of abundance" are appalling. I addressed this "illusion" in the last chapter of Robinson (2021). But I wish to repeat here the last short paragraph from that book.

> If we could somehow meet with the political leaders of today and tomorrow, what plea would we put before them? I would plead, "Please let us preserve and restore, and not plan to replumb or rearrange in some grandiose way!" Maybe this is the plea our environmental movement has always yearned to make through our advocacy organizations. (See Robinson, 2021, Chapter 13, p. 348.)

We should, however, be heartened by the success of our river restoration movement. We are making rapid progress in "alternative" electric power generation (wind and solar) that not only replaces fossil fuel burning, but also hydropower generations. Our lithium-ion battery storage developments, as documented above in this chapter, are already allowing Maine and Massachusetts to quickly progress in this alternative energy development. We should therefore be optimistic concerning the possibilities of river restoration.

As a society, we now recognize the value of dam removals as part of our overall restoration of our watersheds. We have already enjoyed a substantial ecological benefit from our dam removals; we've observed the return of fish runs in substantial numbers. We've begun the redesign of watersheds, such as the Penobscot River, which now has many of its previous dams removed (See Chapter 1). But this was accomplished without lowering hydroelectric generation or weakening flood control. We also observe large increases in recreators paddling, swimming, fishing, and family boating on our restored rivers. They strengthen our political movement for further restoration. For examples, Chapter 9 reviews this political dynamic with respect to the Monongahela River, and Chapter 11 reviews this same dynamic for the Housatonic and the Rogue Rivers. All of these aspects contribute to this significantly large component of our overall environmental movement.

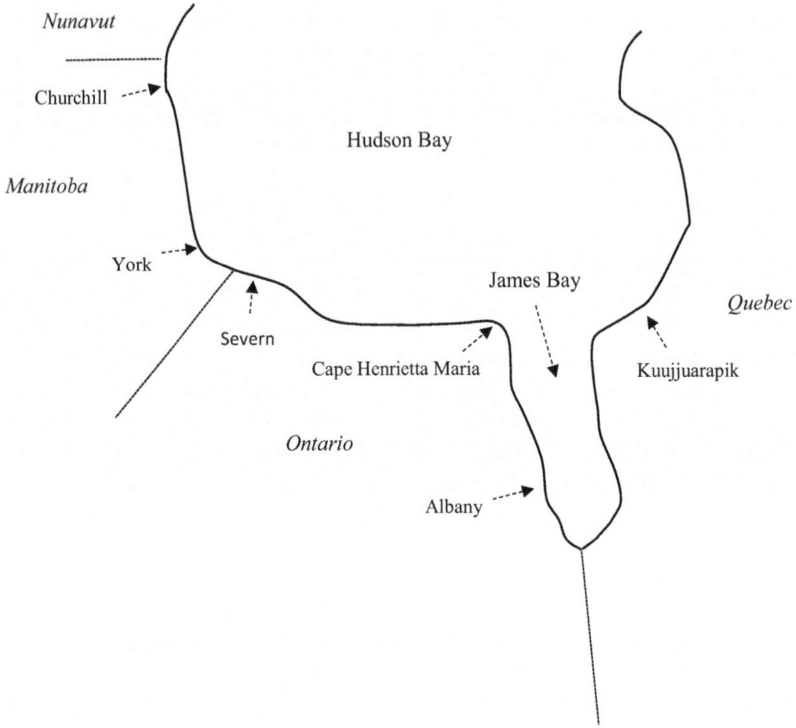

Fig. 13.1 Hudson Bay and James Bay

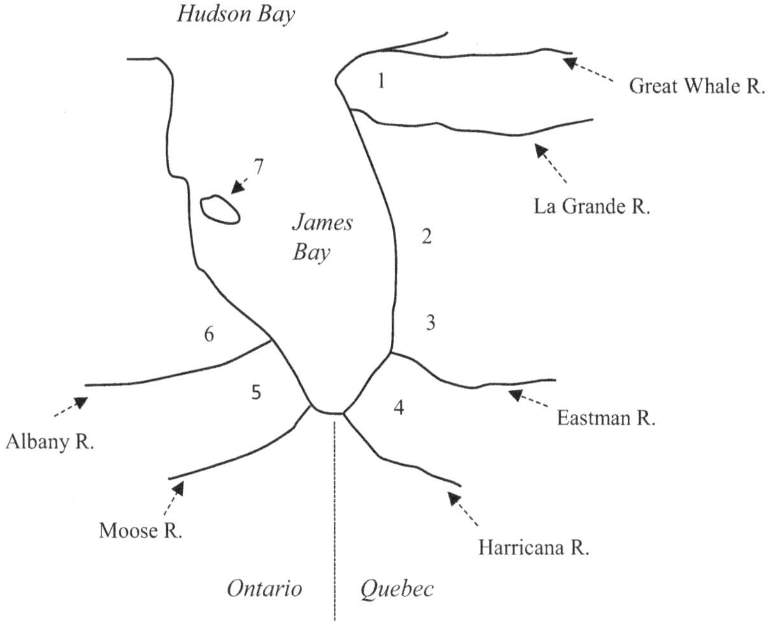

Legend of Towns Along James Bay:

1: Chisasibi
2: Wimindji
3: Eastman
4: Washagamish
4: Moosonee
5: Albany
6: Ashiniski Island

Fig. 13.2 James Bay

References

D'Auria, P. (2023, July 30). Historic Flooding Puts a Spotlight on Vermont's Dams. *Vtdigger*. https://vtdigger.org/2023/07/30/historic-flooding-puts-a-spotlight-on-vermont-dams

Howard, B. C. (2016, June 2). River Revives After Largest Dam Removal in US History. *National Geographic*.

James, I., van der Brug, B., & Greene, S. (2024, March 24). The Klamath River Dams Are Being Removed. Inside the Effort to Restore a Scarred Watershed. *Climate & Environment*. www.latimes.com/environment/story/2024-03-24/klamath-river-restoration

Karlin, R. & Munson, E. (2021, December 17). 147 New York Dams Are "Unsound" and Potentially Dangerous. *Times Union*. www.timesunion.com/projects/2021/ny-dam-hazards/

Price, K. (2024, April 5). As Messy Side Effects of Klamath River Dam Removal Continue, Officials Stress That Short-Term Pain Will Yield Long-Term Gain. *Inside Climate News.*

Rawls, J. (2001). *Collected Papers—John Rawls* (Ed. S. Freeman). Harvard University Press.

Rawls, J. (1987). Idea of an Overlapping Consensus. *Oxford Journal of Legal Studies 7:1–25.* Reprinted in *Collected Papers—John Rawls* (Ed. S. Freeman). Harvard University Press.

Rawls, J. (1980, September). Kantian Constructivism in Moral Theory, *Journal of Philosophy, 77*, 515–572. Reprinted in *Collected Papers – John Rawls* (Ed. S. Freeman). Harvard University Press.

Rawls, J. (1951, April 2). Outline of a Decision Procedure for Ethics. *Philosophical Review, 60*, 177–197. Reprinted in *Collected Papers—John Rawls* (Ed. S. Freeman). Harvard University Press.

Robinson, R. M. (2023). *Environmental Advocacy and Local Restorations.* Palgrave Macmillan.

Robinson, R. M. (2021). *Environmental Organizations and Reasoned Discourse.* Palgrave Macmillan.

Shabman, L., & Stephenson, K. (2007). Environmental Valuation and Decision Making for Water Project Investments and Operations: Lessons from the FERC Experience. *Institute for Water Resources*. www.iwr.usace.army.mil/Portal/70/docs/iwrreports/2007-VSR-01.pdf

Weinstein, E. (2022, February 14). State Mismanaged Aging Dams, Auditor Finds. *vtdigger*. www.vtdigger.org/2022/02/14/state-mismanaged-aging-dams-auditor-finds/

Index

R. M. Robinson, *Restoring America's Rivers*,
https://doi.org/10.1007/978-3-031-81758-8

The manufacturer's authorised representative in the EU is Springer Nature Customer Service Centre GmbH, Europaplatz 3, 69115 Heidelberg, Germany. If you have any concerns regarding our products, please contact ProductSafety@springernature.com

Printed and bound by CPI Group (UK) Ltd, Croydon, CR0 4YY
05/10/2026
02228125-0017